Sensory Ecology, Behaviour, and Evolution

Sensory Ecology, Behaviour, and Evolution

Martin Stevens

BBSRC David Phillips Fellow, Centre for Ecology & Conservation, University of Exeter

OXFORD
UNIVERSITY PRESS

OXFORD

UNIVERSITY PRESS

Great Clarendon Street, Oxford, OX2 6DP,
United Kingdom

Oxford University Press is a department of the University of Oxford.
It furthers the University's objective of excellence in research, scholarship,
and education by publishing worldwide. Oxford is a registered trade mark of
Oxford University Press in the UK and in certain other countries

First published 2013

Impression: 1

British Library Cataloguing in Publication Data

Data available

ISBN 978–0–19–960177–6 (hbk.)
      978–0–19–960178–3 (pbk.)

Printed in the UK by
Bell & Bain Ltd, Glasgow

Contents

Preface

Writing this book has been an illuminating and rewarding journey. Before beginning the project I thought I had a reasonable idea of just how exciting and vibrant the subject of sensory ecology is. I was wrong! It was only as I really got stuck into working on the book that I truly realized just how much incredible and innovative work is being undertaken every year in this field, and the many remarkable examples of natural history that have formed the inspiration for much of this research. I hope I have done the subject at least some of the justice it deserves.

Sensory ecology as a scientific discipline has boundaries that are hard to define. It is perhaps for this reason that few books specifically on sensory ecology actually exist, despite the subject's substantial importance in modern day biology and its rapid growth in the last decade or so. Historically, sensory ecology emerged as discipline that linked conventional ecology with studies of animal behaviour, grounded in the idea that animals need to obtain and use information to behave appropriately in a given environment. This was the approach that Dusenbery's (1992) seminal book, *Sensory Ecology* took, and the idea of information is still central to sensory ecology today. Dusenbery focused on the physics of how different modalities work (for example, sound, light, chemicals), and related this to various ethological examples, such as navigation, searching, and so on. Undoubtedly, this approach is important and the content of that book remains valuable. However, partly thanks to the rise and continued popularity of subjects like behavioural and evolutionary ecology, sensory ecology today deals with all that Dusenbery discussed, but spans across into behavioural ecology and evolutionary processes. For that reason, a modern book on sensory ecology should cover the way that sensory system work, the principles and areas of animal behaviour, through to evolutionary processes. Clearly, to cover all of this in depth would require a massive book, therefore this text should be seen as a broad introduction. The principal aim is not to present an updated version of Dusenbery (1992), but to introduce the subject of sensory ecology, including how sensory systems work and sensory processing, from a more behavioural and evolutionary framework (reflecting how the field has evolved in the last two decades).

I also did not wish to write a book that dealt with each sensory modality one by one in isolation. There are various excellent books devoted to specific sensory modalities and that go into far more detail than I do here. On the contrary, I wanted to produce a book that dealt with the role of sensory systems in a modern behavioural framework, including what sensory systems are used for and how they guide the many behavioural tasks animals must complete during their lives. I also wanted to give the book an evolutionary context, showing how important sensory systems are in processes like diversification and speciation. The various sensory modalities in the book are discussed together, and the aim is to highlight common principles (and differences) across sensory modalities, and what they mean in terms of the way that animals behave and process information. There is much that I have omitted regarding sensory physiology and processing, but my aim was not to present a book focusing only on how sensory systems work. Instead, I have sought to present enough information for the reader to understand important concepts so that the more behavioural and evolutionary issues and discussions could be understood from this basis. Some sensory modalities are presented and covered in more detail, whereas others arise less often or with respect to only some subject areas. This partly reflects the state of play in the subject, with some modalities being worked on from a more behavioural/evolutionary perspective, but also no doubt limitations in my expertise. It is not a suggestion that some sensory systems are more or less important than others, because, as we shall see, the value of any sensory modality depends on the species, environment, and specific behaviour. That said, I acknowledge that this book necessarily reflects some bias in the literature towards studying sensory modalities that are particularly important to humans (especially vision) and consequent relative understudy of some ecologically important modalities (e.g. electroreception and magnetoreception) that we lack.

Generally, this book is more focused on recent work and discoveries, rather than presenting an historical overview of the various topics introduced. The intention here is certainly not to do disservice to older work, which is of course fundamental for new work to build upon. Rather, any bias towards more recent studies is to help illustrate where the field currently stands, to show the diverse and exciting work that is being done today, and to hint at where the important next steps might be. As stated above, this book is intended to be an introduction to sensory ecology. There were various studies that I could not include, and some subjects that are discussed only briefly. For those studies that have been left out, it is not a reflection on their quality, but rather an inevitable outcome when trying to summarize such an diverse and vibrant field. Sometimes I have used well-studied systems or species in several chapters. This is partly intended to help with continuity across the chapters and to highlight the depth of understanding to which researchers have gained from some species and systems. Finally, this book does not explicitly discuss cognition and 'higher' level processes. Clearly these are important in decision making too, yet this goes beyond the sensory scope of this text.

The book is intended to present stand-alone chapters, while also providing a progression from one chapter to the next. It begins with key concepts such as information, and then moves onto the various sensory systems, sensory physiology, and nerve cells, through to important areas in behaviour and communication, and finally to fundamental issues in evolution, such as arms races and divergence. The first chapter discusses how the sensory information available to animals varies greatly across species owing to the sensory systems they have. It introduces the idea that sensory ecology deals with how animals gather and use information from their environment, and that the ecology of a species shapes the form and function of its sensory systems to best acquire and process information and to use it in behaviour.

The following three chapters deal with how sensory systems work. Chapter 2 introduces the different types of sensory system that have evolved to capture information from sound, light, electric, magnetic, mechanical, and chemical stimuli. It describes the overall features of these different sensory modalities, including the sensory receptors involved and what tasks the sensory system is used for. The chapter also describes how some animals have evolved an active sense with an ability to emit signals or calls into the environment and measure changes in the properties of returning reflections to assess the world around them. Chapter 3 focuses on some specific examples in more detail, mainly from vision, hearing, and olfaction to demonstrate the neural mechanisms that exist in sensory processing to encode important information. It covers key principles demonstrating how sensory systems remove redundant information and encode change in the environment, including filtering, feature detectors, receptive fields, adaptation, and parallel processing, and how these processes are made possible by the flexibility of connections between nerve cells and circuits. Chapter 4 discusses how sensory systems face energetic costs, as well as trade-offs in the way that they work to encode different types or features of stimuli. The chapter deals with if and when sensory systems are optimized for one task, as opposed to having features that have evolved to work in many. The second half of the chapter describes how sensory systems do not work in isolation, but rather that animals often integrate information from several modalities and that this can decrease uncertainty about a stimulus or environment.

Chapters 5–8 cover important areas regarding communication in animals. Chapter 5 introduces fundamental concepts relating to signals, cues, and communication in general. It defines what signals and communication are (and the controversy that still surrounds this), and how signals differ from cues. The chapter also argues that to understand the diversity of signals and communication systems in nature fully, we need to carefully consider how signals have evolved in response to the sensory system of the receiver. Chapter 6 introduces the idea of multimodal signals, which involve the use of signal components across two or more sensory modalities. The chapter discusses a framework for understanding multimodal signals and what constitutes one. It covers explanations for the existence and benefits of multimodal signals, including facilitating effective signalling under noisy or variable environments, and allowing receivers to extract different information ('messages') from a signal. Chapter 7 begins by discussing how signalling and communication carry significant costs, not just in terms of energetics, but also through how signals are exploited by eavesdroppers, such as predators, prey, and competitors. The chapter then describes how animals might reduce risks of eavesdropping, including evolving less conspicuous or localizable signals, changes in behaviour, or signalling in a modality that the eavesdropper cannot detect or to which it is insensitive. Finally, the chapter discusses when and how organisms combine multiple functions with the same signal form to perform multiple tasks. The final chapter in this section, Chapter 8, discusses how many signals in nature are dishonest, including various forms of mimicry. This can involve organisms exploiting the communication systems of other species in order to deceive an animal into mistaking it for something else, or exploiting pre-existing biases in the sensory systems of animals to manipulate the receiver's response. The chapter discusses how sensory exploration and mimicry is widely found in mating signals and communication

systems, and how exploitation and sensory biases can lead to exaggeration in signal form.

The final three chapters discuss the role of sensory systems in key evolutionary processes, and how the environment shapes sensory systems and communication. Interactions between species or groups of organisms, both cooperative and antagonistic can be powerful generators of biological diversity. Chapter 9 focuses on two such driving forces: arms races and coevolution. It discusses how predator–prey relationships provide clear examples of arms races, but also how there is little evidence for genuine coevolutionary responses in the sensory systems of the predators to overcome prey defences. In contrast, coevolution seems widespread and diverse in brood and social parasites in birds and insects, and this has lead to extraordinary defences and counter-adaptations in both parasite and host across a range of modalities. Chapter 10 deals with how the environment affects the way that sensory systems work and how organisms interact and communicate. It outlines how organisms use different signal forms in different environments, switch between sensory modalities, or tune their sensory systems to cope best with features of their habitat. The chapter then discusses how different habitats and environments influence the way that information is acquired, how signals are transmitted, and how environmental noise (including those from anthropogenic sources) can interfere with obtaining relevant information. The final chapter (Chapter 11) discusses how sensory and communication systems are involved and are crucial in leading to divergence between populations and speciation. The chapter outlines how we are starting to understand more fully both the molecular mechanisms underlying some of these changes, as well as the selective advantage incurred, and how certain mechanisms can lead to divergence and speciation, including reproductive character displacement, disruptive selection, so-called magic traits, predation pressure, and mate selection, and how these factors operate in different groups and sensory modalities. Finally, the chapter discusses how the environment may cause divergence and speciation in some groups through the process of sensory drive.

Acknowledgements

I owe a huge thanks to many people who have helped at various stages of this book. First, I am hugely grateful to Ian Sherman, Helen Eaton, Lucy Nash, and everyone else at Oxford University Press. Their advice and understanding helped to make this project as enjoyable as it was.

In addition, many people read sections or chapters from this book, were enormously helpful in answering queries or discussing various issues, and gave me valuable suggestions and comments for improvement. Any remaining errors or shortcoming in the book are purely of my own making! Many thanks to: Lina Arenas, Matthew Arnegard, Eleanor Caves, Nick Davies, Esteban Fernandez-Juricic, Tom Flower, Thanh-Lan Gluckman, Holger Goerlitz, James Gould, James Hare, James Higham, Anna Hughes, David Hunt, Sönke Johnsen, Gareth Jones, Simon Laughlin, Martine Maan, Kate Marshall, Amanda Melin, Richard Merrill, Nicholas Roberts, Candy Rowe, Hannah Rowland, Mary Caswell (Cassie) Stoddard, Philip Stoddard, Rose Thorogood, David Tolhurst, Alexandra Török, Jolyon Troscianko, Roswitha Wiltschko, and Tristram Wyatt. I am especially grateful to Graeme Ruxton for reading every chapter of the book (!) and for a range of discussions, especially relating to tricky terminology and definitions.

I am also very grateful to many people who enthusiastically donated images and figures for me to use: Laura Crothers, Christine Dreyer, Damian Elias, Nigel Franks, Madeline Girard, Roger Hanlon, James Hare, Marie Herberstein, Roger Le Guen, Daiqin Li, Martine Maan, Ryo Nakano, Sarah Partan, Erica Bree Rosenblum, Mike Ryan, Leslie Saul-Gershenz, Ole Seehausen, Ulrike Siebeck, Claire Spottiswoode, Devi Stuart-Fox, Keita Tanaka, Marc Théry, I-Min Tso, and Kenneth Yeargan.

I also thank the BBSRC for allowing me the time and freedom to write such a book, and Kate Marshall and Jolyon Troscianko for help with the reference list. Finally, I thank my wife Audrey, for being so encouraging and supportive during the whole process and putting up with me working too hard!

PART 1

Introduction

CHAPTER 1

Sensory Ecology, Information, and Decision-Making

Box 1.1 Key Terms and Definitions

Information: Acquired by organisms from the general environment or from other individuals and reduces uncertainty about a feature or state of the environment, individual, or future events.

Personal Information: Information that individuals gather themselves by directly interacting with the environment.

Proximate Explanation: This is the causal explanation or underlying mechanism explaining *how* something happens.

Socially Acquired Information: Information gathered from other organisms of the same or a different species.

Ultimate Explanation: This is the function or evolutionary advantage explaining *why* something happens.

The world around us is full of sights, sounds, smells, and textures that stimulate our various sensory systems and enable us to interact appropriately with the environment. Our sensory systems are crucial to survival, and this was even more the case when we lived in the wild without modern aids and comforts. We needed them to find food, communicate, navigate, detect predators, and much more. In the modern world, the information our sensory systems provide is every bit as important. To other animals, sensory systems also make the difference between life and death on a daily basis. Sensory systems are the products of hundreds of millions of years of evolution and, given their importance, are breathtakingly diverse across and sometimes even within species. They are often closely linked to the ecology of a species, and have been shaped by an array of selection pressures to enable an animal to perform numerous behaviours. In short, sensory systems are fundamental to survival and reproduction and shape much of evolution and behaviour.

1.1 What is Sensory Ecology?

In the conclusion of *The Origin of Species*, Darwin (1859) presented the description of an entangled bank, with plants of many kinds, birds singing in the trees, insects going about their business, and worms moving through the ground below. In doing so, he illustrated the importance of the environment in the evolution of different types of species, the niches they live in, and the complex interactions between organisms. It is often helpful to think about species in terms of the niches they inhabit, based on a physical separation in space or time. Yet, even organisms living in the same place, at the same time, inhabit different worlds: they live in different sensory environments, bounded by the properties of their sensory organs. For example, a bee using colour vision to search for flowers may be right next to a snake waiting to detect the infrared cues of its prey or an ant following chemical pheromone trails to food. Animals should, and do, only pay attention

Sensory Ecology, Behaviour, and Evolution. First Edition. Martin Stevens.
© Martin Stevens 2013. Published 2013 by Oxford University Press.

to features of the habitat that are important to them. What they can detect and respond to is dictated by their sensory systems and how they are constructed over evolution. Animals can be almost touching each other in space, but be worlds apart in perceptual terms.

Unsurprisingly, animals have evolved a staggering array of sensory organs that are fundamental to survival and reproduction, and shape a great deal of evolution and behaviour. Sensory ecology deals with how animals acquire, process, and use information in their lives, and more recently the role of sensory systems in evolutionary change. Defining almost any subject in biology is difficult because the boundaries between disciplines are frequently blurred. This is certainly the case for sensory ecology, which spans a broad range of ideas and subject areas. On a basic level, it covers everything from the way that sensory systems work (physiology and neurobiology), to the way sensory systems are used (e.g. in behaviour), to the role of sensory systems in evolutionary processes (for example, reproductive isolation).

In many respects, sensory ecology is not a new subject; the classic works on visual ecology by Hailman (1977) and Lythgoe (1979) constitute sensory ecology in the modern sense. However, the subject was perhaps first formally presented as a whole by Dusenbery (1992). At that time, sensory ecology essentially comprised the integration of more 'conventional' ecology with sensory neurophysiology (Bowdan and Wyse 1996). Dusenbery's book focused on the idea that information existed in the environment and that animals needed to obtain and use this information. The book, although a landmark, largely focused on the physics of such potential information, the physiology of the sensory systems, and a few selected ethological examples, such as navigation. Dusenbery (1992) dealt only briefly with many of the evolutionary and ecological implications of these ideas. In the last decade, sensory ecology has come of age. Partly thanks to the rise and continued popularity of subjects like behavioural and evolutionary ecology, sensory ecology now deals with all that Dusenbery discussed, but also spans into behavioural ecology and evolutionary mechanisms. The central issues in sensory ecology are: *i*) how do animals gather and use information from their environment and from other organisms, *ii*) what role

does the ecology of a species have in shaping the form and function of sensory systems to best acquire and process information, and *iii*) how does this influence behaviour and evolution? We will return to the idea of information shortly, because it is an essential (albeit sometimes controversial) concept central to much of sensory ecology. First, however, we discuss some examples of how the information that is available to animals depends on their sensory systems.

1.2 Many Animals Detect and Use Sensory Information Humans Cannot Perceive

Animals often do not perceive the world in the same way that humans do, and we need to be aware of this in studying sensory and behavioural ecology. First, there are entire sensory modalities that humans lack. For example, various animals have a magnetic sense, which they use to navigate over both relatively shorter and longer distances. Likewise, many fish (and some mammals and amphibians) have an electric sense. This can be both passive, involving detecting electric information from the environment (e.g. prey) or active, where the fish emits electricity to the environment and detects the changes in the returning signal. Electric senses are used in many ways, including detecting food, navigation, object detection, and aggressive and courtship interactions. We will discuss how the different senses work and their uses in Chapters 2 and 3.

Equally importantly, even in sensory modalities that humans have, we can only detect certain proportions of the information available to other animals. Here, we focus on examples from two important senses in humans, hearing and vision.

1.2.1 Ultrasonic and Infrasonic Communication

Humans can hear sounds approximately in the region of 15–20,000 Hz (although the ability to hear high frequencies often declines with age). In contrast, many animals can hear sounds at higher frequencies than we can (ultrasonic sounds), and some animals can hear lower frequencies (infrasonic).

Wild house mice (*Mus musculus musculus*) have complex ultrasonic vocalizations (USVs) with frequencies around 30 kHz. These are used by pups to solicit maternal care, but also by males towards females during courtship. Structurally, these vocalizations are similar to songs, such as used by birds, with repeated phrases. Using specialist sound detection equipment, Musolf *et al.* (2010) found that male mice would produce USVs when presented with females or olfactory cues of females, but not towards other males. They then played back sounds of male calls to females and looked at the time females spent associating with different speakers. Females preferred the USVs of males compared to noise, and also preferred the calls of unfamiliar (non-kin) males to those of related males. As yet, it is not clear what aspects of the signal females are choosing and exactly what benefits they may get from this (e.g. male quality or genetic compatibility). Recent work in rats indicates that there are several types of USVs produced, which may play different roles in communication, including in aggressive, feeding, and courtship encounters (Takahashi *et al.* 2010). Beyond rats and mice, ultrasonic communication has also been shown in the calls of some frogs (Feng *et al.* 2006), ground squirrels (Wilson and Hare 2004), and a range of insects, bats, and cetaceans.

Some mammals, such as elephants and certain species of whale, produce infrasonic sounds. African elephants (*Loxodonta africana*) produce low frequency calls with a considerable component below the human hearing range. Playback experiments in the field by Longbauer *et al.* (1991) indicate that elephants respond to these calls even over a distance of several kilometres. Although there is some debate, because low frequency sounds travel further more effectively than higher frequencies, which attenuate quickly, infrasonic calls may allow elephants to communicate and maintain contact between individuals over long distances and for individuals to re-establish contact with a herd. The low frequency vocalizations also have seismic components that are transmitted through the substrate. Field experiments with playbacks of alarm calls to wild elephants indicates that they respond behaviourally to seismic components, including by clumping together more in herds and orientating perpendicular to the playback source (O'Connell-Rodwell *et al.* 2006).

1.2.2 Ultraviolet and Fluorescent Signals in Mate Choice

In humans, colour vision is based upon the relative stimulation of and interactions between three cone types, each tuned to relatively shorter (SW), medium (MW), and longer (LW) wavelengths of light. Although we are highly visual animals, many other species have different, and potentially better, colour vision than us. Birds, for example, have colour vision involving the use of four cone types, including types of SW, MW, and LW cones, but also a class sensitive to ultraviolet light (UV cones). Birds can therefore probably see and discriminate between colours unavailable to us. In fact, the ability to detect UV light is widespread in animals, in both invertebrates and vertebrates.

Some animals, like parrots, have signals that absorb UV light and re-emit it at longer wavelengths (fluorescence). These colour patches are often found adjacent to plumage that reflects UV light strongly. In an experiment with budgerigars (*Melopsittacus undulatus*), Pearn *et al.* (2001) used coloured filters to selectively remove certain wavelengths of light from the ambient spectrum, allowing them to abolish either the UV reflectance of the feathers, and/or the fluorescence of the adjacent patches. They found that fluorescence was used in mate choice, but only when accompanied by the UV reflecting patches. A later experiment by Arnold *et al.* (2002), also with budgies, applied sunblock only to the fluorescent

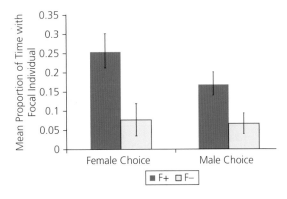

Figure 1.1 When presented with potential mates lacking fluorescence (F−) both male and female budgies spent less time associating with those individuals than individuals with fluorescence (F+). Data from Arnold *et al.* (2002). Error bars are standard errors.

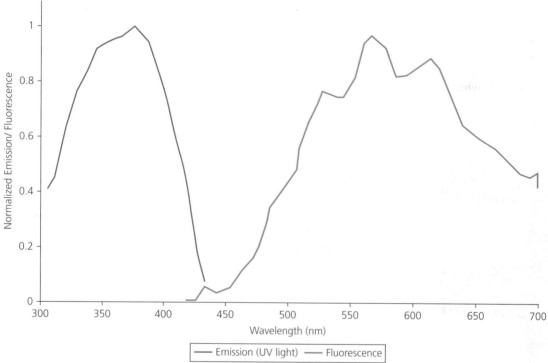

Figure 1.2 An image of a budgerigar under human visible light (left) and under ultraviolet light (right), with an inset showing the region of plumage where there is high ultraviolet reflectance (white area), and low ultraviolet reflectance where ultraviolet wavelengths are absorbed and fluoresced at longer wavelengths (dark area). The plot below shows the emission spectrum of an ultraviolet light shone on a budgie (blue line) and the resulting fluorescence spectrum from the budgie's plumage (green line). Data from Arnold *et al*. (2002).

patches on some individuals. Sunblock absorbs UV, and so also eliminates the light that would normally be used in fluorescence. They found that fluorescence was important in mate choice: both males and females preferred potential mates when they had fluorescence (Figure 1.1).

These two studies are revealing, and although ideally we would want to reduce or enhance the level of fluorescence/UV without eliminating it entirely, and then observe the effects on mate choice, the work indicates that fluorescence alone is not all that matters; rather, it is the dual effect of having adjacent UV-reflecting and fluorescent (UV-absorbing) patches of colour next to each other on the body (Figure 1.2). These two colour patches should stimulate different sets receptors in the birds' eyes, with the UV patches stimulating the UV and SW cones and the fluorescent patches stimulating the MW and LW cones, thus creating strong differences in colour contrast.

Fluorescence is also found in female spiders (Figure 1.3), with ultraviolet coloration seemingly more common in males. Andrews *et al.* (2007) showed that fluorescence is found across many spider families and has probably evolved many times, owing to the presence of different haemolymph fluorophores (which absorb UV light and re-emit it at longer wavelengths) in different families. In the ornate jumping spider (*Cosmophasis umbratica*)

males possess ultraviolet reflecting patches, which juveniles and females lack (Lim and Li 2006). Conversely, females have fluorescent patches that males do not have.

A study by Lim *et al.* (Lim *et al.* 2007) in this species tested the role of UV and fluorescence in mate choice. They used colour filters to remove UV or fluorescent components from potential mates, and then observed the amount of courtship behaviour individuals displayed to either normal or UV/fluorescence-lacking individuals. They showed that males prefer females with fluorescence, and in turn females prefer males with UV (Figure 1.4; see also Lim *et al.* 2008).

Almost all UV sensitive animals are thought to use predominantly UV-A light (315–400 nm). However, some spiders such as the jumping spider (*Phintella vittata*; Figure 1.3) from China, are even sexually dimorphic in the UV-B (280–315 nm) range, with peak reflectance of male coloration at 290 nm (Li *et al.* 2008b). Experiments presenting males to females with different wavebands of light shows females prefer males with UV-B reflectance, but that UV-A reflectance does not affect their decisions (Li *et al.* 2008c).

1.3 Asking Questions about Behaviour

To be able to study the way that sensory systems and information are used, we need to study the behaviour of animals. The field of behavioural ecology emerged from ethology, and is broadly concerned with understanding why animals behave the way they do, from an evolutionary perspective and while considering the ecology of a given species. Behavioural ecology has always had close links with evolutionary theory, with some of the most famous evolutionary theoreticians of modern times (e.g. Hamilton, Maynard-Smith, Trivers) being closely linked with areas and questions common to behavioural ecology. Although there remains some debate, most behavioural ecologists take the approach that the gene is the unit of selection (after Dawkins 1976) although selection commonly acts on the individual organism itself. For a variety of reasons, this is really the only way to understand the broad range of behaviours that animals display, including cases of apparent altruism and group

Figure 1.3 An image of a female *Phintella vittata* jumping spider from China under full spectrum light including ultraviolet wavelengths, showing how the palps of the female glow with fluorescence. Reproduced with permission from Daiqin Li.

Male Female

Figure 1.4 Images of male and female *Cosmophasis umbratica* jumping spiders under human visible light (top), ultraviolet light (middle), and with ultraviolet-induced fluorescence (bottom), showing the strong ultraviolet reflectance in males (but not females), and the strong fluorescence in females (but not males). Reproduced with permission from Matthew Lim.

living (which are really selfishness from a gene's or individual's perspective; Dawkins 1976). In short, individuals are under selection to pass on their genes to the next generation, either directly via their own offspring, or indirectly in the offspring of related individuals (see Davies *et al.* 2012 and Alcock 2009 for full discussion).

1.3.1 Mechanistic and Functional Explanations

The Nobel Prize-winning ethologist Niko Tinbergen specified four famous questions about animal behaviour:

1) What is the mechanism or causal factor underlying a particular behaviour? For example, what genes or hormones lead to a particular behaviour?
2) How does the behaviour develop? For example, is it 'innate' or learnt?
3) What is the function of the behaviour? That is, what is its evolutionary value?
4) What is the evolutionary history of the behaviour?

In reality, the distinction between these questions is not always clear and most modern approaches ask one of two types of question that encompass those above: (1) what is the proximate/causal basis of a behaviour (that is, what is the underlying mechanism), and (2) what is the functional advantage (that is, the evolutionary or ultimate explanation)?

An example illustrates the difference between mechanism and function. Many species of bird seasonally migrate (see Chapter 4). The reasons for this can be varied, but common reasons are to avoid harsh weather during the winter and to exploit different feeding grounds. These are ultimate explanations, describing the functional advantage of migration. Recent work has started to shed light on the molecular basis of migration in birds (its mechanistic basis). Mueller *et al.* (2011) have found evidence that the expression and level of migratory behaviour in blackcaps (*Sylvia atricapilla*) is partly controlled by the *ADCYAP1* gene. This gene encodes a specific polypeptide, which along with its receptors is widely distributed in the brain and elsewhere. It has a broad range of influences on physiological and behavioural functions and metabolism (Mueller *et al.* 2011). Likewise, communication (see Chapters 5–8) is of great importance in many organisms (for instance, to attract mates). Recent advances in research on some animals, such as *Drosophila* flies, have started to reveal the genes involved with reception of communication signals, such as male song, and the molecular changes in females associated with song reception (Immonen and Ritchie 2011). Work with swordtail fish (*Xiphophorus nigrensis*) has also identified genes in females that show different patterns of expression associated with sexual and social stimuli: some genes seem to be involved with interactions with other females whereas others are turned 'on' with interactions with attractive males (Cummings *et al.* 2008b).

Of course, specific genes, or interactions between sets of genes, alone often do not directly encode a specific behaviour. In most cases, interactions between the environment and genetic components lead to a specific phenotype. One of the best-studied examples is song type in birds. In many species the song that an individual produces is partly a result of its genes, but also linked to the songs that it hears (often from its father) during a specific sensitive phase when it is learning its own song. Experiments presenting birds with songs of other species during sensitive phases show that the resulting song is often a mixture of genotypically determined components and learnt components from the tutor.

In many cases, behavioural (or other) traits can reflect phenotypic plasticity. Here, the same genes produce different behaviours depending on features of the environment (for instance due to different expression and regulation). For example, exposure to chemical cues from different types of predator (fish or beetle larvae) causes pacific treefrog (*Pseudacris regilla*) tadpoles to develop different morphologies and foraging behaviour that are better suited to avoid these two different predator types (Benard 2006). Overall, the traits that organisms have, including for behaviour, result from a complex set of interactions between genes, gene expression, the environment, and other aspects of physiology during their lifetime, such as hormones.

1.3.2 Pitfalls and Considerations when Studying Behaviour

When studying behaviour, we need to be careful of several potential traps and pitfalls. One common case is anthropomorphism: giving non-human animals human emotions. For example, saying that an animal 'wants' to have as many offspring as possible is incorrect. Organisms undergo selection over evolutionary time, and this results in specific adaptive behaviours evolving and being maintained that probably have no conscious thought or intent involved. Likewise, it is common, in both popular texts and in scientific manuscripts to talk of animals 'choosing' to behave in a particular way, as if an individual has made a conscious decision. In many cases this is just shorthand for describing cases where over evolution animals that behave in a particular way should be at an advantage. Another common pitfall is teleology. Saying something like 'evolution *aims* to produce individuals that maximize the number of offspring they produce' is wrong. Evolution does not have foresight and does not plan ahead.

Behavioural ecologists often start from the basis that animals should, over evolution, behave optimally. A long history of so-called economic or optimality models of behaviour has aimed to test if animals are optimizing a given behaviour (e.g. foraging), given certain parameters and assumptions. Although it is remarkable how effective very simple models can often be, it is clear that animals often do not often meet the predictions of optimality models. This may simply be because the models themselves are wrong (e.g. they are oversimplified or have incorrect assumptions), but could also reflect a genuine lack of optimization. For example, dunnocks (*Prunella modularis*) are sometimes parasitized by the common cuckoo (*Cuculus canorus*), which lays its eggs in the nests of other species, so that the host parents rear the young instead (see Chapter 9). This is clearly very costly for the hosts: not only do they fail to produce any of their own young (the cuckoo chick ejects them from the nest), but they also rear a chick to which they bear no relation. Many other host species have, over evolution, gained the ability to detect and throw out a cuckoo egg. However, dunnocks do not do this. They simply accept the cuckoo egg even though it looks nothing like their own. Why they do this is not known, but the most common explanation is that of evolutionary lag. The dunnock may be a relatively new host that has not been under selection for enough time for rejection to have evolved. Other reasons why animals may not behave optimally include constraints (e.g. genetic, energetic, morphological, and so on) which prevent a behaviour or trait arising. Finally, animals may be trading-off one behaviour with another if they are detrimental to each other.

1.3.2.1 Beware of 'Just-so' Stories

In a seminal paper, Gould and Lewontin (1979) attacked the so-called 'adaptationist programme', which focuses on how organisms evolve traits (adaptations) that improve their ability to survive and reproduce. Part of their frustration was with the gene-centred view of evolution espoused by Dawkins (1976) and others, but one of their key arguments was that biologists were all too ready to assume that traits were adaptive without considering the alternatives. Gould and Lewontin argued that traits can arise for a range of other reasons, including through random drift, by-products of other selection processes, or that they may have originally evolved in some other function and since have been 'co-opted' for another role (Gould and Vrba 1982). Gould and Lewontin's criticisms were overstated and organisms clearly do have a vast array of adaptations. The gene-centred view is also crucial to understanding evolution in modern biology. However, the fundamental lesson that they presented is important: we must not assume that individuals are perfectly adapted and that every trait we observe is an adaptation for something. Related to this, we must not fall into the trap of creating 'just-so stories'. These are cases when biologists have made up a story that sounds correct to explain a trait or behaviour and then assume it is true without testing the idea properly. We shall consider two examples.

The American artist Abbott Thayer (1909) was one of the pioneers of both natural and military camouflage, and presented many important concepts that have been recently tested and supported regarding the way that camouflage works (Chapter 9). However, he thought that *all* colour patterns on animals were involved in camouflage. The most famous example of this is his painting of flamingoes and spoonbills, which he argued were pink to be camouflaged against the sky at sunset (Figure 1.5)!

Despite Thayer's determination that this was true, to see the fallacy in his argument he had only to ask 'what about during the rest of the day?' and to consider that the birds would more likely appear as dark silhouettes against the sky at this time rather than blend in. Not surprisingly, Thayer's suggestion was strongly criticized by Gould (1991), and even by the ex-US president, Theodore Roosevelt (Nemerov 1997).

Another recent example concerns conspicuous 'eyespots', which are complex circular markings found on many insects and fish (Figure 1.6). These have for over 100 years been assumed by many to mimic the eyes of larger animals, so that a predator would abort an attack for fear of being eaten itself (reviewed by Stevens 2005). The idea has long been a textbook and popular example of defensive coloration, and experimental evidence shows that wingspots on butterflies and moths do halt attacks from birds (Vallin *et al.* 2005, 2007).

Figure 1.5 The artist Abbott Thayer famously asserted that some birds are pink in order to be camouflaged at sunset, despite the obvious flaws in the theory and lack of evidence. From Thayer (1909).

Figure 1.6 'Eyespots' on a range of butterflies and moths like those of the owl butterfly (*Caligo* spp.) are thought by many to mimic vertebrate eyes to scare away predators.

However, such experiments do not test why the markings work. Recent studies with artificial 'moth' prey presented to wild bird predators in the field, marked with spots of different sizes, numbers, and shapes, show that they can be highly effective without mimicking eyes (Stevens *et al.* 2007, 2008a). Spots were more effective in scaring birds when they were larger, of higher number, and greater visual contrast (Figure 1.7). In addition, birds avoided prey with bars and squares as much as prey with circular markings, and more eye-like spots, with a dark 'pupil' and white surrounding 'iris', were no more effective than spots with the inverse arrangement. In short, there is currently no experimental evidence that spots mimic eyes.

Despite the above, many authors still uncritically assume spots mimic eyes. Subjective statements like 'Eyespots on the wings of giant silk moths and other Lepidoptera undoubtedly mimic eyes of mamma-

Figure 1.7 Example stimuli used by Stevens *et al*. (2007, 2008a) to test the idea that eyespots on butterflies mimic the eyes of vertebrates. Stimuli are paper triangles printed with specific markings and pinned to a tree with a dead mealworm attached as an edible component for birds to monitor predation. Stimuli with bars work equally well in preventing predation as circles, and stimuli with circles with a black centre and white surround ('pupil' and 'iris') survive no better than those with the inverse arrangement. Spots of higher contrast enhance survival more than those of low contrast.

lian predators' (Rota and Wagner 2006), and 'An eyed hawk moth presents a remarkable similarity to the fox' (Williams 2010), are still commonplace without justification. The idea that spots mimic eyes is a nice example of how a reasonable, but untested idea can become almost widely accepted over a long period of time without supporting evidence. It does not mean that circular markings never mimic eyes, and evidence may yet show that some do, but we must design appropriate experiments to test our theories and hypotheses before accepting them as true.

1.4 Information

The concept of information is central to much of sensory ecology and has been mentioned various times in this chapter already, but what is 'information'? In terms of behaviour, the natural world has many uncertainties, and behaving in a way that is relevant to what may happen in the future or towards the current/future state of another animal is important. Animals must, therefore, gather information from both the biotic and abiotic environment in order to behave appropriately (Dall *et al.* 2005). Many concepts of information follow the work of Shannon and Weaver (1949) on information in electronic systems, whereby if an event allows a prediction to be made about potential outcomes then it can be considered as carrying information (Laughlin 2011). The key concept here is that information acquisition results in a reduction in uncertainty about something in the world. Studies of animal behaviour have generally been more ambiguous in the way that information has been treated, and the term information is often used informally (Dall *et al.* 2005). A common problem is that while the concept of information is intuitive, it is hard to define what information actually is and how to measure it. On a physiological level, we could define information in terms of, for example, changes in action potential frequency in nerve cells. In many cases, however, this is not possible and we still need to know the initial state of the neurons. One measure of information is the amount of uncertainty that has been removed (Bowdan and Wyse 1996), but the key problem is quantifying what the reduction in uncertainty has been. To do so, we need to know both the initial state of the system and find a way of

quantifying how it has changed. Furthermore, behaviour also depends not just on the amount of information but also the content of that information (Bowdan and Wyse 1996). All of the above means that quantifying information is difficult in many situations, and while the idea of information is conceptually useful it is empirically challenging to measure. We will deal more fully with some of the problems and debates that surround concepts of information with respect to communication in Chapter 5.

Dall *et al.* (2005) argue that two types of information exist. They specify *personal information*, as coming from the direct interactions of an individual with the environment (e.g. during navigation or orientation), and *socially acquired information* that is obtained by observing the behaviour of other animals (such as during foraging or reproduction). Socially acquired information can arise from incidental sources by observation ('cues') or direct communication involving signals, transmitted from either conspecifics or heterospecifics (see Chapter 5).

1.4.1 Information and Decision Making

Here, we discuss two systems, which illustrate the ways that animals gather information and use this to make a decision.

1.4.1.1 Nest Choice in Ants

A range of work by Franks and colleagues has investigated how colonies of the ant *Temnothorax albipennis* (formally *Leptothorax albipennis*) assess the quality of potential nest sites, and make both individual and collective decisions about when to move site. *T. albipennis* colonies are very small (up to a few hundred workers, each about 2.5 mm in length) and in the field live in small crevices in rocks (Franks *et al.* 2002). These crevices are prone to flooding or breaking, forcing colonies to move to a new nest site; a process that is both time-consuming and potentially dangerous. In the laboratory, *T. albipennis* colonies can be housed in between two glass microscope slides and a cardboard interior, allowing the dimensions of the cavity to be manipulated (Figure 1.8).

In addition, the top slide can be removed, damaging the nest and forcing the colony to emigrate to a

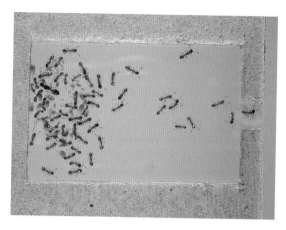

Figure 1.8 A nest of *Temnothorax albipennis* ants housed inside cardboard walls sandwiched between two microscope slides. Reproduced with permission from Nigel Franks.

new nest site. Various features of the emigration circumstances and the nest cavity can be modified and measured, and colonies can be given a choice of nests with different attributes to determine which they will choose (Figure 1.9).

When a decision is made by a scout that a new nest is suitable, they begin to recruit other individuals to it. One way they do this is by the process of 'tandem running', where the scout physically leads an individual to the new nest site (Franks *et al.* 2002; Mallon *et al.* 2001). This is relatively slow, and the following ant needs to remain in regular antennal contact with the abdomen of the leading ant for them to move forward. Ants that have been led can both learn the route and judge the quality of the nest for themselves. Should they find it suitable, they will then begin to recruit others to the nest. This can set up a positive feedback process, whereby if a nest is good then more and more individuals will be led to it. Then, once a certain number of ants are present, a 'quorum threshold' is reached, and the emigration moves more swiftly (Pratt *et al.* 2002). Here, individuals begin physically carrying other adults and brood from the old to the new nest in a process that is about three times more rapid than tandem running (Pratt *et al.* 2002). The quorum threshold is a way of collating the assessments of individual ants to a colony-wide decision. The slow process of tandem running and latencies to recruit other individuals to the site allows other nest sites to be found (Franks *et al.* 2002).

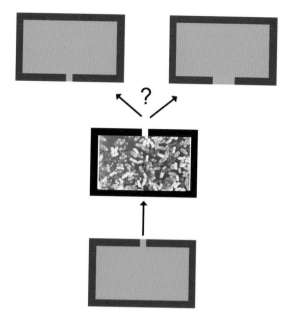

Figure 1.9 An illustration of the set-up used to test nest choice in *Temnothorax albipennis*. Colonies are housed inside microscope slides with an additional red plastic cover to simulate darkness. When the top slide is removed the nest is exposed and the ants have to find a new nest. In some experiments colonies can be given a choice between two potential nest sites with different attributes (in this case different sized entrances) to test how ants both individually and collectively make decisions. Inset image of a *T. albipennis* colony reproduced with permission from Nigel Franks.

Experiments have shown that *T. albipennis* colonies have preferences for specific nest attributes, which may indicate desirable characteristics of the site (Mallon *et al.* 2001). For example, colonies prefer dark cavities (which indicates few gaps in the walls), appropriate cavity dimensions (to house the colony), and relatively small entrances (to prevent access for predators) (Pratt *et al.* 2002). When given a choice between nests with different attributes, ants reliably choose the best one (Franks *et al.* 2003b).

1.4.1.1.1 How do Individuals Judge Nest Quality?

Ants need nests of suitable area in order to house the colony. Mallon and Franks (2000) found that when a scout finds a potential new nest, it lays down individual specific trail pheromones while exploring. Ants typically only begin recruitment after visiting a nest site at least twice, meaning that on subsequent visits they could use the rule of thumb that the less often an ant intersects its previous trail-pheromone path, the larger the cavity area (Mallon and Franks 2000; Mugford *et al.* 2001). Mallon and Franks (2000) presented colonies with a choice of nests, one being half the area of the other. After the scouts had found the nests they removed an acetate sheet that was covering half the floor area of the smaller nest, removing half of the scout's individual pheromone trails. When they did this, about half the colonies now chose the smaller nest, which presumably was judged to be the same size as the larger nest. For ants to use a rule of thumb based on path-intersection frequency, they should keep the length of their initial path constant, regardless of nest characteristics, which they do indeed seem to do (Mugford *et al.* 2001).

1.4.1.1.2 How do Individual Assessments of Nest Quality Translate into a Colony Decision?

Early work showed that nest quality relates inversely proportionately to the time taken for scouts to begin recruiting other workers (Mallon *et al.* 2001; Pratt *et al.* 2002). It was thought for some time that individuals would delay recruiting to nests that were deemed inferior and that this was a key mechanism that would allow other workers to discover better nests and begin recruitment to them faster (e.g. Franks *et al.* 2002). This would allow a better nest to be discovered and chosen. However, recent modelling work indicates that individuals may simply have an internal 'template' of what makes a good nest and simply decide to accept or reject a nest depending on whether it exceeds or falls short of this threshold (Robinson *et al.* 2011). If a nest does not match or exceed the template then the individual will keep searching. This is more parsimonious as it does not require ants to compare nests directly, and longer recruitment times to inferior nests emerge simply as a by-product of scouts rejecting poor nests and continuing to search. The model also seems to fit well with experimental data (Robinson *et al.* 2011). As we will see below, individuals also seem to weigh up current circumstances and sometimes select less desirable nests more rapidly, such as under harsh environmental conditions, and so an issue to resolve is whether and how individual scouts have some flexibility about what their template or threshold actually is.

1.4.1.1.3 How do Environmental Factors Influence Decision-making?

So far, we have discussed experiments where ant colonies show a remarkable ability to accurately select between different nests under relatively benign conditions. However, animals often have to modify behaviour based on information about the external environment. In decision-making, a fundamental trade-off is speed versus accuracy. Animals should make both quick decisions and accurate ones, but rapid decisions often sacrifice accuracy by losing chances to acquire information. This trade-off is illustrated in *T. albipennis*. Franks *et al.* (2003a) forced colonies to emigrate under harsh conditions (either when wind was blown over the destroyed nest, or in the presence of dilute formic acid, which is a cue to the presence of local predatory ants), or benign conditions (no wind or just water respectively). They found that ants began to emigrate (start carrying brood and other individuals) at a lower quorum threshold, used fewer tandem runs, and began moving more quickly under harsh than benign conditions. In some cases under harsh conditions, ants would begin emigrating with a quorum threshold of just one ant. In effect, the ants made a quicker decision and placed more emphasis on individual choice. Franks *et al.* also presented colonies with a choice between one nest type that would normally be chosen reliably, and another nest type that would rarely be accepted under benign conditions (having a smaller cavity height). Under harsh conditions, even though colonies eventually moved towards the better nest, ants sometimes began carrying other individuals and brood to the inferior nest, showing that when they move quickly they can make errors of judgement.

Making a quick decision may seem like a good idea but it creates another problem. In lowering the quorum threshold and beginning carrying earlier, fewer individuals will have discovered the nest and learnt the route. This means fewer individuals to help with the emigration, and somewhat paradoxically could result in a slower emigration (and possibly an error-prone one too). How can this conundrum be solved? The answer seems to lie in the existence of so-called reverse tandem runs (RTRs), where an individual leads another ant from the new back to the old nest (Pratt *et al.* 2002). For a long time the role of these was unknown, but recent work, both mathematical modelling and experiments, has shown that RTRs could speed up emigrations with low quorum thresholds by increasing the number of recruiters to the emigration process (Franks *et al.* 2009; Planqué *et al.* 2007).

Despite the possibility of moving to inferior nests when the emphasis is on speed, colonies can also correct errors later. When colonies housed in intact nests are offered a better alternative, then the colony will sometimes move to the new site (Dornhaus *et al.* 2004). Interestingly, such emigrations only happen when a much higher quorum threshold is reached than those when the old nest is broken (*ca.* 17 ants compared to 8), and the emigration can take days rather than hours. Thus, ants only move to a new nest when there is high consensus about its quality and with an accurate decision; that is, they take their time to gather lots of information.

Finally, in the above experiments ants were presented with nests that they previously had not experienced. However, colonies continuously monitor the environment to determine the presence of other nests. When colonies were allowed to explore an environment containing a good nest site one week before being forced to emigrate, they accepted it and moved more quickly than colonies without this experience (Stroeymeyt *et al.* 2010). Furthermore, when allowed to experience both a good and inferior nest, colonies with prior information emigrated faster and made more accurate decisions than naïve colonies. Therefore, colonies can acquire and store information for future use, and potentially overcome some of the limitations of the speed–accuracy trade-off in doing so (Stroeymeyt *et al.* 2010).

1.4.1.2 Nest Choice and Social Information in Flycatchers

Information can often be acquired from other species. This is well illustrated in studies of nest site choice in northern European birds. Here, flycatchers (*Ficedula albicollis* and *F. hypoleuca*) are migratory and arrive at breeding sites in northern Europe (Sweden and Finland) in the summer. There is a limitation on the availability of suitable nest sites (they require cavities) and good territories, and individuals that arrive earlier in the season tend to

have higher breeding success. Seppänen and Forsman (2007) have conducted experiments showing that later arrivals use resident birds such as great tits and blue tits (chickadees; *Parus major* and *Cyanistes caeruleus*) as a source of social information about the best nest sites and acquire novel arbitrary preferences from other species. Before the arrival of flycatchers, Seppänen and Forsman placed either a circle or a triangle on the nest boxes of the tits, and another empty nest box a few metres away with the other symbol type, giving the impression that all the breeding pairs in an area had specifically chosen nest sites with a specific symbol type. Seppänen and Forsman then put up two more nest boxes when the first male flycatchers arrived, with the two different symbols, and determined which of these box types the arriving females preferred. They found that late-arriving females (but not early arriving females) of both species preferred the boxes with the same symbols to those found on the boxes of the breeding tits (Figure 1.10), even though there was nothing different about those boxes to ones with the other symbol. Such behaviour may be adaptive because the resident tits had longer to gather and process information from the environment than the flycatchers, especially those arriving late. Earlier-arriving birds may have more time and opportunity to gather personal information themselves and so do not need to rely on information from resident birds.

Interestingly, Seppänen *et al.* (2011) have subsequently shown in an experiment of similar design that pied flycatchers do not just copy the choices of tits broadly across individuals, but are more likely to copy the nest choice of tits that are successful at breeding compared to unsuccessful breeders (flycatchers have previously been shown to inspect the nests of resident tits). They found that the likelihood of flycatchers acquiring a preference for a given symbol type was strongly influenced by the number of offspring in the nest of the resident tit that they observed on the day the flycatchers made their nest choice (Figure 1.11). Not only that, but instead of choosing nests with different symbols at random (i.e. not copying) when the resident tits had poor fitness (few offspring), the flycatchers actually chose nests with the opposite symbol. This supports the assumption that animals should only

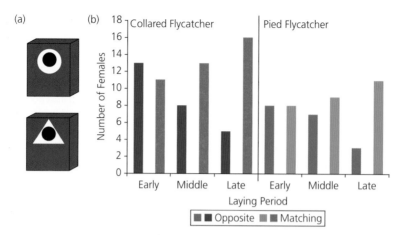

Figure 1.10 a) In the experiments of Seppänen and Forsman (2007), flycatchers observed local tits breeding in nest boxes marked with either a circle or a triangle. The flycatchers could then choose between empty nest boxes with either of these symbols and either copy the resident tits or not. b) The choices of arriving flycatchers in terms of whether they chose to copy the resident tits and select boxes with the same symbols or not. For both flycatcher species, later arriving breeders were more likely to select boxes with the same symbols as the resident tits, but this was not the case for early-arriving breeders. Data from Seppänen and Forsman (2007).

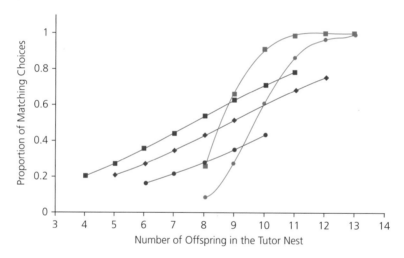

Figure 1.11 Proportion of flycatcher choices in selecting nest boxes with symbols matching those of the resident tits with regards to their breeding success (number of offspring in the nest). Blue and red lines are for the Finnish and Latvian study areas respectively. Flycatchers copied the symbols of resident ('tutor') pairs that were successful, but chose opposite symbols for their nest boxes when the tutors had fewer offspring. Data from Seppänen et al. (2011).

copy individuals that are worth copying when there is variation in the value of different potential individuals that can be observed, and that poor individuals should not be copied. As Seppänen et al. point out, the presence of both copying and active rejection should result in reduced uncertainty about the value of potential nesting sites.

Finally, the experiment by Seppänen et al. (2011) used natural variation in tit offspring number, and so could not completely discount a role of other causative factors that may themselves be correlated with offspring number. However, Forsman and Seppänen (2011) have since replicated the general findings when presenting flycatchers with tit nests containing either four eggs (low clutch size) or 13 eggs (high clutch size and fitness), showing their earlier work to be robust.

1.5 Future Directions

We will see throughout this book that many areas of sensory ecology are in their infancy and are awaiting concerted research programmes. There is still a great deal of confusion about the concept of information and the value of this idea in behavioural studies, and a firmer theoretical framework is needed, in particular regarding how researchers should actually measure and quantify information (and when this is even possible). Otherwise, some scientists and teachers are still guilty of too readily creating and accepting 'just-so stories', and a less subjective approach to understanding behaviour and morphology is important. Finally, embracing new technologies to study non-human perception will illuminate animal behaviour.

1.6 Summary

Sensory ecology is a large and rapidly growing subject that bridges the gap between neurophysiology of the senses, behavioural ecology, and evolutionary processes. Broadly, sensory ecology deals with how animals gather and use information from their environment, how the ecology of a species shapes the form and function of its sensory systems to best acquire and process information, and the impacts of this on behaviour and evolution. Animals acquire information from their environment, including from other organisms, and use this to behave appropriately. However, although information is generally considered to reduce uncertainty about some aspect of the environment or possible events, it is a concept that is hard to define or quantify precisely. Finally, much of the information available in the world to animals is not available to humans (e.g. electricity, magnetic cues, ultraviolet light, and so on), and we should remember that the way one species 'sees' the world could be very different from another.

1.7 Further Reading

Dusenbery's 1992 book remains a landmark text and recommended reading for those who want to get to grips with some of the more physical and mathematical concepts. Although dealing relatively little with concepts of adaptive behaviour and evolutionary processes, the book covers the physics and physiology of sensory ecology, and the framework of information in great detail. For a much fuller introduction to behavioural ecology, Davies *et al.* (2012) and Alcock (2009) are classical texts written with great clarity and authority. Readers should also consult Dall *et al.* (2005) for more discussion on the concept of information from an animal-behaviour perspective.

PART 2

Sensory Processing

CHAPTER 2

Sensing the World

Box 2.1 Key Terms and Definitions

Active Sense: Involves an individual emitting a call or signal (usually sound or electricity), followed by measuring changes in the returning signal to deduce information about the environment.

Electroreception: The ability to detect electric information in the environment, from other individuals, or changes in self-produced electric signals. Electroreception is used in orientation, hunting, and communication.

Magnetoreception: The ability to detect magnetic information in the environment, stemming from the earth's magnetic field, used in orientation and navigation.

Pheromones: Species-wide chemical signals, usually comprising a specific combination of molecules, which elicit innate responses in a receiver.

Signature Mixes: Variable subsets of molecules of an individual's chemical profile that are learnt by other animals.

Sensory systems are crucial to survival and reproduction and vary greatly across animals. The ecology of a species, the habitat it lives in, and the tasks it must perform in its life, dictate what types of sensory modality are most useful. Consequently, different species (and even individuals from the same species) often possess sensory systems that vary in sensitivity to different types of information. The aim of this chapter is to introduce the range of sensory systems used by different animals and their respective uses. Chapters 3 and 4 go into more detail about how sensory information is processed and encoded by nerve cells and circuits, the way that sensory systems are constructed in the face of costs and trade-offs, and how many tasks benefit from information in multiple modalities.

2.1 Signal Detection

Imagine a male moth that needs to detect the pheromones of a female ready to mate. His sensory system needs to detect and correctly identify the pheromone molecules (a signal) against a background of other odour molecules in the environment (noise). However, in many instances, there is not a clear distinction between signal and noise: the two possibilities overlap. A key job of sensory systems is to detect and discriminate a stimulus of importance or interest from background noise. Here, we use the term 'signal' to simply mean a stimulus to which the sensory system should respond, as opposed to the precise definition given in Chapter 5 associated with communication. By noise, we mean something that diminishes the ability of a sensory system to detect or correctly categorize the signal. This could be noise in the environment or 'spontaneous' neural activity that often exists in receptor and nerve cells. In signal detection theory, the response threshold of an individual (either in its sensory processing and/or its behavioural outcome) needs to be placed appropriately to maximize the likelihood of correctly detecting and responding to the signal. There are four potential outcomes (Figure 2.1): *a*) a 'hit', where the animal correctly

Sensory Ecology, Behaviour, and Evolution. First Edition. Martin Stevens.
© Martin Stevens 2013. Published 2013 by Oxford University Press.

detects the signal, *b*) correctly not responding to noise, *c*) a false alarm, where the animal incorrectly responds to noise as a signal, and *d*) a 'miss', where the animal fails to respond to a real signal.

Sensory systems need to be sensitive enough to respond to a weak signal but should not respond too often to similar but irrelevant noise in the environment. More stringent acceptance criteria mean more missed detections, whereas less stringent criteria mean more false alarms. The optimum solution is to set a response criterion that occurs between where the signal and noise overlap, while also accounting for the relative costs and benefits of the four potential outcomes above. For example, failing to identify a predator correctly is likely to be much more costly than sometimes falsely responding when no threat really exists. The above concepts also apply to discriminating similar stimuli (for example when one animal mimics another species or object). This ability is also reduced by the presence of noise (external or neuronal) because this adds uncertainty, especially if the difference between two stimuli is weak.

2.2 Chemical

The ability to detect and process chemical information is perhaps the most widespread sense in animals. There are different types of chemo-sensory

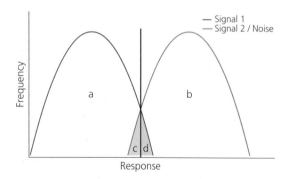

Figure 2.1 Signal detection theory. Animals often have to correctly identify a signal against either background noise or another similar signal with overlapping features. An individual sets a criterion or threshold (black line), either side of which it classifies something as either the signal or noise (or a different signal). The exact location of the threshold depends upon the relative costs and benefits of the potential outcomes (see text).

systems (that are not always separate), but here we focus on olfaction to illustrate key principles of chemical reception. We consider olfaction broadly as the detection of odours in the air or water occurring without physical contact to the object being detected (Eisthen 2002). Other authors consider olfaction as limited to glomerular processing in insects and vertebrates (see below).

Odours are discrete chemical structures, unlike the continuous attributes of wavelength or frequency associated with sound or light stimuli. In both water and air, currents and turbulence disrupt odours into discontinuous and patchy spatial distributions (Webster and Weissburg 2009). Animals therefore often have no direct 'line of sight' to chemical stimuli, and localization can be slower than for other senses. Correspondingly, insects such as male moths orientating towards female sex pheromones are well known for their characteristic zig-zag and wandering flights, as they move into and out of odour plumes. Birds and fish also seem to show these movement types when travelling towards an odour source (DeBose and Nevitt 2008). Animals orientate in the direction of odour plumes based on the direction of odour flow, such as flying upwind towards a source (see Vickers 2006; Willis 2008), and sometimes using bilateral differences in odour concentration (and timing: Gardiner and Atema 2010). Overall, for the detection of chemical signals, the specific compound is not the only consideration, but also the spatial and temporal patterns involved (Atema 1995).

2.2.1 Types and Use of Odours

Odour detection is crucial for many species in mating behaviour, individual recognition, orientation, and prey detection. For example, in insects olfactory receptor genes represent one of the largest families of genes found in this group and have undergone rapid evolutionary change (de Bruyne and Baker 2008). Chemical information greatly influences social behaviour. For example, in lobsters information from urine affects courtship, dominance, and individual recognition (Atema 1995). Odour information is also used over long distances, for example to guide migration towards natal spawning grounds (e.g. in fish).

When chemical signals are used in inter- and intraspecific encounters, we can define two types of stimuli: pheromones and signature mixes. Wyatt (2010) describes pheromones as species-wide chemical signals, usually a specific combination of molecules that elicit an innate response. Pheromones have been found in species from every animal phylum (Wyatt 2010, 2013), and elicit a specific reaction, such as mating behaviour (although this can be context-dependent). For example, all females of a species would possess the same sex pheromone. Although the receptors are often highly specific (see below), this is not a prerequisite for a substance to be a pheromone and some do not have specific receptors. Pheromones include sex pheromones released by females, but also trail and alarm pheromones in social insects and fish, for example.

In contrast to pheromones, signature mixes are variable subsets of molecules of an individual's chemical profile that are learnt by other animals, allowing them to distinguish individuals or colonies (Wyatt 2010). Signature mixes vary across individuals/colonies within a species. Cuticular hydrocarbon (CHC) profiles in social insect colonies, for example, are signature mixes. They can be produced by the organism, or acquired from the environment or other individuals. In ants and various other social insects, CHC blends comprising multiple compounds in different ratios, play an important role in nestmate recognition and aggression towards intruders (see Chapter 9). As with sex pheromone detection, ants can have receptors dedicated specifically to detect CHCs. For example, the carpenter ant (*Camponotus japonicus*) has a type of receptor located on its antennae that responds only to CHC blends of non-nestmates (Ozaki *et al.* 2005). Signature mixes are also widespread in vertebrates, and may be used for things like sex- and individual-specific badges in parental care and inbreeding avoidance.

2.2.2 Detecting Odours

There are many common principles in processing odours across vertebrates and invertebrates (Eisthen 2002; Hildebrand and Shepherd 1997; Masse *et al.* 2009; Su *et al.* 2009; Wyatt 2013), and so here we mainly focus on odour detection in insects (see Wyatt 2013 for information on mammals too, and comparisons with insects). However, note that the gross similarities across vertebrates and invertebrates may not be reflected in things like receptor arrangements and transduction mechanisms in the groups (Kaupp 2010). In Chapter 3, we discuss how odour information is processed and encoded by neuronal mechanisms.

Unlike many other senses, olfaction uses a wide range of receptor types belonging to many gene families (Hansson and Stensmyr 2011; Su *et al.* 2009). In insects, the sensory cells are sensilla (Figure 2.2). These are small structures projecting from the cuticle mostly located on the antennae, with the size and shape varying between species and type (Eisthen 2002; Hansson and Stensmyr 2011; Masson and Mustaparta 1990). Pores in the cuticle allow odour molecules to enter and diffuse into the lymph fluid that bathes the olfactory receptor neurons (ORN). Note that some authors call these olfactory sensory neurons. In vertebrates, ORNs are mainly found in the olfactory epithelia in the nasal chamber, bathed in a mucus lining. ORNs in insects have dendrites that protrude into the lymph fluid. These are the sites of odorant transduction, and where membrane-bound receptors are located. More than one ORN is often found in a single sensillum in insects (Masse *et al.* 2009), and olfactory receptor proteins are exposed in the membranes of ORNs. Binding of an odour molecule to a receptor protein on the membrane (an 'acceptor') leads to a conformational change in the receptor to an active state, causing depolarization of the cell membrane and leading to nerve impulses or hyperpolarization (Eisthen 2002; Masson and Mustaparta 1990).

ORNs normally only express one receptor type, meaning that activity in an ORNs reflects the activity of a specific odour receptor type (Su *et al.* 2009). Receptor proteins are diverse and the number of genes involved varies greatly across species (Bargmann 2006). The mechanisms of transduction and OR protein families differ between insects and chordates, indicating that they have evolved independently (Wyatt 2010). Olfactory receptors seem to act as ligand-gated ion channels but the specific mechanisms of transduction can be complex (see Kaupp 2010; Su *et al.* 2009).

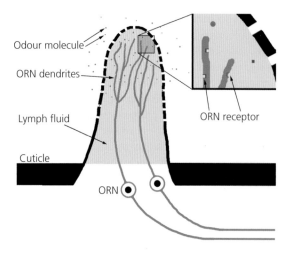

Odour molecule

ORN dendrites

Lymph fluid

Cuticle

ORN receptor

ORN

Figure 2.2 An olfactory sensillum. Odour molecules enter the sensillum through pores, pass through the lymph fluid, and bind with the receptors on the membrane of the olfactory receptor neuron (ORN). Two receptor neurons are shown here, but there are often more than this in one sensillum.

Generally, most olfactory receptors in vertebrates and invertebrates are broadly tuned so that a given odour molecule (e.g. from food) will stimulate several receptors (Masson and Mustaparta 1990; Wyatt 2010). In contrast, olfactory receptor cells that respond to pheromones can be highly specific and activated by a single compound. Broadly tuned receptors tend to be sensitive to structurally similar odorants (Su *et al.* 2009). Different degrees of specialization by different cells reflect the degree of receptor-protein specificity, and how closely a molecule has to match the receptor in order to trigger a response (Reisert and Restrepo 2009). A categorization into 'specialist' cells is sometimes valid for many pheromone systems. However, for non-pheromone processing some cells show more specific responses to certain odour components and others general responses to broader compositions (de Bruyne and Baker 2008; Masson and Mustaparta 1990). In addition, concentrations of odours also determine whether a receptor is 'generalist' or 'specialist' (Kaupp 2010). Therefore, a clear distinction between specialist and generalist is problematic and it is better to think of a continuum of responses (see Chapter 3).

Drosophila has been well studied with regards to the physiology and genetics of olfaction. In *Drosophila*, there are about 50 classes of ORNs (Masse *et al.* 2009) and three types of morphologically distinct sensilla that respond to different things: trichoid (pheromones), basiconic (food odours and CO_2), and coeloconic (food odours, alcohols, ammonia, amines, water vapour) (Kaupp 2010; Wyatt 2013). Some receptors, such as Or82a, are highly specialist, responding to just one odour chemical (Bargmann 2006). Others are activated by as much as 30% of the odours presented. Fruit odours activate about two-thirds of the receptors, owing to the ecology of the flies (Bargmann 2006). Thus, there is a continuum of broadly and narrowly tuned odour receptors in *Drosophila* (Hallem and Carlson 2006). Overall, olfactory sensilla in insects take many different shapes and sizes, but it is not entirely understood why.

2.3 Electricity

Some animals can detect low-voltage electric information: they have an electric sense. This is most widely found in some groups of fish and amphibians, but also in echidna (*Zaglossus bruijnii* and *Tachyglossus aculeatus*; Gregory *et al.* 1989; Pettigrew 1999) and the platypus (*Ornithorhynchus anatinus*; Pettigrew 1999; Scheich *et al.* 1986). Electroreceptors have also been suggested in the star-nosed mole (*Condylura cristata*; Gould *et al.* 1993), and recently in the Guiana dolphin (e.g. *Sotalia guianensis*; Czech-Damal *et al.* 2011), and crayfish (Patullo and Macmillan 2010), and so they could be more widespread than currently known. As air is a poor conductor of electricity, an electric sense is only useful in aquatic or moist environments. Here, we focus on fish because an electric sense has been most studied and is most prevalent in a number of fish lineages. Many species passively detect electric information from their environment. However, other species, in particular some groups of teleost fish, are able to produce electricity themselves and, like echolocating bats, have an active sense. Lineages of electric fish tend to come from habitats where vision is of diminished value, such as where water transparency is poor (for example, due to particulate matter) or when species are predominantly active at night.

2.3.1 Passive Electroreception

All animals produce electricity from the normal actions of muscles and nerves, and some predators are able to sense this to locate their prey. Between the late 1950s and 1970 a steady increase in evidence showed that electroreception existed in some animals (Hopkins 2010). However, comprehensive studies of the function of passive electroreception had not yet been done. Then, a classic experiment by Kalmijn (1971) showed that spotted dogfish (*Scyliorhinus canicula*) can locate a flatfish (*Pleuronectes platessa*) under the sand using just electrical cues. Kalmijn's experiments are wonderfully clear in their simplicity, yet powerful. First, he showed that the shark could locate the flatfish buried in the sand (no visual cues). Second, the shark would also locate a hidden flatfish covered by an agar chamber, which conducts electricity but blocks mechanical cues and some chemical diffusion. However, the shark would not locate the flatfish when the chamber was electrically insulated. Finally, the shark attacked two electrodes hidden in the sand, which only presented electric cues similar to that of a fish.

Electric information in fish is detected by electroreceptors. These are found in jelly-filled pores open to the water and resemble the hair cells of the lateral line system. In cartilaginous fishes, the receptors are arranged into clusters over the head region (the ampullae of Lorenzini). Ampullary electroreceptors are highly sensitive to weak electric fields, and are used in passive electrolocation. They have detection thresholds as weak as 1 μVcm^{-1} in freshwater and 20 μVcm^{-1} in marine species, but possibly lower (Peters *et al.* 2007).

2.3.2 Active Electroreception

Sharks and rays, for example, can only detect electric cues from the environment. However, in 1951 the production of electric signals by fish was clearly demonstrated (Lissmann 1951). Many fish are now well known to produce electricity: they are electrogenic. The largest groups are teleost lineages of weakly electric fish: the Gymnotiformes ('knifefish') from South America, and the Mormyridae ('elephant fish') and Gymnarchidae (Osteoglossiformes) from Africa. By measuring distortions in the electric field generated, weakly electric fish can detect differences in size, distance, shape, and even the material from which an object is made. Electric signals are also used in communication, including mating behaviour. Electric fields are effectively set up instantaneously and do not propagate like sound or light waves. Thus, electric information attenuates (decrease in amplitude) much faster than sound or light and is only useful to within a few body lengths of the individual.

2.3.2.1 Production and Detection of Electric Signals

Electricity is produced by an electric organ, and in most cases this is developmentally and evolutionarily derived from modified muscle fibres (in one lineage, Apteronotus, the electric organ is derived from nerve fibres). The organ is made of a series of electrocytes, which produce synchronous action potentials. The flow of sodium ions into the electrocytes polarizes the skin of the fish, generating an electric field outside the body. Electrocytes are arranged in parallel and series so that their effects are summed, increasing current and voltage, respectively. The effect is to produce electric organ discharges (EODs). Different phases of EODs are produced by varying the direction that sodium ions enter the electrocytes over time (e.g. either in a headward or tailward direction), causing either the head or tail regions to be positively or negatively polarized. Broadly, EODs come in two types (Figure 2.3): 'pulse' fish produce single pulses of EODs separated by large time intervals. These EODs vary between 0.1 ms and 10 ms and have one to five distinct phases. In contrast, 'wave' fish produce a series of quasi-sinusoidal EODs close together in time.

Despite the two main categories of EOD, characteristics vary between species, sex, and even sometimes at different times of day. They also vary with certain behaviours and in different physical and social environments. Note, however, that individual EOD variation stemming from changes in EOD with growth or breeding season, while clearly significant, are often comparatively slight compared to interspecific EOD variation (and EOD variation between sexes in some species). In some fish, the discharge rate of EODs is controlled by pacemaker neurons in the medulla and transmitted to the electrocytes by

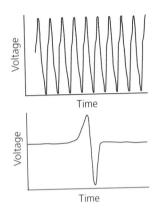

Figure 2.3 Example electric organ discharges from a wave-type electric fish (top: *Apteronotus leptorhynchus*; data from Fugère and Krahe 2010) and a pulse-type electric fish (bottom: *Campylomormyrus compressirostris*; data from Feulner *et al.* 2009b).

spinal relay- and motor neurons. For example, the brown ghost knifefish (*Apteronotus leptorhynchus*) has a pacemaker nucleus comprising about 100 coupled pacemaker neurons that fire simultaneously. These are electrically coupled to spinal relay neurons, followed by spinal motor neurons, which drive EOD frequency (Zakon *et al.* 2002). Recent work has also started to reveal the molecular basis of divergence and diversity in EODs, with changes in EODs across species in both mormyrids and gymnotiforms correlated with parallel changes to some of the same functional regions of the voltage-gated sodium channel α subunit, Nav1.4a (Arnegard *et al.* 2010b; Zakon *et al.* 2006; Zakon *et al.* 2008; see Chapter 11).

In addition to ampullary electroreceptors (to passively detect electric cues), weakly electric fish have tuberous electroreceptors that are specially adapted to detect EODs. Different types of tuberous receptors are often associated with specific tasks (e.g. communication, electrolocation), and encode different features of EODs. In mormyrids, mormyromasts are used in active electrolocation of objects around the fish, whereas knollenorgans are used for detecting the EODs of other individuals in communication (Bell and Grant 1989). Gymnotiforms have just a single type of tuberous receptor for both these tasks, but these can be modified to improve the encoding of either temporal characteristics or differences in the amplitude of EODs.

2.3.2.2 Functions of EODs

Active electrolocation involves using self-produced electric signals to detect, localize, and analyse objects in the environment (see Assad *et al.* 1999; von der Emde 1999). Objects cause distortions in the electric field generated around the animal, leading to changes in the current flowing through the electroreceptors (Figure 2.4): objects with impedance values lower than that of the surrounding water (conductors) 'attract' electrical current because more current flows through the object than the water it replaces. In contrast, non-conductors have higher impedance-values than the water and 'repel' current (von der Emde 1999).

The location of an object is revealed by its effect on receptors at different locations on the body. For example, changes in the response of the receptors at the tail only would indicate an object at the posterior region of the fish. Changes in amplitude reveal information about the impedance values of the object, and more complex features probably require integration of different pieces of electric information (von der Emde *et al.* 1998; von der Emde 1999).

Changes in the temporal properties of EODs in fish can function in communication towards both conspecifics and heterospecifics (e.g. Dunlap *et al.* 2010). Various species of electric fish are sexually dimorphic in their EOD properties, and EODs have increasingly been shown to play a role in mate choice and courtship. For example, males of the pintail knifefish (*Brachyhypopomus pinnicaudatus*)

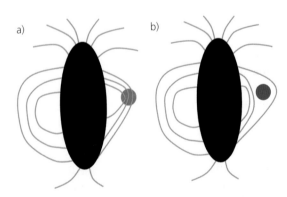

Figure 2.4 In electrolocation, objects in the environment cause distortions in the EODs of fish. Conductors a) attract electrical current towards them, whereas non-conductors b) repel electric field lines.

change their EODs during phases of spawning with females. Females prefer individuals with EODs of longer duration and higher amplitude, and these features correlate positively with male body size (Curtis and Stoddard 2003; Chapter 5). Similar correlations between EOD attributes and body size may also function in dominance relationships between males (e.g. Zakon *et al.* 2002). EOD characteristics can also be important in species recognition (e.g. Fugère and Krahe 2010; see Chapter 11).

Finally, some predatory fish (e.g. catfish such as *Clarias gariepinus*) exploit the electric signals of other fish in order to locate their prey (Hanika and Kramer 2000). They eavesdrop on communication signals of others (see Chapter 7). Other predatory fish, such as the mormyrid *Mormyrops anguilloides*, which feeds on cichlid fish in Lake Malawi, may use synchronized bursts of EODs across individuals during group hunting to maintain group cohesion (Arnegard and Carlson 2005).

2.4 Light

Light is part of the electromagnetic spectrum (it comprises both electrical and magnetic forces). In many respects light behaves like a wave, whereas in others it behaves like a stream of particles (photons). Light is generally more easily understood as a wave when it travels, but more easily understood as a particle when it interacts with matter (see Johnsen 2012). Visible light is the part of the electromagnetic spectrum to which an animal's receptors are sensitive (roughly 380–780 nm for humans, although the upper limit is intensity-dependent). The light that animals are sensitive to differs across, and sometimes even within species, meaning that the range of information available to animals need not be the same. In Chapter 3 we discuss neural mechanisms of vision for encoding important information about stimuli, such as motion and contrast.

2.4.1 Sensing Light

Many microorganisms have light-sensitive molecules used to guide an individual towards or away from a light source (phototaxis). We do not regard this true vision here, however, but instead consider the ability to detect spatial structure in a scene (i.e.

to form an image) as needed for vision. To detect light, animals have photoreceptor cells, which contain a visual pigment. Photopigments comprise a protein (an opsin), bound to a carotenoid chromophore (Kelber *et al.* 2003; Shichida and Matsuyama 2009). The combination of opsin-type and chromophore confers sensitivity of the photopigment to specific wavebands of light. The mechanisms of phototransduction are known in considerable detail (see Arshavsky *et al.* 2002; Porter *et al.* 2012; Shichida and Matsuyama 2009, and references therein). A chromophore absorbs a photon and isomerizes, causing the opsin to change conformation. This activates a G-protein molecule, leading to an enzymatic signalling cascade and an electrical response in the photoreceptor.

Different photoreceptors vary in both the amount and wavelengths of light to which they are sensitive. In vertebrates, rods are distributed widely across the retina and generally sensitive to low levels of illumination (scotopic vision), and usually do not give rise to colour perception. Cone cells generally have low sensitivity and are subdivided into different classes in many animals. They are used in, among other things, colour vision. Almost all photoreceptor cells in animals have membrane specializations to house large quantities of photopigment (Nilsson 2009). These can be based on either modified cilia or microvilli projecting from the cell body, and is often used to create a distinction between ciliary and rhabdomeric photoreceptor cell types (note, however, that this division is becoming less clear and it may be better in future to divide photoreceptors by transduction cascades and opsin families; Lamb 2009). Many vertebrate eyes generally contain the former cell type, and most invertebrate eyes comprise the latter, although again this distinction is not clear-cut and some animals have cells with features of both types. The two cell types also use different types of opsin (Nilsson 2009; Shichida and Matsuyama 2009), although again this may be a simplification (Porter *et al.* 2012).

Different groups of opsins are associated with vision in vertebrates and invertebrates. In vertebrates, there are at least five different classes of opsin genes associated with photoreceptors (although a range of other opsins also exist; see Porter *et al.* 2012): *LWS/MWS* conferring sensitivity in cones to relatively longer wavelengths (peak sensitivity conferred: 490–570 nm), *SWS1* (355–440), and *SWS2* (410–490) conferring sensitivity to shorter and/or ultraviolet (UV) wavelengths in cones, *RhB/Rh2* (480–535 nm) providing sensitivity to medium wavelengths in cones, and *RhA/Rh1* (460–530 nm) found in rods (Bowmaker 2008; Collin *et al.* 2009). Within these groups, tuning of photoreceptor sensitivity to different wavelengths of light is driven by changes in the opsin protein, encoded by the opsin genes (Bowmaker 2008; Collin *et al.* 2009; Osorio and Vorobyev 2005). Of the different types of cone opsin, all four are present in many fish, reptiles, and birds, but one or more have been lost in many mammals, amphibians, and primitive fish. Arthropod visual pigments fall into three families: 'UV', 'blue', and 'green' (Briscoe 2001).

The different wavelengths of light a photoreceptor is sensitive to not only depend on the visual pigment it contains, but also other features of the eye. Specifically, some animals have coloured filters or oil droplets associated with specific receptor types, which tune the sensitivity of the photoreceptor. Diurnal birds, many lizards, fish, and turtles have oil droplets located in the cone inner segments. These selectively absorb light before it reaches the photopigment and modifies the light spectrum that is available for the receptor to respond to (Bowmaker 1980; Goldsmith 1990; Hart *et al.* 2000). This may enhance colour discrimination by reducing the spectral overlap between cone types (Vorobyev 2003). Coloured filters in some stomatopods and butterflies are based on tiered retinae, where light passes through a set of distal receptor cells before reaching the proximal cells (Kelber *et al.* 2003). Many invertebrates also use screening pigments. The effect of these mechanisms is to tune the receptor to specific wavebands of light, for example for enhanced colour discrimination and/or in line with prevailing light conditions (see Chapter 10).

2.4.2 Forming Images

To detect the direction of light and to form an image, information needs to be gathered from an arrangement of photoreceptors in an eye. The eyes found in animals are varied (Figure 2.5). However, very broadly, there are camera-type eyes (such as in

Figure 2.5 Types of eye. Both birds and preying mantis have excellent vision used in foraging, movement, and communication. Birds have a camera-type eye (top-right), where light passes through the cornea and a single lens to a concave retina, which contains the receptor cells. In the preying mantis compound eye (middle-right), light passes through thousands of lenses (facets) into the ommatidia, each of which contains a limited number of receptor cells. Compound eyes come in two broad types: apposition compound eyes (middle right) comprise many ommatidia, each with a set of photoreceptors receiving light from only one lens, whereas superposition compound eyes (bottom-right) involve light passing through a large number of lenses and onto photoreceptors on a convex retina. Note that various other types of eye also exist as well as variations between these, including simple pinhole-type eyes with a small opening, and eyes where light is reflected from a mirror-like surface onto a retina.

vertebrates) that have a single lens in front of an array (retina) of photoreceptors where the retina is concave, and compound eyes, with hundreds or thousands of lenses (or facets) with a convex array (Goldsmith 1990; Land and Nilsson 2012). There are also eyes that are somewhere between these types, and eyes without lenses. Although eyes of some type are found in the majority of animals, it is unclear how many times eyes have evolved (Lamb *et al.* 2009).

Insect compound eyes comprise a mosaic of ommatidia. Compound eyes are broadly of two main types. Apposition compound eyes involve each set of (normally eight to nine) photoreceptors ('rhabdomeres') receiving light from only one lens.

In contrast, superposition compound eyes work more like camera eyes (except with a convex rather than concave retina), with light passing through a zone from a large number of lenses and focused onto a range of photoreceptors. The latter has increased sensitivity and tends to occur in nocturnal insects (Kelber and Roth 2006), although a range of insect species that have undergone a transition from a diurnal to a crepuscular or nocturnal lifestyle retain apposition eyes modified for dim light (Frederiksen and Warrant 2008; Greiner 2006). In vertebrate camera-type eyes, the retina is a mosaic of different photoreceptors, with the relative proportions of photoreceptor types often varying across different regions of the retina. This reaches an extreme in a fovea or visual streak, which are characterized by a dense concentrations of specific receptor cells. In humans, the fovea involves high densities of longwave- ('red') and mediumwave- ('green') sensitive cones and no rods, and provides high visual acuity.

2.4.3 Achromatic and Colour Vision

2.4.3.1 Luminance

A fundamental aspect of vision is encoding of achromatic (brightness) information. This can be termed 'luminance' when quantified with respect to the receiver's vision, and varies in species owing to different luminance encoding mechanisms. Luminance vision is fundamental to many aspects of sight because much information in the world is encoded by differences in brightness (for example, the boundaries between objects). It is therefore often used in spatial vision, motion perception, and detection of small objects (Osorio and Vorobyev 2005). Correspondingly, the receptors used for luminance vision tend to be the most abundant type in the eye. In humans, luminance vision is largely based on the additive combination of the longwave (LW) and mediumwave (MW) cone outputs. In contrast, birds, fish, and various other non-mammalian vertebrates seem to use an additional photoreceptor type, the double cones (two receptor cells linked together), which have broad sensitivity (Kelber *et al.* 2003; Osorio and Vorobyev 2005). Bees and some other insects, for example, use their longwave-length receptors for luminance.

2.4.3.2 Colour

Colour vision is the ability to detect differences in two light spectra based on their composite wavelengths, independent of light intensity (Kelber *et al.* 2003; Kelber and Osorio 2010). An animal's perception of colour does not depend on the absolute stimulation of the receptors, but rather the relative stimulation of two or more receptor types of different sensitivity and their neural interactions. As Newton (1718) realized, light itself is not coloured, but colour perception (like other forms of vision) stems from the sensory system and brain. In principle, the number of receptor types involved in colour vision dictates the range of colours an animal could potentially see. Trichromatic animals have colour vision involving three receptor types, whereas tetrachromatic vision involves four receptor types. Much of colour vision depends on opponent processing of photoreceptor outputs. Here, antagonistic colour pairs are processed in a separate neural channel. In humans, blue–yellow and red–green, represent opponent neural pathways (Chatterjee and Callaway 2003; Chichilnisky and Wandell 1999). Colour opponency is the product of excitatory and inhibitory connections between the receptor types, with either member of the opponent pair perceived depending upon excitation or inhibition of the pathway. In some primates for example, cells (e.g. ganglion cells; see Chapter 3) in the visual system have been found that are stimulated by light of one part of the spectrum and inhibited by light in another part of the spectrum. Outside of primates, evidence for multiple opponent channels has been found, for example, from behavioural experiments with birds (Osorio *et al.* 1999) and neurophysiological studies of horizontal and ganglion cells in turtles (e.g. Ammermüller *et al.* 1995; Ammermüller *et al.* 1998; Twig and Perlman 2004; Ventura *et al.* 1999; Ventura *et al.* 2001). Colour-opponent mechanisms in insect vision have been studied in some species, but not much is yet known about how coding of colour channels works or the variation that exists (Briscoe and Chittka 2001).

There is a great variety in the types and numbers of receptors potentially used in colour vision across animals (Bowmaker 2008; Kelber *et al.* 2003; Figure 2.6), although explaining this diversity has proved diffi-

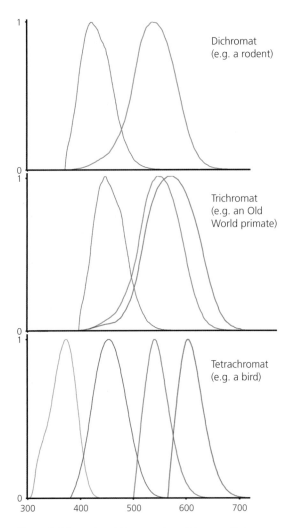

Figure 2.6 Spectral sensitivity curves for a dichromat with two cone types (e.g. a rodent), a trichromat with three cone types (e.g. an Old World primate), and a tetrachromat with four cone types (e.g. a bird) used in colour vision.

bees use three receptor types in colour vision. Stomatopod crustaceans ('mantis shrimp') have beyond ten different types of spectrally distinct photoreceptors (Cronin and Marshall 1989). Some animals are also sexually dimorphic for the types of receptor and colour vision that they may have; for example, some butterflies. Many butterflies have four or even five receptor types, although this does not mean that they use them all in colour vision. Instead, different combinations of photoreceptors may be used in different tasks (Kelber and Osorio 2010). For example, the Japanese yellow swallowtail butterfly (*Papilio xuthus*) has eight varieties of photoreceptors with different spectral tuning, sensitive to various parts of the spectrum (Koshitaka *et al.* 2008). Behavioural studies suggest that when foraging for nectar, it uses tetrachromatic colour vision, with UV, blue, green, and red classes of receptor. Some fish also seem to go from mono- to di- to trichromatic vision depending upon light levels. There will almost certainly be more flexibility in colour vision uncovered across and within species as work into the processing of colour beyond the initial interactions and comparisons of receptor types progresses.

Finally, many conventional assumptions are unlikely to be hard rules. For example, the assumption that colour vision only operates during the day has been shown to be wrong in an increasing number of species that use colour vision at night (Kelber and Roth 2006), and receptors like double cones, often assumed to be involved in luminance, may be used in colour vision in some fish (Pignatelli *et al.* 2010).

2.4.4 Polarization

Polarization describes the way that as a wave of light propagates its electric field vibrates, with the axis of vibration being called the electric field vector (e-vector) angle. When light comprises photons with the same e-vector angles, it is called plane-polarized or linearly polarized (for a full description of polarization see Johnsen 2012). Many animals, including birds, reptiles, amphibians, fish, and especially various invertebrates from cephalopods to insects, may be capable of perceiving polarized light (reviewed in Cronin *et al.* 2003; Cronin and Marshall 2011; Marshall and Cronin 2011; Nilsson and Warrant 1999; Wehner 2001). In unpolarized

cult (see Chapter 4). For example, many mammals are dichromatic and some are trichromatic, whereas many birds and some reptiles and fish are potentially tetrachromatic.

Note that simply identifying a given number of receptors (e.g. four) does not mean that the animal has colour vision comprising that number of dimensions (e.g. is tetrachromatic). Ultimately, relevant neural comparisons are needed between each receptor type. The number of receptor types also varies greatly across invertebrates. Some insects, such as light, all e-vectors are represented randomly over time with the light having, on average, no e-vector direction. Natural light can be partially polarized due to the effects of scattering, where certain e-vectors are more common. Circular polarization occurs when the e-vector rotates as the wave travels. Rotations of the e-vector in a clockwise or counter-clockwise direction, seen from the direction of the receptor, constitute right- and left-handed circularly polarized light.

2.4.4.1 Detecting Polarized Light

In order to be able to detect and discriminate polarized light, animals need to have appropriate visual mechanisms (Horváth and Varjú 2004; Johnsen 2012). The probability of light being absorbed by a photoreceptor is greatest if the e-vector of the light aligns (is parallel) with the excitable dipole of the visual pigment molecule, and lowest if the light has an e-vector perpendicular to the axis of the pigment molecule (i.e. absorption probability depends on the angle between the pigment bond and the polarization of light). This means that visual pigment molecules are automatically sensitive to polarization (Cronin *et al.* 2003; Nilsson and Warrant 1999). For the photoreceptor (which contains many pigment molecules) to have polarization sensitivity, there needs to be general/partial alignment of preferred absorption angles of the pigment molecules in the receptor. In addition, the receptor cells need to be orientated in the eye. In vertebrate rods and cones the photoreceptors are insensitive to e-vector direction and the basis for polarization sensitivity is uncertain, although it may stem from the double cones (Johnsen 2012). However, in the rhabdomere of invertebrate photoreceptors, microvilli can be aligned in parallel and kept straight so the animal could achieve sensitivity to a specific e-vector (Cronin *et al.* 2003; Marshall and Cronin 2011; Nilsson and Warrant 1999). In cephalopods like cuttlefish and octopus, microvilli are also arranged in an orthogonal structure that makes them sensitive to linear polarization (Mäthger *et al.* 2009; Moody and Parriss 1961; Shashar and Cronin 1996; Shashar *et al.* 1996; Talbot and Marshall 2010, 2011).

A range of relatively simple tasks and detection of polarization levels can be achieved using just one class of polarization-sensitive receptors. However, true polarization vision is often described as the ability to discriminate between stimuli based on angle and degree of polarization alone, independently of colour or intensity information (Bernard and Wehner 1977; Johnsen 2012; Marshall *et al.* 1999; Nilsson and Warrant 1999). To achieve this fully, animals would need three classes of receptor sensitive to different e-vectors of light (as with colour vision that requires two or more receptor types sensitive to different wavelengths of light), and opponent processing channels between these receptors (Labhart and Meyer 2002; Marshall and Cronin 2011; Nilsson and Warrant 1999). Polarization opponent interneurons have been found, for example, in the optic lobe of crickets, receiving antagonistic input from orthogonally arranged photoreceptors sensitive to polarization (Labhart 1988). A number of studies have investigated the receptors and their properties that seem to be used in polarization vision in, for example, insects (e.g. Labhart 1980; Wehner *et al.* 1975), and shown that sets of ommatidia in stomatopod crusteceans (mantis shrimp) compound eyes can discriminate between linearly polarized light and use this to guide behaviour (e.g. Marshall 1988; Marshall *et al.* 1999), and even between right- and left-handed circular polarizations of light (Chiou *et al.* 2008; Roberts *et al.* 2009). Recent work also shows that some cuttlefish can have very refined abilities to discriminate e-vector angles down to just 1° (Temple *et al.* 2012).

2.4.4.2 Functions of Polarization Sensitivity and Vision

It has been known for some time that insects such as bees use patterns of polarization in the sky for orientation and navigation (e.g. Brunner and Labhart 1987; von Philipsborn and Labhart 1990), as do some vertebrates such as birds (see Chapter 4). Orientation in both vertebrates and invertebrates (on land and underwater) may exploit polarized light patterns relative to the direction of the sun (Labhart and Meyer 2002; Lythgoe and Hemmings 1967). Much work on polarization sensitivity and vision has been and still is being undertaken in orientation and navigation (e.g. Henze and Labhart 2007; Homberg *et al.* 2011; Kraft *et al.* 2011; Lerner *et al.* 2011). However, animals may also use polarization in other tasks. This includes communication. For example, Marshall *et al.* (1999), Chiou *et al.* (2012), and others have shown that some mantis shrimp have body patterns of polarization, most notably on the antennal scales, and polarized light may also be used in signalling in *Heliconius* butterflies (Sweeney *et al.* 2003), and cephalopods (Mäthger *et al.* 2009). Some beetles may also possess circular-polarization discrimination abilities and use this in communication (Brady and Cummings 2010). Polarization vision could also be useful in hunting. For example, cuttlefish (*Sepia officinalis*) may use polarization patterns to detect silvery fish (Shashar *et al.* 2000), and to detect transparent prey (Shashar *et al.* 1998), although these ideas are still largely lacking in experimental support (see Johnsen *et al.* 2011; Marshall and Cronin 2011).

2.4.6 Infrared

Some snakes (pit vipers, pythons, and some boas) are well known for using infrared radiation to detect and strike at prey and for orientation. Pit vipers have heat-sensitive nerve fibres found in pit organs on the head (Goris 2011; Newman and Hartline 1982; Safer and Grace 2004). Information from the pit organs is relayed to the optic tectum of the midbrain, which is well known for its role in vision (Newman and Hartline 1982). Vampire bats (*Desmodus rotundus*) also have high numbers of thermal receptors, located at the central nose leaf. These are capable of detecting small temperature differences and probably underlie their ability to acquire blood meals from other animals (Kürten *et al.* 1984).

The pit organ of snakes comprises a cavity with a thin membrane (Figure 2.7), close to the surface of which are thousands of nerve endings that respond to very small changes in temperature, with even values as low as 0.003° C discussed (e.g. Goris 2011; Newman and Hartline 1982). The structure of the pit organ is grossly similar to camera-type eyes, with a narrow opening so that the patch of illuminated membrane allows direction to be deduced (Goris 2011). Note that the pit organs can detect changes in temperature relating to either cool objects against a warm background or vice versa (Van Dyke and Grace 2010).

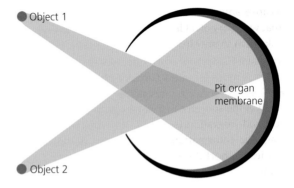

Figure 2.7 The pit organ found in some snakes is superficially similar to a camera-type eye, with an opening through which infrared information passes and stimulates nerve cells lying close to the surface of a thin membrane layer at the back of the pit organ. The location of stimulation can reveal the direction of the object.

The gene *TRPA1* underlies a temperature-sensitive calcium ion channel (TRPA1) on the sensory nerve fibres in the membrane (Gracheva *et al.* 2010). Geng *et al.* (2011) showed that this gene is under strong positive selection in snake species with pit organs, but not in species without the organs. Furthermore, 11 amino acid sites in the TRPA1 protein sequences have diverged in pit-bearing snakes, but are conserved in other species. Parallel amino acid changes seem to be responsible for modification across different groups of snakes (Yokoyama *et al.* 2011). Therefore, evolutionary modification of infrared sense organs in pit-bearing snakes is associated with changes in the *TRPA1* gene.

2.5 Magnetic

Sensitivity to magnetic fields (magnetoreception) has been found in many animals, including birds (Wiltschko and Wiltschko 2007), mammals such as bats (Holland *et al.* 2006) and mole rats (Marhold and Wiltschko 1997; Němec *et al.* 2001), insects such as bees (e.g. Schmitt and Esch 1993), beetles (e.g. Vácha *et al.* 2008), and flies (e.g. Gegear *et al.* 2010), turtles (e.g. Lohmann and Lohmann 1996a), fish (e.g. Hellinger and Hoffmann 2009; Quinn 1980), lobsters (e.g. Boles and Lohmann 2003), and some amphibians (e.g. Phillips and Borland 1992), and cartilaginous fish. The magnetic sense is the only known sensory modality that is not used in communication because animals cannot generate magnetic signals (that we know of, at least). It is also not used in other tasks such as prey location, and is apparently largely restricted to orientation and navigation, although it may play a role in calibrating internal clocks. Much work on the use of magnetic cues has been undertaken in the subject of relatively long-range orientation and migration (see Chapter 4), but the use of magnetic information is also important in many species of short distances, such as moving through home ranges or in nest construction (e.g. De Jong 1982; Jacklyn 1992). Although much progress has been made in understanding magnetoreception, many questions remain about how it works in different species. One problem is that, for most senses, the stimulus being detected requires identifiable receptors exposed to the external environment, whereas because magnetic fields pass through animal tissue, the receptors could be based anywhere and dispersed widely in the body (Johnsen and Lohmann 2005). Therefore, the sensory receptors involved in magnetoreception are relatively poorly understood. Much of what we know comes from experiments manipulating aspects of magnetic fields and analysing changes in behaviour. However, in the last decade we have started to gain a much better understanding of how magnetoreception works.

2.5.1 The Source and Use of Magnetic Cues

The basis of magnetic-guided behaviour is the magnetic field of the earth. How exactly this field is created is not well understood, but probably stems largely from motion in the earth's molten iron core due to the coriolis effect, with a smaller (and variable) component produced inductively by ions in the jet streams (see Gould 2010). Field lines leave the southern magnetic pole and re-enter at the northern hemisphere. At the magnetic equator, the field lines are parallel to the earth's surface and the inclination angle is $0°$, whereas at other locations the field lines intersect the earth at different angles, becoming steeper, and then $90°$ at the poles (Lohmann *et al.* 2007; Wiltschko and Wiltschko 2005, 2007, 2010). The intensity of the magnetic field is also highest at the poles and lowest at the magnetic equator. Information from the earth's magnetic field is not entirely consistent because the earth's magnetic field varies over time, and due to local anomalies caused by material in the earth's upper crust and electromagnetic radiation from the sun (although how much of this variation is significant enough to be a problem for animals is unclear). Perhaps for this reason, animals often seem to use their magnetic sense along with information from other modalities, such as vision (see Chapter 4). However, magnetic information is present both day and night, irrespective of weather and season (Lohmann *et al.* 2007).

Birds are the most studied group with a magnetic sense, with over 20 species demonstrated to use magnetic information (see Wiltschko and Wiltschko 2007, 2010). Most work has been undertaken on migratory birds. These become active at the time of year when they would normally be migrating ('migratory restlessness'), and the direction that they 'aim' to go in can be measured with an Emlen funnel: a funnel-shaped cage lined with paper that records the scratch marks of birds when they move. Experiments involving altering the direction of the magnetic field around the funnel results in birds adjusting the direction of their movements in line with this modification (Wiltschko and Wiltschko 2007, 2010; Figure 2.8). However, a magnetic sense also seems to be important over shorter distances (e.g. homing behaviour in pigeons), and even on a local scale (such as search behaviour in chickens; see Wiltschko and Wiltschko 2010).

The main potential use of magnetic cues in animals is in gaining information about a specific location (a map) or of a direction (a compass). Compass information is uncontroversial and enables animals to head in a particular direction (e.g. north or southeast, for example). However, magnetic cues need to be used in conjunction with other information in order to reach a specific goal. The inclination of the magnetic field lines can provide information about direction (i.e. an 'inclination compass') such as 'poleward' or 'equatorward'. Birds, for example, seem to use this type of compass, rather than one sensitive to polarity (e.g. magnetic north versus south), whereas rodents, salmon, and some invertebrates seem to use a polarity compass sensitive to the polarity of the field lines (Wiltschko and Wiltschko 2005).

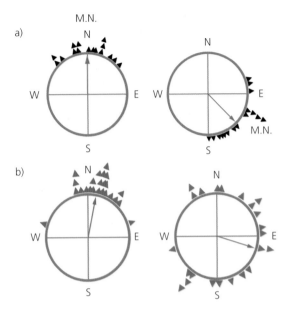

Figure 2.8 a) European robins (*Erithacus rubecula*) orientate with respect to the local magnetic field (left), and switch their orientation behaviour accordingly when magnetic north is moved 120°. b) Orientation of robins towards the magnetic field is accurate when under monochromatic light of 565 nm ('green' light; left), but is abolished when under 635 nm light ('red' light; right). Arrows are mean vectors across all individuals (of standardized length). Data from Wiltschko and Wiltschko (2007).

In contrast, magnetic maps allow individuals to go to a specific destination and compute geographical location (Lohmann *et al.* 2007). Various studies have demonstrated the use of magnetic maps by capturing animals and testing them inside apparatus with magnetic coils. The experimenter simulates a false present location in terms of magnetic cues and the animal orientates towards a position that corresponds to the shifted location of the capture site in accordance with this. Magnetic maps would allow true navigation, whereby an individual could determine its position relative to a goal and find the way there based on magnetic field information (Boles and Lohmann 2003). This could involve using changes in the angle of field lines and variations in intensity of the magnetic field (Gould 2010; Lohmann *et al.* 2007). Magnetic maps could be used in either long-distance migratory pathways, such as in sea turtles and some birds, or to navigate to specific target areas in a habitat, such as turtles, newts, and lobsters (see review by Lohmann *et al.* 2007).

However, whether animals can navigate based on magnetic field information alone is currently unclear.

2.5.2 Sensing Magnetic Cues

The proposed mechanisms for magnetoreception in animals are complex (see Johnsen and Lohmann 2005 for a summary), but two leading theories exist that invoke either the presence of magnetic substances or light-dependent chemical processes. The mechanisms animals use, even in relatively closely related groups, seems to vary widely and different mechanisms may be found even in the same individuals (Gould 2008). Other mechanisms based on induction occur in sea animals, especially cartilaginous fish, but this is not well studied or understood (Wiltschko and Wiltschko 2005).

The first main mechanism involves the use of substances that are permanently magnetic and interact with the earth's magnetic field (Kirschvink and Gould 1981). Magnetite, a form of iron oxide, has magnetic properties that depend on the size and shape of the particles (see Gould 2008, 2010; Wiltschko and Wiltschko 2005, 2007, 2010). When of a certain size, particles can act as permanent magnets with the spins of adjacent atoms aligned. This magnetic field interacts with that of the earth's, creating a pressure to align these two fields that could be measured by secondary receptors such as stretch- or mechanoreceptors or hair cells (Gould 2008; Johnsen and Lohmann 2005). Alternatively, animals could use smaller superparamagnetic particles where the spins of unpaired electrons align themselves with the direction of the magnetic field (Gould 2008). Magnetite-based organs have been found in fish, birds, reptiles, amphibians, mammals, and insects. In birds, for example, magnetite is found in the orbital and nasal cavities and in the skin of the upper beak (Wiltschko *et al.* 2006; Wiltschko and Wiltschko 2007). In bees, it is found in the abdomen (Gould *et al.* 1978).

Recently, evidence has accumulated that some birds, amphibians, and insects have a magnetic sense that requires an input from vision. This second main mechanism is termed the 'radical-pair' interaction, developed by Ritz *et al.* (2000). Changes in the excited spin state of photopigment molecules occur in response to light. The magnetic field alters the transition between spin states (singlet radical pairs with anti-parallel spin change to triplet pairs of parallel spin), and the products of triplets are chemically different from singlets: comparing triplet yields from different directions could be measured between receptor molecules of different orientation in order to get compass information (Biro 2010; Johnsen and Lohmann 2005; Wiltschko and Wiltschko 2005, 2007, 2010). Owing to the arrangement of photoreceptors in the eye, radical-pair processes may generate different patterns over the retina. The photopigment molecules involved are probably cryptochromes and these have been found in the eyes of some magnetoreceptive birds. Recent work in robins (*Erithacus rubecula*) and chickens (*Gallus gallus*) has identified the ultraviolet sensitive cone type as having cryptochrome molecules that fulfil the requirements potentially to underlie the radical-pair model (Nießner *et al.* 2011). Other research has also shown that photochemical reactions in vitro can provide magnetic sensitivity to even weak magnetic fields such as of the earth (Maeda *et al.* 2008), supporting the feasibility of the radical-pair theory.

In the radical-pair process, assessing magnetic fields is light dependent, owing to the process of absorbing a photon. Early work showed that magnetic orientation in salamanders is light-dependent and depends on specific wavebands (Phillips and Borland 1992). Magnetoreception in *Drosophila* is also light-dependent, and depends on a cryptochrome photoreceptor molecule mediated by UV-blue light (e.g. Gegear *et al.* 2010). Robins become unable to use magnetic cues based on a radical-pair process when under light spectra lacking wavelengths below 568 nm (Figure 2.8b) (see Wiltschko and Wiltschko 2005, 2007), and recent work has shown that under relatively longer red-light orientation based on magnetite receptors takes over (Wiltschko *et al.* 2011). Orientation in darkness also seems to depend upon the magnetite receptors in the upper beak and not on a radical-pair process (Stapput *et al.* 2008). For the radical-pair mechanism, light of shorter wavelengths is needed. Light-dependent magnetoreception in birds also seems to depend on input from one eye (e.g. Wiltschko *et al.* 2002; usually the right, although this may not to be a rule across all birds; see Wiltschko and Wiltschko

2010), although recent work indicates that the presence of light alone may not be sufficient but also access to information about visual contrast or contours (i.e. an image is needed; Stapput *et al.* 2010). The radical-pair mechanism involving photopigments may primarily be involved in directional information, whereas magnetite-based receptors provide information about magnetic field intensity, giving different inputs into navigation (Wiltschko *et al.* 2011). Thus, the radical-pair mechanism seems to work only as a compass and not as a map. However, there remains much to learn and this simple dichotomy is unlikely to hold as a rule.

2.6 Mechanical

The ability to detect vibrations, either from incidental sources or those used in communication is widespread in insects and other invertebrates (Cocroft and Rodríguez 2005; Hill 2009) and vertebrates. The use of vibrations as a sensory modality has, however, been somewhat understudied. Vibrations can be detected after transmission through a variety of substrates, from the ground and leaf litter through to plant structures and spider webs. These substrates have very different transmission properties (see Chapter 10). In communication, both vertebrates and invertebrates produce vibrations by drumming appendages against the substrate. In arthropods, this is often extended to more subtle tremulations by rocking or vibrating body parts, and even to the use of specific structures such as tymbal organs in some insects (Hill 2009). Vibratory information is often used in mating behaviour and parent–offspring communication, but also in other interactions. For example, it is even used in signals between beetle pupae to prevent accidental damage by conspecific larvae living in the same patch of soil (e.g. Kojima *et al.* 2012). Some caterpillars can also discriminate between the vibratory signals of conspecifics and predators, and distinguish these from vibrations caused by wind and rain (Guedes *et al.* 2012).

In arthropods, detection of mechanical information usually occurs via innervated hairs extending from the cuticle (Dusenbery 1992). Insects tend to have subgenual organs on their legs, and spiders have slit sense organs (Hill 2009). The compound slit sense organ (metatarsal lyriform organ) found in spiders is sensitive to vibrations through the tarsus, with vibrations producing compression of the slits (Barth and Geethabali 1982). In addition, slit sensilla on the pre-tarsus on the leg, such as in the spider *Cupiennius salei*, are also substrate-borne vibration detectors, stimulated by compression, and seemingly important in prey detection and capture (Speck and Barth 1982).

In vertebrates, sensory hair cells are often involved in detecting mechanical stimuli, with movement of the cilia causing excitation or inhibition (Dusenbery 1992). In fish, the lateral line provides information about flow patterns around the animal, objects, and other organisms in the environment (see Bleckmann 2008). Lateral line cells comprise neuromasts that are either on the surface of the animal (superficial neuromasts) or within canals (canal neuromasts). They are distributed around the head, tail, and along the side of the fish. The number of neuromasts and the properties of the lateral line system vary across fish species. Superficial neuromasts respond to water motion, whereas canal neuromasts respond to pressure variations and gradients between neighbouring cells (Bleckmann 2008; Dusenbery 1992). Increases in pressure arise when displacement of water caused by the fish's movement is restricted by a nearby object, with specific receptor arrangements providing information about size, location, motion, and direction of the object. Like the electric sense, the lateral line system is probably only useful over short distances.

The lateral line system is also used by surface-feeding fish in detecting waves generated by moving prey (Bleckmann 2008). Information is also important in effective shoaling behaviour in fish. For example, by treating groups of firehead tetras (*Hemigrammus bleheri*) with aminoglycoside antibiotics (which damaged the superficial neuromasts), Faucher *et al*. (2010) showed that groups of fish were no longer able to maintain a shoal (larger distances between individuals, lack of coordination, and collisions occurred).

2.7 Sound

Sound information is considered here as airborne and distinguished from similar stimuli like substrate-borne vibrations, even though the two may occur together and even involve the same receptors. It is used in communication, prey detection, and in some species, even in orientation via comparing self-generated sound signals with returning echoes (echolocation).

Sound is produced by mechanical disturbance in a medium (e.g. water or air), and propagates as waves from the source. Sound waves can be characterized by their velocity, phase, frequency, and wavelength (with frequency and wavelength inversely proportional). Sound travels more quickly in water than air, and can propagate over a longer distance. As sound waves propagate through a medium they cause changes in pressure and disturbance in molecules at a given point. Very close to a sound source, air molecules experience large movements that comprise a significant amount of the total intensity but only small deviations in pressure occur ('near field'), whereas away from the sound source ('far field'), sound propagates as a pressure wave (Greenfield 2002; Hoy and Robert 1996). Most sounds cover a wide range of different frequencies, as opposed to those that are of just one frequency (pure tone). In Chapter 3 we cover the neural basis of sound reception and localization in bats and owls.

2.7.1 Sound Detection in Insects

Receptor organs to detect sounds close to a source are often called near-field detectors (Hoy and Robert 1996). In insects, these include the Johnston's organ of mosquitoes, which responds to rapid wing beats during mating (Arthur *et al*. 2010) and cercal organs of cockroaches. Near-field receptors lack eardrums and are usually restricted in function to short-range detection and to relatively low frequencies of around 75–100 Hz (Hoy and Robert 1996). They are often used in courtship and social behaviour in Diptera and Hymenoptera (Greenfield 2002).

Far-field receptors in insects are often tympanal hearing organs, which can be sensitive to a range of high frequencies, from 2–100 kHz or higher (Hoy and Robert 1996). These ears comprise a tympanal membrane (an 'eardrum' of thinned cuticle), an air-filled sac or tracheal expansion, and sensory neurons attached to the chamber or the tympanal membrane (Greenfield 2002; Hoy and Robert 1996).

The sensory receptors are a type of chordotonal organ, which are organs that generally function in detecting mechanical deformation. Airborne sound waves cause the membrane to vibrate in response to pressure differences between the air chamber and the external environment, or to internal pressure differences linked to the tracheal system. This is detected by the chordotonal organ, causing a response in the sensory neuron. Tympanal organs in insects are diverse and sensitive to a wide range of frequencies and often the direction of sound (e.g. Forrest *et al.* 1997; Hedwig and Meyer 1994; Schöneich and Hedwig 2010; Windmill *et al.* 2007; Yack *et al.* 2007). They are found in at least seven orders of insect and located on various different positions on the body, including the head, wings, and legs (Hoy and Robert 1996). This indicates different evolutionary origins of the organs across and within groups, and their use in a wide range of tasks. For example, various nocturnal insects such as moths have tympanal organs tuned to ultrasonic frequencies, presumably as an adaptation to detect the echolocation calls of bats (see Chapter 9). Many moths have co-opted this function to use hearing organs in communication, such as mate choice and male–male competition, and have evolved a range of sound production mechanisms (Conner 1999).

2.7.2 Sound Detection in Vertebrates

In vertebrates, hearing is associated with the inner ear, comprising fluid-filled chambers and hair cells. In mammals, sound is captured by the outer ear and then causes vibrations of the tympanum (eardrum), which are transmitted via the bones of the middle ear to the inner ear, where they pass along the basilar membrane in the cochlea (Zwislocki 1981). The basilar membrane is covered with hair cells that are stimulated by the vibrating tympanum, with different hair cells responding to the different frequencies of sound. Vibrations propagate on the basilar membrane from the cochlear base towards the apex. Hair cells sensitive to high frequencies are located near the base, with sensitivity to low-frequency sounds (which travel further) at the cochlear apex (Ren 2002; Ruggero 1992). This conifers a tonotopic arrangement, whereby hair cells on different parts of the basilar membrane respond to different sets of frequencies. Each hair cell generally excites several sensory neurons.

The general situation above is broadly similar in birds (Knudsen and Gentner 2010; Köppl 2011) although there are differences in anatomy and sound vibration transmission between the different regions of the ear. Like mammals, auditory nerve fibres in birds are tuned to sensitivities that correspond to the location on the basilar membrane that they innervate (Konishi 1970). In frogs, species with tympanic ears have two hearing organs, the basilar papilla and the amphibian papilla, of which only the latter seems to have a tonotopic organization (Van Dijk *et al.* 2011). Hearing organs seem to have evolved independently in mammals and amphibians and perhaps multiple times in reptiles, but how similar the mechanisms of sound processing across these groups are is not yet clear (Grothe *et al.* 2010). Species also vary in the sensitivity of their hearing to sound levels and different frequencies of sound (Figure 2.9).

2.7.3 Echolocation

Some animals, especially bats and toothed whales (e.g. dolphins and porpoises) have evolved complex echolocation ('biosonar'), also found in a few other groups in more rudimentary forms (e.g. cave swiftlets, oilbirds; Jones 2005). Echolocation may also occur in some small nocturnal mammals, such as shrews (Siemers *et al.* 2009), and can even be learnt by humans (Schenkman and Nilsson 2010), albeit without the sophisticated neural mechanisms and pathways found in bats and dolphins but rather using brain regions normally devoted to vision (Thaler *et al.* 2011). Echolocation, like electrolocation, is an active sense (Nelson and MacIver 2006). It involves emitting sound into the environment and measuring changes in the returning echoes. Echolocation is primarily used for gaining information about position relative to other objects (spatial orientation), and in some species, for detecting and identifying prey.

Most of the approximately 1000+ species of echolocating bats live on insects, although many species have diverged to live on other sources, such as nectar, pollen, vertebrate prey (lizards, frogs, fishes), and even

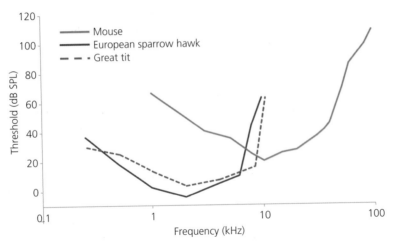

Figure 2.9 Threshold sensitivity curves for two birds: the European sparrow hawk (*Accipiter nisus*), and great tit (*Parus major*), and for the mouse *Peromyscus leucopus*. Data from Klump *et al.* (1986) and Ralls (1967). The graph shows the sensitivity of hearing to sounds of different frequencies played at different sound pressure levels. Greater sensitivity means that the animal can detect quieter sounds (lower sound-pressure level). The two birds have similar tuning curves, but the mouse is sensitive to higher frequencies of sound.

blood (Neuweiler 1989, 1990). Echolocating bats emit rapid sequences of high-frequency (11–212 kHz for the dominant frequency, but higher for some harmonics) sound pulses, broadcast through either their mouth or nostrils, and receive sounds with their large external pinnae (Jones 2005; Jones and Holderied 2007; Neuweiler 1989, 1990). Bats call at very high intensities, even beyond 140 dB peak sound pressure level at 10 cm for some insectivorous species (Surlykke and Kalko 2008). This helps to maximize detection range because sound has to both travel to and return from the target, and for small objects only a small proportion of the sound is returned. Many bats also contract muscles in the inner ear to avoid deafening themselves to outgoing calls (Jones and Holderied 2007) or have neural mechanisms that prevent the bat's own calls from stimulating its hearing pathways. However, not all call attributes have evolved for prey detection and orientation. Small divergences in the frequencies of echolocation calls of closely related bat species can facilitate intraspecific communication (see Jones and Siemers 2010; Schuchmann and Siemers 2010a,b). The way that echolocation is used in bats and the neuronal mechanisms are discussed in Chapter 3.

Bats are the most studied of echolocating animals, but an increasingly large number of the 70+ toothed-whale species are being tested for echolocation (e.g. Kyhn *et al.* 2010; Nachtigall and Supin 2008). Dolphins and porpoise produce sound from the nasal passages and transmit this through a waxy 'acoustic melon' on the forehead (Norris and Harvey 1974). The melon directs sound into a forwards beam (as occurs in bats; Surlykke *et al.* 2009), but also acts as an impedance-matching device. It has a low sound-velocity core and a high velocity periphery to convert the sound propagation speed to match that of seawater by increasing the speed closer to the animal's surface (Norris and Harvey 1974). This means that less energy is lost at the transition from animal to seawater than would be the case without the melon. The lower jaw is used in dolphins to receive echoes and transmit these to the inner ear (Brill *et al.* 2006).

The evolution of echolocation, especially in bats, has received much recent attention (see for example Jones and Teeling 2006; Jones and Holderied 2007; Teeling 2009). The gene *FoxP2* is involved in the development and control of mammalian vocalizations, including human speech (Teeling 2009; Teramitsu and White 2008). Interestingly, this gene seems to be conserved in most mammals but is variable in bats (Li *et al.* 2007). Perhaps even more exciting, however, is the role of the gene *Prestin*. This drives the motility of the outer hair cells in the cochlea of mammals, being responsible for amplifying sound waves and having a major role in conferring sensitivity and frequency selectivity (Jones 2010; Teeling 2009; Zheng *et al.* 2000). Li *et al.* (2008a) showed that mutations in *Prestin* may be associated with the evolution of constant-frequency echolocation calls, which often require precise tuning of auditory systems (see Chapter 3). Furthermore, molecular phylogenetic analysis of the *Prestin* gene in a range of species of bats and whales shows that echolocating dolphins clustered alongside echolocating bats in the tree (Li *et al.* 2010; Liu *et al.* 2010). In contrast, large-scale molecular analyses show that all bats form a monophyletic group separate from whales, suggesting that toothed whales and echolocating bats have undergone convergent changes in the *Prestin* gene that are associated with the evolution of echolocation (Jones 2010).

2.8 Future Directions

Although more work is always needed, we are gaining a good understanding of the type, function, and range of sensory receptors used in most senses. The main exception is magnetoreception, and a major goal is to understand better how animals detect and discriminate magnetic information and how the receptors work. Across all senses, more information on the physiology underlying receptor tuning and specificity would be valuable. In addition, recent work in echolocation, olfaction, electric fish, vision, and infrared sensors has shown how molecular changes underlie tuning, diversification, and convergence in sensory receptors. This will be a major avenue of future work and has enormous potential in illuminating processes of evolution and mechanisms underlying sensory systems.

2.9 Summary

Sensory systems have evolved to detect important cues and stimuli from the environment, and to discriminate these from noise (environmental or physiological) or other irrelevant stimuli. There are many different types of sensory system evolved to capture information from sound, light, electric, magnetic, mechanical, heat, and chemical stimuli. They involve converting information into nervous signals. The sensory worlds of animals vary greatly. Not only do different species possess different sensory modalities, but also within a given modality, the tuning varies across and even within species. The features of sensory systems that animals possess relates to the environment in which they live and important tasks they must complete in their lives (e.g. finding mates and food, navigating, avoiding predators). Nonetheless, there remain many common principles about how sensory systems work across species. Finally, some animals have evolved an active sense with the ability to emit signals (acoustic or electric) into the environment and measure changes in the properties of returning echoes or distortions in electric fields.

2.10 Further Reading

Bradbury and Vehrencamp's (2011) recently revised book on communication covers many of the sensory modalities discussed here in great detail, including much more information about the physical basis of different modalities and the types of sensory systems found in animals. Dusenbery (1992) also covers different sensory modalities, with emphasis on the physical properties of the types of information available. Greenfield (2002) covers sensory systems in arthropods in detail. Otherwise, many excellent textbooks are available that are devoted to specific sensory modalities. Regarding recent books, readers should consult Johnsen (2012) for an excellent (and entertaining) discussion of optics as relevant for biologists, and Land and Nilsson (2012) for discussion on animal eyes. Wyatt (2013) discusses olfaction in considerable depth across taxa, and Hill (2008) discusses vibratory communication.

CHAPTER 3

Encoding Information

Box 3.1 Key Terms and Definitions

Across-Fibre Pattern: The coding of a wide range of variable stimuli (such as volatile compounds in food odours) can be achieved by the relative activity of combinations of nerve cells rather than having a dedicated pathway for each component. Across-fibre patterns reflect the processing of different combinations of odorants, each of which can stimulate a range of receptors to different degrees.

Doppler Shift: Motion induced changes in frequency occur when a receiver moves towards (positive Doppler shifts, resulting in increased frequency) or away from (negative Doppler shifts, resulting in reduced frequencies) a sound wave.

Filter: These cut out irrelevant or unimportant information and pass or enhance important information.

Fovea: An area or property of a sensory system with a disproportionate weighting towards certain stimulus features of particular interest, for example specific frequencies of sound encoded by large numbers of hair cells on some vertebrate basilar membranes, or high densities of cone cells in a specific region of the retina of some animals.

Harmonic: Most sounds comprise more than just one frequency, including the fundamental (or first harmonic, being the principle frequency), and harmonic frequencies that are integer multiples of the fundamental frequency.

Lateral Inhibition: Connections between cells mutually inhibit each other when active, allowing sensory systems to enhance information about contrast.

Labelled Line: Some pieces of particularly important sensory information, such as pheromones in olfaction, are processed in dedicated pathways with specific receptors and corresponding nerve pathways.

Parallel Processing: Different aspects of information of a stimulus (such as timing and intensity in sound reception) may be processed separately in different neural pathways, and brought together later.

Receptive Field: On a basic level, this is the area of parameter/sensory space that influences the activity of a neuron. Interactions between nerve cells (such as via lateral inhibition) can create regions of importance that stimulate or inhibit a cell. A common type is centre-surround receptive fields, with a stimulus in the central region causing excitation, and a stimulus in the surrounding area causing inhibition.

In Chapter 2, we discussed how sensory receptors respond to signals and cues from the environment. Here, we focus on some examples, mainly from hearing, vision, and olfaction, to highlight important principles about how sensory information is processed and encoded by nerve circuits. In Chapter 1, we defined information as a reduction in uncertainty about the environment or an event. A fundamental question is how sensory systems enhance relevant information about a stimulus, including its strength, temporal structure, and composition. In many instances, sensory systems act like filters. They pass and enhance important information but cut out background noise or irrelevant information. Correspondingly, sensory systems respond primarily to change because this tells the animal something new. There is little point in encoding stability because this reveals no new information.

3.1 Contrast and Receptive Fields in Vision

3.1.1 Lateral Inhibition, Contrast, and Edge Detection

The (compound) lateral eye of the American horseshoe crab (*Limulus polyphemus*) has been a model system to investigate key processes such as lateral inhibition, filtering, and circadian rhythms affecting vision (Barlow 2009). Each *Limulus* ommatidium has a visual field, whereby light falling on this region excites the retinula cells (10–20 of them) and this passes on to the following eccentric cells. Pioneering work by Hartline and others (1956, 1957) showed that light falling on one area of the eye inhibits the outputs of neighbouring cells. When a single ommatidium is illuminated, but not the neighbouring cells, bursts of action potentials can be recorded from the eccentric cell. However, when light is shone on both the measured ommatidium and the surrounding region, the response of the eccentric cell is reduced. The inhibition between

---

Box 3.2 Transmission of Information Along and Between Nerve Cells

The following is a brief description of how information is transmitted along and between nerve cells. The aim is to provide basic background information to help understand the concepts in this chapter. Nerve cells (neurons) come in several types and vary greatly in physiology and anatomy. Broadly, sensory neurons respond to changes in the environment (here, we are concerned with the external environment), and transfer this information to the central nervous system. Motorneurons act on muscles to bring about responses and movements, and interneurons connect between neurons. A typical nerve cell is made up of a cell body, the axon (an elongated section of highly variable length along which nerve signals travel), dendrites (cell processes that branch out and receive signals from other neurons), and synapses (that transmit signals to other neurons).

Neurons generate and transmit electric information that conducts along the axon. One of the most common ways they do this is via action potentials. Action potentials involve changes in the membrane potential of a neuron, driven by a combination of electrical potential and ion concentration gradients. At resting potential, a gradient exists that is in the region of -60 to -80 mV, maintained by open potassium (K^+) channels. When a stimulus arrives at the neuron, it may depolarize. Here, voltage-gated sodium ion (Na^+) channels open, leading to an influx of Na^+ ions into the cell across the concentration gradient. If the influx of Na^+ causes the membrane potential to depolarize to a critical threshold then an action potential occurs (with a membrane potential of around $+40$ mV). The influx of Na^+ lasts for approximately 1 millisecond and the neuron starts to repolarize when potassium (K^+) ion channels open, allowing K^+ to leave the cell. After an action potential the neuron may briefly hyperpolarize, called the refractory period, owing to Na^+ channel inactivation. The refractory period places a limit on the number of action potentials that can occur in a given time and prevents the nerve impulse from travelling backwards. After an action potential, ion pumps restore the balance of ions inside and outside the cell to that before the action potential using ATP. Action potentials are all-or-nothing, and so their amplitude is relatively constant. Information (for example in terms of stimulus strength) is therefore encoded in the number and frequency of impulses.

Not all neurons produce action potentials. Others produce a graded response (especially sensory receptor cells). Here, stimulus intensity is encoded by the degree of de- or hyperpolarization. This allows greater information about stimulus strength because the response is continuous, whereas action potentials are all or nothing, with the upper frequency of action potentials limited by the refractory period to around 500 Hz. However, action potentials travel without decrement (i.e. they do not lose intensity as they travel along an axon), whereas graded potentials decrement with distance, meaning that they are only really valuable over relatively short

continued

Box 3.2 *Continued*

---

distances. Graded potentials can be either depolarizing (such as in insect photoreceptors) or hyperpolarizing (for example vertebrate rods and cones), whereas action potentials always involve depolarization.

Nerve-cell impulses are transmitted between neurons across synapses, which can be electrical or chemical. In electrical synapses, special channels called connexons connect pre- and post-synaptic cell membranes allowing current to flow through the cells. In chemical synapses, which are more common, a nerve impulse arrives at the pre-synaptic membrane. This depolarizes the membrane and causes an influx of calcium (Ca^{2+}) ions that triggers the binding of vesicles containing neurotransmitter to the pre-synaptic membrane (Figure Box 3.1). The neurotransmitter is released into the synaptic cleft and binds with receptor molecules in the post-synaptic membrane. Following this, depending on the properties of the receptor molecules, the post-synaptic membrane can either depolarize (for example, by allowing in Na^+ ions) or hyperpolarize (for example by allowing K^+ ions to leave or Cl^- ions to enter the cell). Consequently, the transmitted signal can be either an excitatory post-synaptic potential (EPSP) or an inhibitory post-synaptic potential (IPSP).

The use of synapses has several important implications and allows signals to be amplified, summed and subtracted. This can occur spatially, whereby two or more synaptic connections impinge on one post-synaptic neuron, or temporally with repeated signals from the same pre-synaptic neuron (Figure Box 3.2). Summation can produce an enhanced outcome if the combined effect of the pre-synaptic signals causes greater depolarization in the post-synaptic neuron and a greater EPSP. This can allow weak signals from more than one location to be amplified, or can be used to set a requirement that signals must to arrive from two different locations simultaneously before a responses occurs. Alternatively, IPSPs produced by some synapses could subtract from the excitation supplied from other synapses.

Finally, it is worth considering that neuronal signalling is not perfect: repeated presentations of the same stimuli can result in quite different neural responses (bursts of action potentials). Therefore, although the refractory period can limit response rate, variability in responses can also limit distinguishable/identifiable stimuli are.

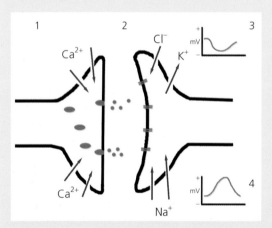

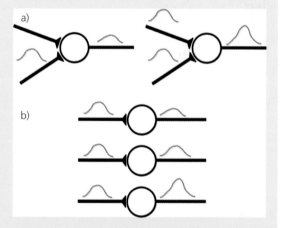

Figure Box 3.1 An illustration of signal transmission at a chemical synapse. At stage 1 a nerve impulse arrives causing depolarization of the pre-synaptic nerve cell, resulting in an influx of calcium ions. This causes vesicles containing neurotransmitter to fuse with the cell membrane and release neurotransmitter into the synaptic cleft (2), which then bind with the receptor molecules on the post-synaptic cell membrane. If an IPSP is produced (3), either chloride ions enter the cell or potassium ions leave the cell, causing hyperpolarization. If an EPSP is formed (4), sodium ions enter the cell, causing depolarization.

Figure Box 3.2 a) Illustrates spatial summation. Here, signals arrive at the post-synaptic neuron from two cells. If both cells signal at the same time then the response in the post-synaptic cell can be greater. b) Illustrates temporal summation. Repeated signals arriving from the same pre-synaptic cell can cause sequential increases in the size of the post-synaptic cell response.

neighbouring cells is mutual: lateral inhibition. This arises because each eccentric cell is connected to its neighbours. Broadly, the level of inhibition given to a specific cell is greater when *i*) the intensity of the illuminating light in the surrounding region is higher, *ii*) the area around the ommatidium under illumination is larger (although this effect declines with each increase in area; Barlow 1969), and *iii*) when the location of the illuminated region is close to the ommatidium being measured (Hartline *et al.* 1956). When two neighbouring ommatidia are both illuminated, the degree of inhibition experienced by each cell is linearly related to the degree of activity in the other, which is in turn directly influenced by the level of illumination (Hartline and Ratliff 1957).

Lateral inhibition in *Limulus* means that the outputs of ommatidia are not independent but are strongly influenced by activity in neighbouring regions. The discovery of lateral inhibition in *Limulus* has far-reaching implications. It shows how relatively simple processing mechanisms can act as filters, removing redundant information. In this case, the effect is to increase contrast because the outputs of neighbouring eccentric cells will differ most when cells on one side are illuminated and cells on the other side are in the dark (Figure 3.1). This is also a rudimentary edge or boundary detector because changes in contrast tend to be associated with the boundaries of objects or edges of patterns (Elder and Sachs 2004; Van Deemter and Buf 2000). It is not the overall light intensity across the eye that matters here, but the spatial change in illumination.

Encoding objects by edge and boundary information reduces the amount of data to be processed but retains key information about shape and location. Researchers have often aimed to determine how edge detection may work in animals, including in those with more complex visual systems. Some of the proposed algorithms are based on encoding sharp changes or gradients in intensity at different spatial scales (e.g. Marr and Hildreth 1980), and others are based on phase changes in local 'energy' detected by filters (e.g. Morrone and Owens 1987; Morrone and Burr 1988). The exact 'algorithms' used in edge detection in vertebrates are debated, and more complex cells are likely involved than in *Limulus*, but the underlying prin-

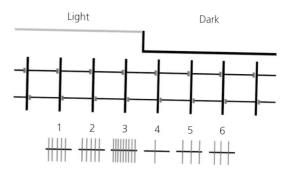

Figure 3.1 Lateral inhibition in *Limulus* eccentric cells. Cells 1–3 are illuminated and are therefore stimulated, whereas cells 4–6 are in the dark. Cell 3 is highly stimulated by the light and also receives little inhibition from cell 4 adjacent to it. In contrast, cell 4 is in the dark but receives substantial inhibition from cell 3. Therefore, the output of cell 3 is substantially greater than cell 4, enhancing the contrast between light and dark.

ciples are probably based on similar mechanisms of lateral inhibition.

3.1.2 Receptive Fields

Vertebrate retinae have three sequentially connected cell types: receptor cells (which produce a graded response), bipolar cells, and ganglion cells (which produce action potentials), with connections between them made by amacrine and horizontal cells (Ewert and Hock 1972). Interactions between these cells can give rise to so-called receptive fields. Analysis of ganglion cells in cats (Kuffler 1953) and frogs (Barlow 1953) showed that receptive fields often have a common arrangement: a 'centre-surround' organization (Figure 3.2). Illumination on cells of a central region is excitatory, whereas illumination on the surrounding area inhibits the ganglion cell. This is an 'on-centre off-surround' arrangement but 'off-centre on-surround' receptive fields also occur (Wandell 1995). In the central region of a receptive field, signals pass directly from the receptors to the bipolar cells and then on to the ganglion cells. Receptor signals from the surround region go to the ganglion cells via the horizontal, bipolar, and amacrine cells. Bipolar cells also have receptive fields. Although there is variation, receptive fields of this type are common in vertebrate visual systems.

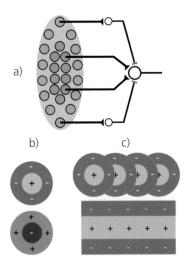

Figure 3.2 a) An example of a receptive field, such as of a retinal bipolar or ganglion cell. The cells on the left are photoreceptors, with ones in the centre corresponding to the excitatory region stimulating the receptive field cell. Cells from the surrounding region inhibit the receptive field cell. Modified from Kandel *et al.* (1995). b) Example 'on-centre off-surround' cells with an excitatory centre and inhibitory surround, and 'off-centre on-surround' with the opposite arrangement. c) Sets of receptive fields arranged in certain ways may be used to create line and edge detectors.

An important feature of receptive fields is inhibition. A stimulus on the central region increases the number of impulses from the receptive field cell, but the same stimulus falling on the surrounding region causes inhibition. This broadly means that the stimulus size that causes most excitation is approximately proportional to the size of the excitatory region of the receptive field (Ewert and Hock 1972; Lythgoe 1979). In *Limulus*, eccentric cell receptive fields do not respond to large patches of light or dark, but provide information on local contrast in a visual scene rather than absolute intensity. In the mammalian retina, the input from a receptor into a bipolar cell is inhibited based on responses of multiple cells around, and the effect is to remove average light levels.

3.1.3 Movement and Feature Detection

Toads (*Bufo* spp.) respond with prey-capture behaviour to a small horizontally elongated object moving from left to right in front of it (i.e. an object elongated parallel to the direction of movement). However, they do not attempt to capture objects elongated in

the horizontal plane (i.e. elongated perpendicular to the direction of movement) (Figure 3.3). They therefore have a visual system that encodes prey-like objects such as worms and insects (Beck and Ewert 1979; Ewert and Hock 1972; Ewert *et al.* 1978).

Toads have retinal ganglion cells with centre-surround receptive fields, some of which are sensitive to movement. Particular features of prey (e.g. size, orientation, and contrast) are encoded by different cells (Ewert and Hock 1972). Ganglion cells alone are not sufficient to categorize worm-like stimuli. In particular, certain types of retinal ganglion cells (class R2) along with cells in the brain (such as in the T5 neurons in the optic tectum and TH3 neurons from the thalamus) are most sensitive to worm-like moving stimuli owing to interactions between them (Borchers and Ewert 1979).

Prey-catching behaviour is elicited by local motion detectors in the optic tectum that respond to motion in only small areas of the visual field but do not respond to global motion over the majority of the visual field (Satou and Shiraishi 1991). This allows the toad to detect and respond to moving prey, but not to large-scale motion induced by its own movement or large background movements. Again, local motion detectors in the optic tectum can be explained by their receptive field properties because objects that are too large do not cause responses, being big enough to stimulate both the excitatory and inhibitory regions of the receptive field (Satou and Shiraishi 1991).

Prey-capture behaviour in toads is another example of how sensory systems often act as filters, passing information about worm-like stimuli but not anti-worm like stimuli by virtue of receptive fields.

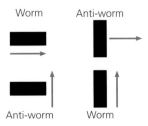

Figure 3.3 Toads will respond with prey-capture behaviour towards worm/insect like stimuli that are elongated parallel to the direction of movement (worm). However, they will not respond (or show defensive behaviour) towards stimuli elongated perpendicular to the direction of movement (anti-worm).

The toad's visual system also acts as a feature detector for biologically relevant stimuli (see also Chapter 4). Capturing worm-like prey is of great importance in survival, and so selection has favoured specific processes in vision to encode these stimuli.

Movement detection in animals goes way beyond prey capture. Motion detection is essential to guide movement in orientation and to respond to the actions and signals of others. A simple model of understanding motion detection is the Reichardt model (Figure 3.4). Here, neighbouring cells in the eye respond in sequence to a moving stimulus in a specific direction, with the output of one receptor delayed. If the two signals arrive at a motion detector (often called an elementary motion detector) at the same time, this produces a response and the animal can detect motion in a given direction (Krekelberg 2008). Some animals seem to have arrays of many elementary motion detectors. Understanding how visual systems encode the speed and direction of movement can be complex. Although it is just a model, the basic principle of the Reichardt detector is used in many models today to explain motion processing across various animal groups (Burr and Thompson 2011), although in many cases there is no direct evidence that a Reichardt-type process in particular underlies motion perception.

In mammals, such as rabbits, motion is encoded by various cells including direction-sensitive ganglion cells and interneurons that impinge on these called amacrine 'starbust' cells ('starbust' because of their almost circular radiations of dendrites; Fried and Masland 2007). The amacrine cells provide the inhibition to make other cells direction-sensitive, but they also have directional sensitivity themselves. This stems from a receptive field with an excitatory region and an inhibitory region where amacrine cells reciprocally inhibit each other (Lee and Zhou 2006).

3.2 Sound Localization in Barn Owls

For over 40 years, the barn owl (*Tyto alba*) has been a model system to study sound detection and localization and the neural mechanisms involved (Takahashi 2010). Payne and Drury (1958) and Payne (1971) demonstrated that barn owls (*Tyto alba*) locate and capture prey (such as rodents) in the dark based on sound. First, they showed that owls could successfully capture a running mouse on a substrate of dry leaves in complete darkness. Second, the owl would attack a piece of moving paper dragged through the leaves, discounting odour or infrared cues. Third, blocking of either ear meant that the owl would misdirect its attacks short of the location of the mouse, but accuracy was restored when the earplugs were removed. Finally, replacing the substrate with sand, which eliminated sounds from the mouse running, resulted in a loss of detection by the owls or of missing the prey. Tying a piece of paper or dry leaf to the mouse's tail, however, caused the owl to attack the leaf/paper. Next, the use of speakers allowed Payne (1971) and Konishi (1973) to show that owls respond most to frequencies of about 6–9 kHz, and that they make fewest localization errors for sounds between 4–8 kHz (Knudsen *et al.* 1979).

3.2.1 Morphological Adaptations

How do barn owls locate and capture prey in the dark? First, they have a number of morphological adaptations. The disk-like face of the owl is covered with acoustically transparent feathers that do not reduce the intensity of arriving sounds and it acts as a sound collector (Payne 1971). In addition, the ear-flaps are asymmetrically positioned on either side of the body. The ear openings are approximately the same size, but the left ear is located higher than the right ear and an asymmetry in the facial ruff (feathers) means that the left ear is more sensitive to sounds from below (Konishi 1973). Plugging either ear, which results in little change in the time sounds

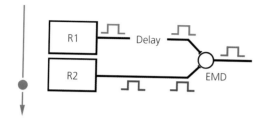

Figure 3.4 A simplified version of the Reichardt model for motion detection. A moving stimulus passes receptor 1 (R1) first, followed by receptor 2 (R2), causing a signal in both pathways. The pathway from R1 is delayed, so that signals from both R1 and R2 arrive at the elementary motion detector (EMD) cell at the same time, triggering a response. Because the pathway from R1 is delayed, only motion passing from R1 to R2 will cause a response, and so the model is direction specific.

arrive at the ears but lowers the sound intensity at the plugged ear, results in systematic errors in the vertical plane (elevation) but only minor errors in azimuth (horizontal plane) (Knudsen *et al.* 1979; Knudsen and Konishi 1979; Konishi 1973). In the inner ear, owls also have a longer cochlea than many other birds, and the increased length of the basilar membrane may allow a greater range of frequencies to be detected and discriminated (Payne 1971). Like other vertebrates, owls have the characteristic pattern of sensitivity on the basilar membrane, with progressive changes in frequency sensitivity occurring along its length. However, recordings of primary auditory neurons show that owls have an expanded area of sensitivity, with approximately half of the length of the membrane being most sensitive to frequencies of 5–10 kHz (Köppl *et al.* 1993). This disproportionate representation of frequencies is akin to the idea of a fovea in visual systems, and corresponds to frequencies that the owl relies on most in sound localization (Knudsen and Konishi 1979).

3.2.2 Time and Intensity Information

The owl locates its prey in both the horizontal (azimuth) and vertical (elevation) planes. Sounds from different angles in the horizontal plane arrive at different times at the two ears. Correspondingly, localization in the horizontal plane is based on interaural time differences (Knudsen and Konishi 1979). The hair cells are 'phase-locked' and produce a response only to specific points on (phases of) the sound wave. This allows very fine discrimination and accuracy in assessing timing differences because the auditory system can compare the phases of a sound wave arriving at each ear.

Louder sounds cause responses in more hair cells and their sensory neurons. The intensity of sounds reaching the owl's ears depends upon the angle the sound is coming from (Payne 1971). This is significant owing to the asymmetry of the left and right ears. The right ear is most sensitive to sounds from above the horizontal plane, and the left ear to sounds below this (Knudsen *et al.* 1979). This provides information about differences in intensity between the two ears, which is used in determining elevation of a sound source. Therefore, interaural time differ-

ences (ITD) encode the position of a sound source in the horizontal plane (azimuth), and interaural level differences (ILD) in sound pressure level or intensity encode sound source location in the vertical plane (elevation) (Takahashi 2010).

In the auditory brainstem, timing and intensity are processed by parallel pathways (Figure 3.5), which are then combined later into a spatial map. Intensity is processed via the nucleus angularis, and then to the posterior lateral lemniscus where differences in intensity between the left and right ear are encoded. This forms a map of intensity differences at different frequencies. Time (phase) for each ear is encoded by the nucleus magnocellularis (NM), with interaural time differences (phase differences) then encoded in the nucleus laminaris (NL), followed by the anterior lateral lemniscus. The two pathways are then combined in the central nucleus of the auditory midbrain (inferior colliculus).

3.2.3 Coincidence Detection

There is a problem with using time differences to localize a sound source. The ears of the owl are close together and differences in the timing of sound reaching one ear and the other are minute; for example, in the range of microseconds. However, action potentials last for around a millisecond. So how can the owl encode timing differences that are so small? The answer lies in 'coincidence detection'. It takes longer for a nerve signal to arrive at a given point if the length of the axon is increased or other delays

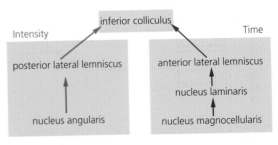

Figure 3.5 Parallel processing of time and intensity in barn owl hearing. Interaural time and intensity differences are processed separately in different pathways, and then brought together in the inferior colliculus, where a spatial map is formed, combining time and intensity differences.

(such as synapses) are added. This means that a series of delay sensitive neurons, which respond only when they receive stimulation from both left and right sides at the same time can be created, each encoding specific time differences (Figure 3.6). In the owl, information about the time a sound arrives at each ear (interaural time difference) is processed by neurons of the nucleus magnocellularis (NM) and nucleus laminaris (NL). Axons from the NM have different lengths and act as delay lines, and neurons in the NL respond to different delay combinations from axons from the left and right NM. These are coincidence detectors and only respond

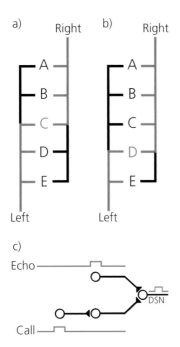

Figure 3.6 Example models of coincidence detection, which combine inputs separated in time. a) and b) illustrate how coincidence detection could work in barn owl hearing, with an array of delay-sensitive neurons, each of which only responds when signals from both the left and right ear arrive at the same time. In a) a signal arrives at both ears at the same time, travels down the two axons and causes cell C to respond. In b) a sound arrives at the right ear first, and consequently the signal travels further before it coincides with the sound signal from the left ear, causing cell D to respond. c) is another example of a coincidence detector, as proposed by Sullivan (1982b) with respect to echolocation in bats. Here, the bat emits a call and retains a neural copy of the signal, which is delayed (e.g. by longer axons or synapses) before reaching the delay sensitive neuron (DSN). If the signal from the returning echo arrives at the same time as the call signal, then the delay sensitive neuron will respond.

to simultaneous stimulation from axons from both the left and right sides. They are capable of encoding time differences of less than a few tens of microseconds (Ashida *et al.* 2007; Carr and Konishi 1990; Carr and Boudreau 1993; Peña *et al.* 1996). The NL therefore creates a map of interaural time differences.

3.2.4 Auditory Spatial Map

Knudsen and Konishi (1978b) found neurons with receptive field properties in the external auditory nucleus (located in the inferior colliculus) that shows how time and intensity differences are used. These cells respond to combinations of elevation and azimuth (specific combinations of ITD and ILD); that is, the neurons respond to sound stimuli in a particular area and provide a map of auditory space. The receptive fields are characterized by an excitatory centre and antagonistic surround (Knudsen and Konishi 1978a). Sounds falling on the outer portion of the receptive field are inhibitory, whereas sounds falling in the centre are excitatory. The receptive fields enhance spatial contrast independent of absolute stimulus intensity, in much the same way as in vision (see above). The difference with features of receptive fields in vision, however, is that visual centre-surround receptive fields can be based on lateral inhibition of neighbouring receptors because spatial information falls directly onto the retina, whereas in hearing further neural architecture is needed to bring together information from different ears and so the map is a computed one (Knudsen and Konishi 1978a). Note that visual and auditory information is combined during stimulus processing in barn owls to increase accuracy in detection and localization (see Chapter 4).

3.3 Echolocation

Having discussed the ability of barn owls to locate sound sources, we now discuss echolocation (mainly in bats) as an active sense (see Chapter 2), and what this can tell us about sensory processing. In their extreme forms, bat echolocation calls can be either short (about 1–10 ms) frequency modulated (FM) calls comprising downward sweeps in frequency, or relatively long (10–100 ms) largely constant frequency

(CF) calls (Neuweiler 1990; although the calls of many bat species comprise a mixture of these). Note that the category of CF calls should not be mistaken to mean that only one frequency is present in the call; in fact, bats with CF calls often use a fundamental frequency (first harmonic), plus a range of other harmonics. The key thing with CF calls is that all the harmonics are of approximately constant frequency. Likewise, bats with FM calls can either mainly use calls with one harmonic or use a mixture of harmonics. In bats with long CF components, the fundamental sometimes has less energy than the other harmonics, and some species use this in making comparisons between the emitted call and the harmonics of the returning echoes (see below).

3.3.1 Information from Echolocation

The properties of returning echoes enable bats to determine a range of important features about their environment and potential prey. Different species of bat forage in varied habitat types and niches, and this affects the way that echolocation works (see Neuweiler 1989, 1990). For instance, fast-flying species tend to forage for insects above vegetation, and here the environment is largely acoustically quiet (little acoustic clutter from other objects) and with low densities of prey. These species need to have echolocation calls that cover large distances with high energy in order to detect prey and to cope with sound decay due to spreading and attenuation. Some bats (e.g. *Macrophyllum macrophyllum*) emit more intense calls in habitats that are less cluttered and this increases detection ranges (Brinkløv *et al.* 2010). Other species of bat forage in or close to vegetation and have to deal with an array of returning echoes ('echo clutter') from the environment. One of their main problems is that echo clutter may mask the returning echo of a prey animal. Many species in this type of foraging environment emit a CF call with frequency modulated (FM) components (Neuweiler 1984) or short FM calls to reduce overlap between prey echoes and background clutter (e.g. Kalko and Schnitzler 1993), or differences in echoes between harmonics (Bates *et al.* 2011). Below, we discuss some of these important attributes and how they are processed by the bat auditory system.

3.3.1.1 Target Direction

Information in the horizontal plane (azimuth) is given by differences in intensity arriving at each ear (interaural intensity difference). Bats combine effective directional hearing with a wide range of adaptations in their pinnae to facilitate sound detection and localization. Movements of the pinnae can also help in determining elevation based on the direction the pinnae face and direction-dependent spectral properties of a sound.

3.3.1.2 Target Size

Echo intensity gives information about target size. Larger targets reflect more sound, leading to greater intensity of returning echoes. However, intensity is also a function of range, with more distant targets reflecting lower intensity. Therefore, the combination of echo delay and intensity encodes target size. Recent work has also shown that the spatial width of the returning echoes ('sonar aperture') can contribute to size perception in bats, and that there are neurons in the auditory midbrain and cortex that respond most to echoes from objects with specific sonar apertures (Goerlitz *et al.* 2012; Heinrich *et al.* 2011).

3.3.1.3 Target Identification

Differences in the spectrum of the echo can give information about surface texture. FM calls in bats seem especially useful in cluttered environments to separate targets from background vegetation, by gaining many spectral samples of different frequencies of the environment (Jones and Holderied 2007). The echoes returning from potential prey can be quite different in characteristics from echoes returning from vegetation, as can echoes from special leaf structures from some plants, used to direct pollinating bats to flowers (see Chapter 5). CF calls can also be useful in detecting prey movements (acoustic 'glints') against clutter.

3.3.1.4 Target Range (Distance)

Range or distance to an object is encoded by the time delay between the emitted signal and the returning echo. Longer delays indicate greater distances. Generally, FM calls are thought to give most information about range because they give multiple samples at different frequencies and are very short (and therefore temporally precise) (Feng 2011; Neuweiler 1990). CF calls are generally not so helpful because they are often longer in duration than FM calls and CF calls could overlap with returning echoes.

Sullivan (1982a) found the auditory cortex of the bat *Myotis lucifugus* contains delay-sensitive neurons that may encode specific delays corresponding to returning echoes by coincidence detection. This is an example of a sequential delay, encoding time between an emitted call and a returning echo, as opposed to an interaural delay encoding differences in time that a sound arrives at two ears, as in barn owls and various other species. Sullivan (1982a) suggested that the neurons were involved in encoding distance, and proposed a model based on the idea that the input to neurons from the emitted call is delayed relative to the echo, with the two signals arriving together in time at a delay-sensitive neuron (Sullivan 1982b; Figure 3.6c). Different delays would correspond to specific distances and would be encoded by different delay-sensitive neurons.

The inferior colliculus and auditory cortex of horseshoe (Rhinolophidae) and mustache (*Pteronotus parnelli*) bats have neurons that respond to pairs of stimuli of specific frequencies separated by a time delay. These two species are distantly related and so provide nice examples of convergent evolution in sensory processing. Neurons encoding different time delays are found in different locations in the array. In the FM–FM area of the auditory cortex, neurons respond specifically to delays between the first harmonic of the emitted call and different harmonics of a returning echo (Neuweiler 1990). Neurons with different delay sensitivities are arranged topographically in different locations, in a map of time delays encoding different ranges. Time delay neurons seem to be found in all echolocating bats and encode times that are consistent with distances of prey-capture behaviour. However, the exact arrangement varies across species.

3.3.1.5 Velocity and Target Flutter (in CF bats)

Velocity can be encoded by the phenomenon of Doppler shift. This describes changes in the frequency of sounds arriving at a receiver, owing to the motion of the receiver (or the object being detected) towards or away from the sound source. A useful analogy is to imagine you are on a beach with the waves from the ocean coming towards you at a constant frequency. If you run directly towards the waves then the frequency that they reach you would be higher (positive Doppler shift), whereas if you run away from the waves the frequency would be lower (negative Doppler shift). Changes in frequency caused by Doppler shift can be useful to bats because as the bat flies more quickly it will encounter its returning echoes faster, creating a greater positive Doppler shift. The same principle applies to even finer changes in frequency caused by the movement of insect wings. As an insect beats its wings, they move towards and away from the source of the echolocation call. As they do so they cause small positive and negative Doppler shifts (Figure 3.7). The exact speed and movement of the wing beats creates characteristic oscillations frequency owing to this Doppler shift effect. Bats of the CF-type use this to detect flying insects against the background, and even identify them based on the oscillations (Lazure and Fenton 2010; Schnitzler and Denzinger 2011).

In bats that are able to detect and analyse Doppler shifts, the species are of the CF type because, being of the same frequency, it is easier to measure motion-induced changes in echo frequency. In those species, the auditory system contains high specialization to the frequencies of CF echoes, not just of the species but often of the individual bat, and even to a level that is highly tuned to detect Doppler shifts caused by insect wing beats (Neuweiler 1990; Schnitzler and Denzinger 2011). Specialization begins with an

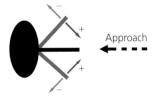

Figure 3.7 Insect-wing beats cause characteristic oscillations in Doppler shift that are detected by echolocating bats. As the insect moves its wing upwards and downwards towards its midline, it creates positive shifts in received frequency with respect to a bat approaching from the right. Conversely, when the wings are moved away from the midline (and bat), then negative Doppler shifts are created.

auditory fovea (Bruns and Schmieszek 1980; Schuller and Pollack 1979) in the cochlea, comprising an expanded frequency representation on the basilar membrane. This specialization continues throughout the auditory pathway, including nuclei in the lateral lemniscus and inferior colliculus, and of the auditory cortex (Bruns 1976a; Kössl and Vater 1985; Neuweiler 1984, 1989, 1990; Suga and Jen 1975, 1977). Auditory neurons can be highly sensitive to prey flutter, encoding modulations as small as 10 Hz (Neuweiler 1989).

The greater horseshoe bat (*Rhinolophus ferrumequinum*) has a (FM–)CF–FM call with a long (50 ms) CF component of about 83 kHz (Suga *et al.* 1976). Specialization at the inferior colliculus (auditory midbrain) to frequencies of 83–86 kHz stems from morphological specializations at an enhanced area of the cochlea (Bruns 1976a,b; Bruns and Schmieszek 1980). The auditory nerve fibres and cochlear nuclear neurons are also sharply tuned to 83 kHz (Schnitzler *et al.* 1976; Suga *et al.* 1976). About 50% of the neurons in the inferior colliculus are tuned to frequencies between 78–88 kHz (Suga *et al.* 1976), with just over 16% of all neurons devoted to frequencies between 83–84.5 kHz (Schuller and Pollack 1979). The over-representation of these frequencies is maintained into the auditory cortex in the CF area, which is even more sensitive to frequency changes (Bruns and Schmieszek 1980).

As with the greater horseshoe bat, the mustache bat uses the second harmonic of its CF call component, with frequencies of 61–62 kHz in Doppler shift analysis and, again like the horseshoe bat, has an inner ear highly tuned to these frequencies and to other harmonics (Kössl and Vater 1985; Suga *et al.* 1975). Up to 50% of the cochlea length is occupied by a relatively small range of frequencies (50–74 kHz) and there is very sharp tuning within this region to specific frequencies (Kössl and Vater 1985; Neuweiler 1984, 1989, 1990; Suga *et al.* 1975; Suga and Jen 1977). In addition, Doppler-shifted echoes of the CF component are encoded by narrowly tuned neurons that produce responses to the combination of the bats emitted call and the returning echo (Suga *et al.* 1975). These highly tuned neurons can detect Doppler shift frequency changes as small as 0.01%, enabling the bat to encode information about wingbeat frequency

(Suga and Jen 1977). Measurements from neurons in the inferior colliculus show that it dedicates 50% of its neurons to a band of frequencies just 350–450 Hz wide, occurring around 61–63 kHz depending on the individual (Pollak and Bodenhamer 1981). Again, this tuning seems to be at least partly established by the specialization at the cochlea.

A problem with having an auditory system that is so finely tuned to small differences in frequency is that the bat can perceive Doppler shifts (increase in echo frequency) created by its own velocity when flying. Bats like the mustache and greater horseshoe bat with CF–FM calls use Doppler shift compensation (see review by Schnitzler and Denzinger 2011). By adjusting the frequency of their calls during motion, they can keep the frequencies of the returning echoes within the range of their acoustic fovea. Positive Doppler shifts, such as those experienced during flight, cause bats to lower the frequency of their calls by an amount approximately the same as the Doppler shift (Pollak and Bodenhamer 1981; Suga *et al.* 1976), so that the echo always returns at the frequency where auditory sensitivity is highest. Overall, the horseshoe and mustache bats represent excellent examples of convergent evolution, whereby sophisticated echolocation calls and processing have evolved in these two species, each using slightly different mechanisms for Doppler shift analysis and compensation.

3.4 Olfactory Processing

In Chapter 2 we discussed how olfactory receptor neurons (ORNs) in animals respond to different odours, with some being narrowly tuned and others broadly. The way that information is encoded by the olfactory system reveals how processing of sensory information relates to the type of stimuli detected. Olfactory system must encode potentially many thousands of odorants. In addition, odours often occur in a noisy environment, and so the olfactory system must enhance contrast between relevant odours and the background (Smith and Getz 1994). How is this achieved?

Axons of the ORNs go to the antennal lobe in insects and the olfactory bulb in vertebrates. A common feature of olfaction in both vertebrates

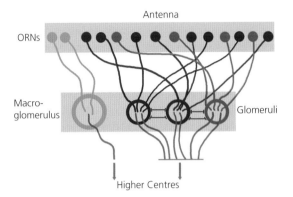

Figure 3.8 In insects, processing of odour information begins with odour reception in the olfactory receptor neurons (ORNs) in sensillae on the antenna. Signals then pass to the glomeruli, with some interactions between the glomeruli by interneurons. ORNs of the same type pass to the same glomerulus, before heading onto higher centres via projection neurons. Processing of each pheromone component (molecule) in males often occurs in an enlarged macroglomerulus, with the number of glomeruli in the macroglomerular complex reflecting the number of pheromone components.

and invertebrates is that processing occurs via glomeruli (Figure 3.8). Glomeruli contain synapses between the olfactory receptor neurons and projection interneurons. Different patterns of activity in glomeruli occur with different chemical stimuli. ORNs that have the same receptor type converge on the same glomerulus, so that activity in an individual glomerulus reflects the responses of a specific ORN type (Bargmann 2006; Christensen 2005; Hansson and Stensmyr 2011; Wyatt 2013). There are typically 50–300 glomeruli in arthropods, and in the cockroach there is a convergence of about 1000:1 from ORNs to glomeruli (Hildebrand and Shepherd 1997). Projection neurons from the glomeruli then pass to higher centres, where the outputs are integrated with information from other sensory systems, as well as aspects of experience and learning (Bargmann 2006; Christensen 2005; Su *et al.* 2009).

3.4.1 Gain Control and Stimulus Amplification

Projection neurons also respond to sensory input from a single glomerulus, relaying information from a specific type of receptor (Bargmann 2006;

Masse *et al.* 2009; Su *et al.* 2009). In the glomeruli there is a high level of convergence of ORNs to projection neurons (50:1 in *Drosophila*; Su *et al.* 2009). Convergence of many ORNs to one projection neuron allows the amplification of weak signals from the peripheral sensory system by summating information from ORNs of the same type (Bargmann 2006; Masse *et al.* 2009; Su *et al.* 2009). The averaging of strong signals also increases the signal to noise ratio. In *Drosophila*, the amplification of signals seems to occur by a strong synaptic release of neurotransmitter between the ORNs and the projection neurons (Su *et al.* 2009). There is strong neurotransmitter release even with weak pre-synaptic ORN activity, causing strong activation of projection neurons. In contrast, strong pre-synaptic ORN activity may lead to depletion of neurotransmitter and little amplification (Su *et al.* 2009). This explains how non-linear amplification of ORN activity occurs in projection neurons, where weak signals are amplified more than strong signals. Flies may detect concentrations of odours over eight orders of magnitude with their ORNs, and non-linear amplification could prevent saturation of projection neurons (Masse *et al.* 2009; Su *et al.* 2009). Amplifying weak ORN responses more than strong ones in the projection neurons acts as a 'gain control' mechanism (Masse *et al.* 2009).

Although different glomeruli encode responses from different ORN types, in some cases projection neurons may integrate information from several sources. There is growing evidence of tuning of glomeruli due to interactions and lateral inhibition between glomeruli via local interneurons (Bargmann 2006; Christensen 2005; Masse *et al.* 2009; Su *et al.* 2009). Lateral inhibition across glomeruli, related to overall ORN activity across a whole antenna, may also prevent saturation because inhibition is pre-synaptic to the projection neurons (i.e. on the ORN axon terminals) (Masse *et al.* 2009; Su *et al.* 2009). Information about odour identities is therefore encoded by relative activities in different glomeruli. High overall levels of ORN activity reduce projection neuron activity, and the coding of odours reflects a balance between amplification mechanisms both within and across glomeruli (Masse *et al.* 2009).

3.4.2 Encoding Different Odours

Olfactory processing has many mechanisms to enhance contrast and sensitivity, but how are different types of stimuli encoded? First, important stimuli are sometimes given a dedicated pathway. For example, in the American cockroach (*Periplaneta americana*), males have two specialist types of receptors, each responding respectively to the key components of the female pheromone (periplanone A and B; Sass 1983). Subsequently, there is a male-only enlarged macroglomerulus specifically for processing of each of these sex pheromones (Hildebrand and Shepherd 1997). This is a 'labelled line', where a specific odour stimulus has a dedicated pathway, beginning with narrowly tuned receptor cells (see Chapter 2) through to encoding mechanisms in the nervous system (Masson and Mustaparta 1990; Smith and Getz 1994). Other specializations exist in species to encode stimuli of high importance, such as host–plant chemical cues (e.g. Kalberer *et al.* 2010), and perhaps recognition templates in social insects (Ozaki *et al.* 2005), although more work is needed to test this.

Labelled lines are, however, not efficient for encoding many odours, and common stimuli such as food comprise odours with hundreds of different volatile chemicals. Most odours are also only relevant or around an animal for a limited time. Encoding all the potential odours that an animal may encounter in its life via labelled lines would be almost impossible (Smith and Getz 1994). Olfactory coding of non-pheromonal odours can be understood by an 'across-fibre pattern' model (Getz and Akers 1997; Masson and Mustaparta 1990). In the American cockroach, Selzer (1984) presented individuals with different stimuli corresponding to food odours (e.g. pentanol, hexanol, dodecanol, citronella) and made recordings of olfactory receptor cells from the antennae. Different cells responded more to certain odour stimuli: some cells were relatively specialist in response to specific odour components, whereas others had a broader response to wider combinations of stimuli. In an across-fibre pattern, the activity in a neuron and across glomeruli reflects the different components present. Specific combinations and ratios of compounds generate specific neural patterns (Masse *et al.* 2009; Masson and Mustaparta 1990). The pattern of spatial and temporal activity across many receptors codes for odorant identity (Smith and Getz 1994; Su *et al.* 2009).

Pheromones have a crucial role in the life history of a species, for example the use of sex pheromones in mating. They are also constrained within a species. This explains why responses to pheromones are often innate rather than learnt, and why sensory systems invest in high numbers of dedicated sensory cells and in labelled-line processing. Cues that are highly diverse and variable (e.g. food odours) cannot be specified along specialist pathways and therefore need across-fibre coding based on the responses of different combinations of receptors. In reality, however, across-fibre and labelled-line systems are probably two extremes of a continuum (Smith and Getz 1994).

3.5 Common Principles across Species and Modalities

In this chapter we have concentrated on a limited number of examples to illustrate key principles of how sensory systems encode salient information. We should be careful of over-simplification because differences and exceptions always occur between species and taxa. However, it is nonetheless apparent that many concepts apply broadly across very different groups of animals and sensory systems. For example, Hartline *et al.* (1956) noted that in *Limulus* the output of receptor cells declines gradually when continuously illuminated due to adaptation. Likewise, adaptation occurs in ORNs, enabling them to function over a broader dynamic range and detect odours above background levels (Su *et al.* 2009). Thus we come back to one of the main messages of this chapter: sensory systems respond to change because this tells the animal something new about the environment. Adaptation to continuous stimuli acts as a filter, only encoding relevant changes.

We do not need to look far for other common principles. Analogous to echolocation in bats, the sensory systems of electric fish are often tuned to the frequency of electric organ discharges (EODs) of individuals (Zakon *et al.* 2002). Bats sometimes change the timing and/or frequency of their echo-location calls (or in some instances cease calling altogether), to avoid interference ('jamming') from the calls of other individuals around them (Chiu *et al.* 2008; Jarvis *et al.* 2010). Many electric fish also have a jamming avoidance response (JAR). This involves individuals changing their EOD charac-teristics to avoid overlap with individuals around them with similar EODs. We discussed above how gain control is a mechanism in olfactory process-ing, but this is also found in active senses. Dol-phins and porpoise adjust both the amplitude of their calls and the sensitivity of their hearing in line with distances to targets (Au and Benoit-Bird 2003; Linnenschmidt *et al.* 2012; Nachtigall and Supin 2008). This compensates for relative loss of sound with propagation through the water and keeps echo intensity at optimal levels for the audi-tory system independent of target distance and size. Bats have similar gain control mechanisms (e.g. Hartley 1992; Kick and Simmons 1984).

Adaptations for heightened sensitivity and processing to relevant stimulus attributes also occur widely. Barn owls and bats have an acoustic fovea to enable high sensitivity to important sound frequen-cies. In human retinae, for example, the longwave- and mediumwave-sensitive cones are most densely distributed in a central region of the retina called the fovea. The high density of just two cones provides high acuity vision. Changes in sensitivity also occur on a temporal scale. In *Limulus*, for example, modi-fications in the sensitivity of the visual system are partly controlled by circadian rhythms to ensure effective vision under low light levels at night (dark adaptation) by reducing spontaneous noise and increasing receptor sensitivity (Barlow 2009, 1983; Kaplan and Barlow 1980; Pieprzyk *et al.* 2003).

The parallel processing found in barn owls for timing and intensity differences exists elsewhere too. For example, the fish lateral-line system has superficial neuromasts that respond to water motion, and canal neuromasts that respond to pres-sure gradients (Chapter 2). It seems that outputs from superficial and canal neuromasts are proc-essed in two parallel pathways to the brain, provid-ing information about surface velocity and pressure gradients separately (Bleckmann 2008). Receptive fields and maps, like those found in vision and hearing, are also found in lateral-line processing in fish and amphibians (Bleckmann 2008), and in the processing of infrared information in snakes (Hart-line *et al.* 1978; Newman and Hartline 1982). There-fore, while we must be careful of overgeneralizations, looking for common principles can reveal much about how sensory systems have evolved to encode information.

3.6 Future Directions

More work in different groups of animals is need to determine the degree of across-fibre patterns and labelled-line processing that is found in different olfactory systems and how much of a continuum exists. In general, more comparisons across modali-ties would also be valuable in establishing how wide-spread key concepts such as fovea, receptive fields, parallel processing, and so on are, and how the neu-ral basis differs across modalities. In particular, such work should relate these findings to the ecology of the species and key behaviours or specific tasks (e.g. prey capture) that may be central to processing.

3.7 Summary

Sensory systems need to encode important informa-tion from irrelevant background noise. In many cases, they respond to and encode changes in the environment as this tells the animal something new. The importance of filtering out irrelevant informa-tion and enhancing important features is demon-strated by numerous mechanisms that exist at various stages of sensory processing and in many common approaches across taxa and sensory modalities. Principles such as filtering, feature detectors, receptive fields, adaptation, fovea, and parallel processing are found widely. Such process-ing is made possible by the high flexibility in the way they nerve cells can behave, and crucially the way that sets of nerve cells are connected. This does not mean that the neural mechanisms are the same across species in processing, but the range of com-mon principles shows how sensory systems have evolved to encode different pieces of information about important stimuli.

3.8 Further Reading

Simmons and Young's classic text (2010) covers a great deal about how nervous systems bring about behaviour, including many of the examples in this chapter (and others) in more detail. Readers are also encouraged to consult recent neurobiology textbooks to gain a fuller understanding of how sensory and nerve cells work, beyond the simplified overview presented here in Box 3.2. Snowden *et al.* (2006) covers many of the issues regarding visual processing of contrast, colour, and motion in more detail.

CHAPTER 4

Sensory Systems: Trade-Offs, Costs, and Sensory Integration

Box 4.1 Key Terms and Definitions

Feature Detector: Some cells in sensory systems combine information (from nerve cells and receptors) in such a way as to only respond to specific combinations of cues, which correspond to a particular stimulus type, such as a prey item.

Mono-, Di-, Tri-, and Tetrachromacy: Colour vision depends upon the relative stimulation of more than one receptor class, sensitive to different wavelengths of light, and their neural interactions. Monochromatic animals have just one receptor type, or do not make comparisons between receptor types, and so do not have colour vision. Di-, tri-, and tetrachromatic animals have colour vision involving the use of two, three, or four receptor types respectively.

Sensory (Multimodal) Integration: Combining information from two or more senses can decrease the level of uncertainty about a stimulus. This can reduce detection times and increase the accuracy of localization. In some cases, it is also important in correctly identifying and classifying a stimulus.

In Chapters 2 and 3 we discussed the range of sensory systems that exist, and how these systems have evolved to effectively gather and encode information. It is clear that sensory systems are remarkably effective in what they do and crucial to survival and reproduction. However, they require considerable investment. In addition, sensory systems normally have to function in many tasks, and these can sometimes require different and conflicting features of sensory processing. Therefore, sensory systems often face trade-offs. In this chapter we discuss the costs involved in producing and maintaining sensory systems, and the types of trade-offs that can exist. We will ask if and when sensory systems have evolved or been tuned to perform a specific important task in a species, as opposed to being generally used for a wide range of functions. Finally, we will discuss how integration of information from multiple sensory modalities may overcome trade-offs and allow animals to reduce uncertainly in detecting and recognizing important cues and stimuli.

4.1 Energetic Costs of Sensory Systems

Sensory systems require considerable energetic investment to encode and transmit information. Therefore, they are often selected against when, for example, a population moves into an environment where a sensory system is no longer important. Selection against costly sensory apparatus that are of reduced value explains, for instance, the loss of vision in cave-dwelling animals. Moreover, in populations of *Drosophila* flies that have been kept in captivity, individuals show decreases in eye size over time, presumably because there is less need for eyes in captivity and selection favours saving energy (Tan *et al.* 2005). The cost of sensory systems can place constraints on their evolution and they therefore need to be efficient.

One of the primary costs in maintaining an effective sensory system is the energy required (Laughlin 2001; Niven and Laughlin 2008). The process of pumping sodium and potassium ions across neural cell membranes (see Chapter 3) requires substantial amounts of ATP when analysed over the whole nervous system. In fact, 50% of the total energy expended by mammalian brains seems to be driven by cell signalling, especially due to sodium and potassium pumps, which are highly concentrated in nerve cells (Laughlin 2001). ATP consumption by blowfly photoreceptors and large monopolar cells has been measured at about 10% of ATP consumption of a resting fly in bright light, again, largely due to sodium and potassium pumps (Laughlin et al. 1998). Olfactory processing by glomeruli also requires considerable energetic investment (Nawroth et al. 2007).

Sensory systems should therefore evolve under selection pressure for efficiency (see Balasubramanian et al. 2001; Fain et al. 2010; Laughlin 2011; Niven and Laughlin 2008). In Drosophila, for example, the use of certain potassium channel types in photoreceptors can reduce metabolic costs (Niven et al. 2003). Animals can also use mechanisms in the short term during their life to reduce energy consumption. The spider Dinopis subrufus hunts at night by building small silk 'nets' that it throws over prey that pass beneath it. It therefore needs excellent nocturnal vision, and has large posterior median eyes that are highly sensitive in low light conditions (Blest and Land 1977). At dusk, individuals rapidly (in 1–2 hours) synthesize new photoreceptor membranes, which are quickly broken down again at dawn (Blest 1978; Blest et al. 1978a,b). Other spiders that are active at night also cycle membranes, but to a lesser extent (e.g. Blest and Day 1978; Blest et al. 1980; Grusch et al. 1997). Similar changes occur in crab retinula cells (e.g. Stowe 1980, 1981), and other arthropods. Blest (1978) suggested a possible reason for this extreme breakdown of membrane could be that since the spider is not active during the day, and does not need its vision so much then, the metabolic cost of maintaining large amounts of membrane would be greater than the cost of continuously cycling the membrane between day and night. Blest did not favour this idea, but more recent work (above) showing how energetically costly vision is suggests that this could be a valid reason for the membrane breakdown.

Weakly electric fish not only need to detect electric information from the environment but also produce electricity to measure changes in the field generated. The production of electric organ discharges (EODs) is energetically costly (Salazar and Stoddard 2008; Stoddard and Salazar 2011; see also Chapter 7). Under high oxygen levels in the water, there may be no measured relationship between metabolic rate and EOD, but when oxygen levels are low, fish reduce EOD amplitude, presumably to save energy (Reardon et al. 2011). Circadian rhythms, involving lower EOD activity during the day, also seem to have evolved at least in part to reduce energy costs (Salazar and Stoddard 2008).

4.2 When Are Sensory Systems Optimized for One Task Rather than Being Generalized for Many?

Clearly, most sensory systems function in many tasks. However, we can still ask if and when some sensory systems are optimized for one specific function (e.g. finding a specific food type or mates). The problem with optimizing performance to one task is that it may reduce performance in other functions. Therefore, there is likely to be a trade-off involved that limits specialization. To begin with, we will discuss tuning of photoreceptor sensitivity and colour vision in animals, because this is perhaps where the above issues have been discussed most.

4.2.1 Dimensions of Colour Vision and Visual Pigment Sensitivity Tuning

Animals have an amazing diversity of photoreceptor types, and in the tuning of these receptors to different wavelengths of light. How has such variation been driven, and do these differences relate to specific aspects of an animals' life history?

4.2.1.1 Vertebrates

In Chapter 2 we discussed how five different classes of opsin gene are found in vertebrates and produce photopigments sensitive to different wavelengths of light. Four opsin pigment types are found in cone cells involved with colour vision: LWS/MWS, RhB/Rh2, SWS2, SWS1 genes produce pigments that are sensitive to relatively longwave (LW), mediumwave (MW), shortwave (SW), and ultra-shortwave light respectively (Bowmaker 2008; Collin *et al.* 2009).

4.2.1.1.1 Birds

Birds are a diverse and widespread group, yet their visual pigments (and the coloured oil droplets that filter light before it reaches the visual pigment) seem to be relatively constrained across species, habitat, and life history (Hart and Hunt 2007). All diurnal birds studied so far have four cone types and the sensitivity of these to different wavelengths of light varies little across species. This could be because light levels in terrestrial systems do not vary as much as in aquatic environments, reducing the need for tuning of sensitivity to ambient conditions, in contrast to many aquatic animals (see Chapter 10). In addition, the spacing of the cone types across the bird visual spectrum may be optimal to discriminate between the wide range of different colours that birds encounter in nature, involved with food sources, orientating, and mating, across a wide range of habitats and light conditions (Hunt *et al.* 2009). The main exceptions are penguins (that have shortwave-shifted vision to aid foraging in water), and nocturnal species (where col-our discrimination is reduced to enhance night vision) (Hart and Hunt 2007).

There is, however, one major exception. Visual pigments encoded by the *SWS1* opsins in birds are sensitive to very short (violet) and ultraviolet (UV) light. Broadly, birds can be placed into one of two groups based on the sensitivity of their SWS1 cone type. Most small songbirds have a cone that is maximally sensitive around 355–380 nm (ultraviolet light: 'UVS' group), whereas many non-songbird species have the ancestral violet cone type ('VS' group) sensitive mostly to between 402–463 nm (Hart 2001a; Hart and Hunt 2007; Ödeen and Håstad 2003; Figure 4.1). VS species are still sensitive to UV light but their peak sensitivity is shifted to longer wavelengths. Changes between the VS and UVS systems seem to have occurred more often than traditionally believed; nearly ten times in passerine birds alone (Ödeen *et al.* 2011), implying a selective advantage of these different systems. At present, however, the selective pressure for variation in UV sensitivity in this cone type is unknown.

There is little direct evidence that avian vision seems to be tuned to improve discrimination of specific signals in birds, such as begging displays, flower colours, or plumage coloration. Instead, cor-

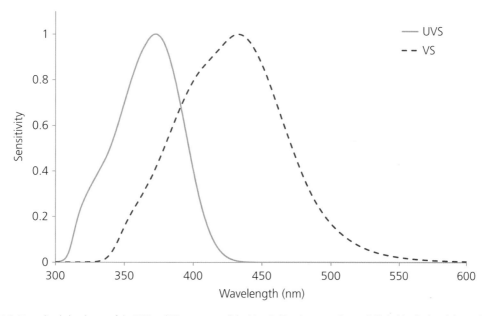

Figure 4.1 Normalized absorbance of the UVS and VS cone types of the blue tit (*Cyanistes caeruleus*; solid light blue line) and the peafowl (*Pavo cristatus*; dashed dark blue line) respectively (data from Hart *et al.* 2000; Hart 2002).

relations between visual sensitivity and tuning probably result from changes in the signals to match the pre-existing sensitivity of the vision (Avilés and Soler 2009; Coyle *et al.* 2012; Mullen and Pohland 2008; Ödeen and Håstad 2010; but see Ödeen *et al.* 2012), although in practice it is hard to know whether the signal or the vision changes first. However, it is important to remember that considerably less than 100 species of bird have had their spectral sensitivity carefully analysed out of > 9000 species. There is also more to colour vision than the spectral sensitivity of the receptors. The relative proportions of receptor types in the retina and their distribution also matter, and crucially how the receptor outputs are coded and weighted in subsequent neural processing. Little is known about what variation in neural processing exists in bird colour vision, but there is some evidence that oil droplets and receptor numbers vary across species and that may reflect variation in diet and feeding behaviour, and habitat type (Hart 2001b), although more work is needed.

4.2.1.1.2 Fish

Does the situation in birds reflect that of vertebrates as a whole? The answer seems to be 'no'. First, fish can vary in both the number of receptor types and in their sensitivity tuning much more than birds, and their vision seems to be more flexible across species. Cichlid fish, for example, may have highly variable colour vision across and even within species (Sabbah *et al.* 2010). It is not clear why, but this could be linked to the importance of visual communication and foraging. Fish are also well known to vary in the tuning of their photoreceptors with respect to light conditions, which vary much more in water than in terrestrial environments (see Chapter 10). Despite this, there is still little direct evidence that changes in vision in fish reflects optimization for a specific task, although it is possible that shifts in sensitivity along with light conditions are driven by silhouette detection (e.g. detecting the form of prey or predators). However, some deep-sea dragon fish have photoreceptors that are highly tuned to detect the longwave light that they produce from their bioluminescent organs (Denton *et al.* 1970; Douglas *et al.* 1998, 1999; see Chapter 7; Figure 4.2). Likewise, other fish have a good match between their short mediumwave bioluminescence and their visual sensitivity (Fernandez and Tsuji 1976), and may match the emission properties of other species if this

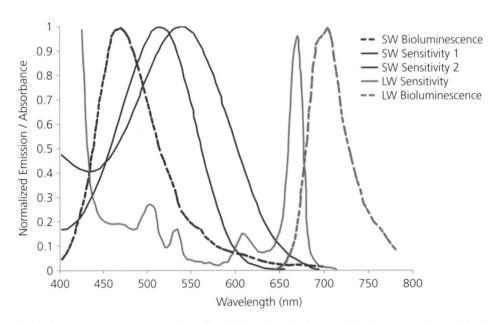

Figure 4.2 Bioluminescent emission spectra (broken lines) of the SW (blue line) and LW (red line) bioluminescence-producing photophores in *Malacosteus niger*, and the spectral sensitivity of the different receptor types. Data from Douglas *et al.* (2000).

facilitates prey detection. The use of biolumines- cence in such species may function in communica- tion. These instances potentially provide good examples of correlation between a sensory tuning and a specific visual signal.

4.2.1.1.3 Mammals

Mammals also vary greatly in vision between spe- cies, and can have just one cone type (no colour vision: monochromatic), or two or three cone types involved in colour vision (dichromatic or trichom- atic). Early mammals moved to a nocturnal exist- ence, requiring high sensitivity over colour vision. As a consequence, they lost two of the four ancestral cone visual pigment classes and became dichro- matic (Hunt *et al.* 2009). Subsequently, some (but not all) nocturnal and marine groups also lost another of the pigment classes and become monochromatic, whereas many primates (including humans) re- evolved trichromacy. In mammals, the SWS1 pig- ments can give rise to peak sensitivities ranging from the ultraviolet at 360 nm (such as in some mice and rats) to 440 nm (as in some squirrels; Hunt *et al.* 2009). It is unclear why some species of rodents have retained UV sensitivity, although there have been suggestions that it may allow individuals to detect recent UV reflecting urine markings (Chávez *et al.* 2003), or facilitate food detection (although this is currently not supported by experiments; Honkavaara *et al.* 2008). There is less variation in sensitivity in the LWS pigment. When shifts do occur in dichromatic species, it is unclear why, and such changes seem not to confer any great benefits in discriminating between objects (Jacobs 2009).

Therefore, across mammals as a whole, broad changes in the dimensions of colour vision seem to occur in line with large-scale changes in life history (e.g. mono- or dichromacy in nocturnal mammals), but there is little evidence that this has occurred for specific tasks. In addition, the reasons underlying tuning of photoreceptors are poorly understood. Nevertheless, there is one major exception: pri- mates. Trichromacy can be found in all of the pri- mate lineages: the New World monkeys from Central and South America (platyrrhines), Old World monkeys and apes from Africa and Asia (catarrhines), and the prosimians (lemurs, lorises, and tarsiers) (Hunt *et al.* 2009; Jacobs 1996, 2009).

Colour vision stems from the *SWS1* gene, which is autosomal, and the *LWS* gene found on the X chro- mosome. Trichromacy in catarrhines occurs because, in addition to the *SWS1* gene, the *LWS* gene is dupli- cated, producing two pigments, one sensitive to relatively more LW light and one to more MW light (Jacobs 2009). The sensitivity of Old World monkeys seems very similar across species (Jacobs and Deegan 1999). In platyrrhines and prosimians, how- ever, the *LWS* gene is polymorphic, with different alleles producing pigments with different sensitiv- ity values (Jacobs 1996). In platyrrhines, there are frequently three alleles of the *LWS* gene, each con- ferring different sensitivities. Because females have two X chromosomes, they can be either trichromatic (heterozygous for *LWS* with three different pairs of alleles possible) or dichromatic (homozygous for one of the three *LWS* alleles), whereas all males are dichromats (one of the three alleles) (Jacobs 2009). However, in howler monkeys, males are also tri- chromatic because a similar *LWS* gene duplication has occurred as in catarrhines (Hunt *et al.* 2009; Jacobs 2009).

The major factor promoting trichromacy in pri- mates seems to have been the enhanced ability to detect ripe fruit and leaves based on red–green con- trast. There is convincing modelling evidence that trichromatic colour vision is good at encoding ripe fruit and fresh leaves in terms of high contrast against foliage backgrounds (e.g. Dominy and Lucas 2001; Osorio and Vorobyev 1996; Osorio *et al.* 2004; Párraga *et al.* 2002; Riba-Hernandez *et al.* 2004; Sum- ner and Mollon 2000). In addition, behavioural experiments support this hypothesis. For example, captive trichromatic tamarin monkeys (*Saguinus* spp.) are more efficient at selecting food targets emu- lating ripe fruits than dichromats (Smith *et al.* 2003). Similar findings have been found in Geoffroy's mar- mosets (*Callithrix geoffroyi*) (Caine and Mundy 2000). However, not all studies have found supportive evi- dence (e.g. Dominy *et al.* 2003; Hiramatsu *et al.* 2008; Vogel *et al.* 2006) and other factors such as different behaviours in di- and trichromats may complicate things. In addition, there are likely to be other advan- tages of trichromatic colour vision (which may have arisen after routine trichromacy was established), such as discriminating between the red faces of indi- viduals that may indicate dominance or fertility cycles (e.g. Fernandez and Morris 2007; Higham *et al.* 2011).

It is not yet clear whether polymorphism in tri- and dichromacy in platyrrhines reflects an ongoing evolutionary transition towards full trichromacy, or whether there is a heterozygote advantage to females. Alternatively, at the moment there is increasing evidence that polymorphism may exist because, although trichromats have an advantage in finding fruit, dichromats are better at finding colour-camouflaged prey (Caine *et al.* 2010; Melin *et al.* 2007; Melin *et al.* 2010; Saito *et al.* 2005; Smith *et al.* 2012; see Chapter 9). This may allow different colour vision phenotypes to exploit different food resources and foraging niches. Overall, however, in primates tri- and dichromatic colour vision may be related to specific behavioural tasks in some groups.

4.2.1.2 Invertebrates

4.2.1.2.1 *Butterflies*

Butterflies also have a wide range of receptor types (commonly between three and five but sometimes more) with different sensitivities, and this can even vary between sexes. It is often not clear if all receptors are used in colour vision, or even in the same behavioural tasks. Generally, the basis for this diversity has been puzzling, and as with vertebrates there was little evidence of links to specific behavioural tasks. However, Bernard and Remington (1991) suggested that tuning of photoreceptor sensitivities in some butterflies is well suited for tasks such as discriminating the wing colours of different species and ovipositioning behaviour in females. The small white butterfly (*Pieris rapae crucivora*) is sexually dimorphic with respect to the sensitivity tuning of one of the shortwave-sensitive receptor types (Arikawa *et al.* 2005). Because females but not males have UV wing reflectance, this change may enable males to discriminate between individuals better. Another subspecies of *Pieris rapae*, where both sexes are more monomorphic, seems to lack these changes in sensitivity.

Swihart (1972) also suggested that diversity in photoreceptors and neural coding in *Heliconius erato* may be linked to the importance of colour in this species. Recent work on *H. erato* has shown that

it has evolved a second UV opsin in addition to the other UV, blue, and longwave opsins already known (Briscoe *et al.* 2010). The two UV opsins in *H. erato* confer sensitivity to 355 nm and 398 nm, and may allow a greater degree of discrimination of the characteristic UV–yellow wing patches found on these butterflies (Bybee *et al.* 2012). Visual signals in *Heliconius* are used in mating and in preventing hybridization (see Chapter 11). It seems that this additional UV opsin is found only in *Heliconius* and not in related groups, and may explain why the yellow wing colours of *Heliconius* are so rich in UV (Briscoe *et al.* 2010). The evidence is convincing for a correlation between the visual system and the communication signal, but one of the remaining challenges is to determine whether the change in coloration or the change in visual system occurred first.

4.2.1.2.2 *Bees*

Uniformly spaced receptors in honeybees may be ideal for encoding flower colours, and there does seem to be a match between the reflectance from flowers and the sensitivity of some bee visual systems (Chittka and Menzel 1992). This benefits the pollinators as they can learn to discriminate rewarding from unrewarding flowers more effectively, and it benefits the flowers because pollinators are more likely to visit the same flowers, transferring pollen to individuals of the same species. As Chittka and Menzel (1992) point out, insect colour vision needed to exist in some form before the flowers evolved colours, but whether the insect colour vision was subsequently tuned over evolution to discriminate flowers better is unclear. Therefore, it remains uncertain whether colour vision in pollinators is tuned to flower detection, or whether flower colours have evolved to exploit pre-existing attributes of the visual system.

4.2.1.2.3 *Fireflies*

Fireflies produce flashes of bioluminescence in communication, especially in mating (see review by Lewis and Cratsley 2008). The colours of the bioluminescent emissions are species specific, yet it seems that fireflies respond not specifically to the colour of the emissions but rather to the pattern of flashes (Cronin *et al.* 2000a; Lall and Worthy 2000; but see

Booth *et al.* 2004). Lall *et al.* (1980b) showed that species of *Photuris* and *Photinus* fireflies produce different bioluminescent emission spectra depending on the time of day. They analysed 55 species and found that generally, twilight-active species produce yellow emissions of between 560–580 nm, whereas nocturnal species tend to produce green emissions around 550–560 nm. The study also suggested that the relatively LW-shifted emission spectra of the twilight species is an adaptation to increase contrast against green foliage when light levels are still relatively high (see also Cronin *et al.* 2000a). Coloured pigments in the eyes act as filters, removing certain wavelengths of light and tuning visual sensitivity in different species (Lall *et al.* 1988), in addition to differences existing in the actual visual pigments (Cronin *et al.* 2000a). There is often a very close match between the peak emission spectra of species and the peak spectral sensitivity of their respective visual pigment, sometimes as little as a two nm difference (Cronin *et al.* 2000a; Lall *et al.* 1980a,b, 1982, 1988; Lall 1981a,b).

Similar matches in emission spectra and sensitivity have been found in species of Japanese fireflies (Egughi *et al.* 1984). It is likely that when bioluminescence first evolved in fireflies it was green in order to coincide with the spectral sensitivity of the visual system, but subsequent tuning by filters have then acted to increase the match between emission and vision. Similar results exist in click beetles that also have variation in emission spectra across species and often a match between the bioluminescent emission and the visual system (Lall *et al.* 2000, 2010). Therefore, in fireflies and other similar groups we have perhaps the best example of a close correspondence between visual sensitivity and a specific signal.

Despite many uncertainties, we can make the following broad conclusions about diversity in tuning of receptors and the numbers of receptors involved in colour vision across taxa: *i*) large-scale changes in life-history can have a significant impact; for example, switching from a diurnal to nocturnal lifestyle favours a reduction in the number of receptor types involved in colour vision in favour of enhanced luminance vision; *ii*) large changes in ambient light spectra in different habitats (e.g. shallow to deep-sea habitats) can lead to tuning of visual pigments to match ambient light; *iii*) in most cases, visual sensitivity and receptor number seems driven by the need to perform many tasks rather than being optimized for one; *iv*) some instances of tuning of visual pigments and changes in the number of receptors present can evolve under selection for very important tasks (finding food or mates) but this seems rare. However, these general conclusions may change as more species are studied. In addition, although it is often argued that visual sensitivity and receptor numbers used in colour vision reflects the need that most animals have to perform a range of functions under different environments, substantial diversity in the number and tuning of photoreceptors across (and sometimes even within) species still exist. At present it is poorly understood why this is the case. In addition, examples are being found that potentially link attributes of vision with a specific aspect of life history or task. Clearly, there is much work to be done.

4.2.2 Specializations in Other Sensory Organs

There are instances of specialization of sensory systems to specific tasks in other modalities, but as with visual sensitivity, more often of generalization. Good examples of how both the morphology and tuning of sensory systems relates to one task occur in parasitic flies. As discussed in Chapter 7, females of the parasitoid *Ormia ochracea* have a hearing organ that is highly convergent with that of its main cricket hosts (Robert *et al.* 1992). The morphology of the hearing organ is much like that of the tympanic organ found in female crickets to find singing males. Moreover, the organ is tuned to frequencies of 4–6 kHz, which closely matches the calls of males. Finding a host is essential for parasitoids, and selection has favoured a sensory system that is highly specialized for this. In fish, the auditory sensitivity of the Lusitanian toadfish (*Halobatrachus didactylus*) also seems well tuned to calls from conspecifics, although it is also capable of detecting and discriminating between the calls of other species in the environment and of predators such as dolphin (Vasconcelos *et al.* 2011). In olfaction, highly tuned receptors to detect pheromones occur (see Chapters 2 and 3), and again this reflects the crucial importance of finding a mate. Conversely, receptors that respond to more than one odour-molecule type, particularly for substances like foods are widespread. Therefore, different mechanisms may be used for different stimulus types, some specialist, others generalist. Overall, the general picture across modalities is that sensory systems are used in many tasks and are broadly tuned, rather than having become specialized for one task alone. When instances of tuning and high specialization do arise, they tend to occur in cases where there is very strong selection pressure with regards to a particular task of great importance, where increases performance in that task are sufficient to overcome any costs of reduced function in other tasks, and in systems that involve a specific stimulus of consistent properties. However, when and why specialization evolves requires more work, both theoretical and empirical. More often perhaps, specialization to specific types of stimuli for certain tasks occurs in post-receptor processing, as evidenced by examples such as prey capture in toads, barn owls, and bats (Chapter 3).

4.3 Trade-Offs in Processing Different Components of Stimuli

There are many trade-offs in the way that sensory systems acquire information. Varied components and features of stimuli sometimes require different processing mechanisms. Here, we discuss some examples with respect to vision and hearing.

4.3.1 Vision

In vision, larger photoreceptors allow increased light capture and therefore sensitivity. However, this can come at a cost of spatial resolution because there are fewer photoreceptors in a given area of the retina to make comparisons between. Furthermore, in order to evolve an eye that is capable of forming an image, as opposed to simply detecting the direction of light, each photoreceptor needs to receive light from a restricted range of angles, meaning that they receive less light (Nilsson 2009). Thus, there is a trade-off between sensitivity and spatial resolution. In addition, methods of increasing sensitivity can affect other features of stimulus processing. For

example, prey-capture behaviour in toads (see Chapter 3) is diminished under lower light levels with toads being slower to respond to worm-like stimuli. Low light levels require increased sensitivity with longer retinal integration times (temporal summation by ganglion cells) but this comes at a cost of reduced speed of response (Aho et al. 1993). Thus, toads are more likely to detect a moving prey item, but less likely to capture it. Enhanced sensitivity can also affect colour vision. Lind and Kelber (2009) have shown that parrots that remain active into twilight have a retina with relatively greater proportions of rods than cones (increasing sensitivity under low light levels) than species that stop activity at dusk. However, because twilight-active species have fewer cone responses to pool over a given retinal integration area, they lose effective colour vision more quickly under declining light levels.

Many aspects of eye form and receptor mechanisms are linked to ecology. For example, comparative studies of 20 species of dipteran fly shows that species (and sexes) have different photoreceptor sensitivities and response times depending upon their ecology (Laughlin and Weckström 1993). Species with fast flight patterns, for instance, for mate searching and chase-behaviour, have rapid phototransduction mechanisms and high temporal resolution for responding to fast changes in light and dark (Weckström et al. 1991). There can also be differences between the sexes. For example, male houseflies (*Musca domestica*) have receptors that are 60% faster than those of females, enabling them to rapidly chase females in flight and undertake quick manoeuvres (Hornstein et al. 2000). In contrast, slow flying species that operate under low light levels have slower photoreceptor mechanisms (Laughlin 1996). Therefore, nocturnal species with slow flight patterns favour overall sensitivity and light detection over temporal resolution, whereas fast-flying diurnal species have rapid vision capable of capturing fast changes in stimuli and surroundings. Slow-flying species also reduce energetic costs by not having a rapidly functioning visual system, and this may be a major reason why slow-flying species do not have fast photoreceptors (Laughlin and Weckström 1993).

It is often assumed that during foraging for food, animals such as birds cannot detect predators while their head is down. This is thought to be due to limitations in being able to simultaneously detect objects from below and in front of the animal (e.g. food), and objects above the head and to the side (e.g. predators) without using scanning head movements. Based on this logic, it is often widely thought that many animals must perform regular vigilance behaviour (scanning) in between foraging bouts. However, the above assumption need not be so straightforward, especially in some taxa (Fernández-Juricic *et al.* 2004). First, differences in visual systems across species may affect the performance of individuals in different tasks. For example, European starlings (*Sturnus vulgaris*) have better visual acuity than house sparrows (*Passer domesticus*), and starlings are better at spotting potential predators at a distance than are sparrows, potentially due to the differences in acuity (Tisdale and Fernandez-Juricic 2009). Constraints on sensory systems, along with behavioural traits, may also limit predator detection. Foraging techniques across species (e.g. probing, pecking, or hunting) may influence or reflect aspects of vision, such as binocular overlap and relative blind areas behind or around the head (Fernandez-Juricic *et al.* 2010; O'Rourke *et al.* 2010a). This could interact with other behaviours, such as scanning for predators or prey with head movements (Fernandez-Juricic *et al.* 2011; Gall and Fernandez-Juricic 2010; O'Rourke *et al.* 2010a, b). For example, in species with larger blind spots, individuals may need to scan the environment more with rotations of the head and eye movements in order to detect predators, limiting things like foraging time (Guillemain *et al.* 2002). However, food detection and predator scanning need not trade-off if the visual field is wide enough to detect predators while at the same time foraging for objects on the ground (such as in some sparrows and finches; Fernández-Juricic *et al.* 2008). In addition, areas of the retina with high visual acuity (such as visual 'streaks' of high retinal ganglion density across the retina) may be oriented in such a way that would facilitate both predator detection and foraging (such as in some geese; Fernández-Juricic *et al.* 2011). Overall, therefore,

assumptions that trade-offs exist can be oversimplified without considering multiple properties of animal eyes and behaviour, which may allow for conventionally conflicting tasks to be combined with reduced detriment to one another (see Fernández-Juricic 2012 for discussion of many of the above issues).

4.3.2 Hearing

In auditory processing, there is widely thought to be a trade-off between temporal and frequency resolution. For example, across animals like birds and mammals, the basilar membrane is arranged where hair cells in different sections respond to different frequencies of sound (Chapter 2). Each region of the membrane is considered to act as a band-pass filter, only responding to sounds of a specific frequency range. Smaller filter bandwidths allow finer discrimination to be made between sound frequencies. However, narrower bandwidths are thought to lower temporal resolution because they cannot detect fast changes in intensity (Okanoya and Dooling 1990). Demonstrating that this trade-off exists has been difficult, but recent work with birds provides some evidence. One way to test temporal resolution is to present a study animal with two sounds ('clicks') separated by a specified time interval, and then measure the degree of response in the auditory system (such as action potentials in the auditory nerve or brainstem) to the second sound. Animals with greater temporal resolution show greater ability to respond to the second sound stimulus presented closer together in time than individuals with poorer temporal resolution. Henry *et al.* (2011) showed that temporal resolution was greater in house sparrows (*Passer domesticus*) than in Carolina chickadees (*Poecile carolinensis*) and breasted nuthatches (*Sitta carolinensis*). This is consistent with previous findings showing that house sparrows have lower frequency resolution (broader auditory filters) than the other two species. In addition, individual chickadees are variable in their frequency resolution, and those individuals with lower frequency resolution have better temporal resolution (Henry *et al.* 2011). It is possible that different species vary on which side of the trade-off they fall based on the characteristics of their song and ecology.

Another potential trade-off relates to the idea that sensory systems face constraints on simultaneously processing more than one aspect of information (processing different 'streams' of information). The pallid bat (*Antrozous pallidus*) uses echolocation for orientation but detects prey based on the sounds the prey make. At times, returning echoes and prey sounds overlap temporally, potentially reducing the bat's ability to distinguish between them and to process the different streams of information separately. Barber *et al.* (2003) trained bats to fly between a wire array (maze) using echolocation and to locate a food reward revealed by a speaker playing sounds. They played the sounds just before the bats started to negotiate the array, forcing them to process both lines of information within a 300–500 ms time-window. Bats presented with both tasks simultaneously were less effective at navigating through the array and less accurate in locating the food reward than when dealing with either task alone. This effect was more marked when the sound from the speaker was longer in duration, increasing overlap with echoes. In addition, bats also increased the intervals between their calls, potentially to reduce the overlap between returning echoes and the food reward sound. Processing of different streams or channels of information is likely to be widespread in other modalities too. For example, different populations of neurons have been identified in electric fish that may encode different features of prey (e.g. Pothmann *et al.* 2012). There have been relatively few studies of trade-offs in simultaneously processing multiple streams of information, but this is an important area for future work.

4.4 Integrating the Senses

The relative advantages of different sensory systems are not the same. For example, vision is rapid because light travels quickly, whereas olfaction is relatively slow and persistent because it relies on the transmission of odour molecules across space. Odours can also be blown around and mixed up by wind and water movements, potentially enabling olfactory sensing over large temporal and physical scales but also making odour cues/signals patchy in distribution. However, sight is obscured by objects in front in the observer,

whereas odour molecules easily pass by objects and sound can refract/reflect around objects (although this is frequency-dependent). Light availability for vision is also variable depending upon the time of day, whereas magnetic cues are ever present (subject to certain local anomalies and drifts in the earth's magnetic field). These differences mean that information from one sensory modality alone can be incomplete or unreliable. Although different streams of information can compete for attention and reduce performance within a modality, combining information from two or more senses (sensory or multimodal integration) can improve performance. Integrating information from different sensory modalities can also be used to 'calibrate' each system. Overall, many behavioural tasks require or benefit from multimodal information.

4.4.1 Sensory Integration

Sensory or multimodal integration can help an animal in detecting, localizing, and correctly identifying a stimulus. In effect, sensory integration reduces uncertainty and provides the individual with more information because each sensory modality can provide an independent assessment of the environment or a stimulus, and because noise in each modality may be unrelated (Munoz and Blumstein 2012). Multimodal integration is thought to help coordinate head and eye movements towards a stimulus and allow faster reaction times to important events (Colonius and Diederich 2004). This is most often investigated in two modalities (bimodal), but sometimes three (e.g. Diederich and Colonius 2004). In many instances, information in each system is redundant because it is the same in each modality. However, this is still useful and can enhance the response, especially if information in either or both modalities is weak, because uncertainty is reduced. Alternatively, information in either modality may be non-redundant. Here, each sense provides different 'messages', and the response of the individual may be different to both senses together than it is to either modality alone (see Chapter 6 for more information on responses to multimodal signals in communication).

4.4.1.1 Hearing and Vision

Much work on sensory integration has investigated detection and orientation with respect to stimuli in vision and hearing. In Chapter 3 we discussed the auditory processing mechanisms that barn owls (*Tyto alba*) use to locate prey. Although the owl's ability to locate prey in darkness based on sound alone is remarkable, in nature there is often enough light for vision to play a role too. In fact, barn owls have excellent vision and use this in hunting (Takahashi 2010). When an owl detects a potential prey item, it quickly turns its head to face it. Whitchurch and Takahashi (2006) tested the ability of owls to orientate towards a stimulus based on hearing alone, vision alone, or the two combined. When presented with a stimulus in only one modality, the time to orientate was related to the stimulus strength. Movements of the head or eyes were quicker in response to acoustic cues, but more accurate in response to visual cues. Responses to simultaneous audiovisual information, however, combined these benefits, being both accurate and fast. These results supported previous findings in humans, also showing that information in one sense can produce a fast response (hearing) and information in a second sense (vision) can increase accuracy, with both senses combined being fast and accurate (Corneil *et al.* 2002). The benefit of combining senses in owls was greatest when the auditory and visual information was simultaneous or when the visual information occurred first, and when the signal-to-noise ratio of the auditory target was low (Whitchurch and Takahashi 2006).

In the barn owl's optic tectum there are neurons that respond more to the combination of auditory and visual information than to stimuli presented in either modality alone (Zahar *et al.* 2009). This is multisensory enhancement. Recently, responses of neurons in the entopallium area of the barn owl forebrain (which is connected from the optic tectum along the tectofugal pathway) have been found to correspond to stimuli encoded by visual and auditory information from locations that coincide in space and that are coincident in time (Reches and Gutfreund 2009; Reches *et al.* 2010). In many cases tested so far, multisensory integration produces greater neural responses to novel or unexpected stimuli. This is consistent with processing of unimodal information, whereby sensory systems predominantly respond to changes (see Chapter 3). In mammals, sensory integration between sound and visual cues occurs partly in the superior colliculus (e.g. Frens and Van Opstal 1998; King and Palmer 1985; Stanford *et al.* 2005). Again, some neurons respond more to the combination of sound and visual cues together, provided that they are coincident in space. Often, the relative level of enhancement (neuronal and behavioural) is greater when either or both cues are weak than when the cues are strong already (Stanford *et al.* 2005). This is logical, because there is less uncertainty about stronger cues in the first place, and so less need for multimodal enhancement.

Generally, work investigating integration of auditory and visual cues shows the following results (from Whitchurch and Takahashi 2006): *i*) information from both modalities together improves localization and reaction times to a stimulus when stimulus information is aligned in space, compared to when information from the two senses is misaligned; *ii*) misalignment between visual and sound cues decreases accuracy in the task compared to either modality alone; *iii*) performance is enhanced when auditory and visual cues occur simultaneously in time; and *iv*) multisensory enhancement is more likely to occur when the cues from either modality alone are of low amplitude or signal-to-noise ratios.

4.4.1.2 Vision and Olfaction

In *Drosophilia*, localization of an odour source (e.g. food) is driven by odour and visual cues, rather than chemical information alone. Combined, these cues affect motor patterns such as flight control (Duistermars and Frye 2008; Frye and Dickinson 2004; Stewart *et al.* 2010). Odour cues increase responsiveness to visual information, enabling flies to travel along an odour plume to the source (Chow and Frye 2008). This may enable individuals to locate an odour source while accounting for ambiguities that can arise due to things like wind-induced disturbance of plumes that make it hard to know the precise direction to travel to the source. Localization of the exact point of an odour source can be difficult, and so may be facilitated by visual information (Frye *et al.* 2003). Male moths also often require visual information when locating females based on pheromones, for example using optic flow patterns across their eyes as they fly. Such relationships vary across groups, however. For example, male cockroaches seem less affected by a lack of visual information than male moths in locating females (Willis *et al.* 2011).

4.4.2 Sensory Integration and Stimulus Identification

Sensory integration could also be important for object classification to correctly identify a stimulus and distinguish it from other possibilities. For example, female *Oophaga pumilio* frogs lay their eggs in small water pools and then revisit the pools to feed each tadpole with an unfertilized egg. In response to their mother, the tadpole swims to the surface and begs by vibrating its body. However, this behaviour should only be presented to mother frogs because other visitors to the pools include predators, to which the tadpole should dive to reduce capture risk. Tadpoles swim more when presented with visual cues or both visual and chemical cues of adult *O. pumilio*, but they only beg in response to visual, chemical, and tactile cues together (Stynoski and Noble 2012). However, they do not beg in response to visual, chemical, and tactile cues from similar sized frogs of another species, or to cues of a potential predator (a spider). Therefore, the combination of cues from three different modalities allows the tadpoles to correctly identify beneficial (mother) from antagonistic or neutral stimuli. Similarly, antipredator responses in tadpoles of other species to potential fish predators are greater when a combination of cues are present (chemical, tactile, and visual), but weak if only presented with subsets of these cues (Stauffer and Semlitsch 1993).

In some cases, researchers have found the nerve cells that help animals to correctly identify a stimulus based on multimodal information. Infrared (IR) detection is important in some snakes for prey and object detection (see Chapter 2). The processing of information from IR receptors has a similar pattern to that of visual information, with a spatial map (Chapter 3) being formed in the optic tectum (New-

man and Hartline 1982). Both visual and IR maps correspond well, and cells in the optic tectum have been identified with specific responses to combinations of IR and visual stimulation. Hartline *et al.* (1978) found cells with receptive fields in the tectum that respond to: *i*) stimuli in either modality ('OR' cells) or to both combined and *ii*) neurons that only respond when information is present in both modalities ('AND' cells). Further work showed the presence of four more cell types (Figure 4.3) that responded: *iii*) weakly to vision alone, or strongly to vision plus IR (IR-enhanced visual cell); *iv*) weakly to IR alone, but strongly to IR plus vision (vision-enhanced IR cell); *v*) to vision alone, but were inhibited by IR (IR-depressed visual cell); and *vi*) to IR alone, but were inhibited by vision (vision-depressed IR cell) (Newman and Hartline 1982; Figure 4.3).

The OR and AND cells may be primarily useful in prey detection, because they respond to an event that occurs in the visual or IR field of the snake. Furthermore, if the stimulus in either or both senses is weak, then the AND cells produce an enhanced response to relevant stimuli, increasing detection. The AND cells may also approximate to feature

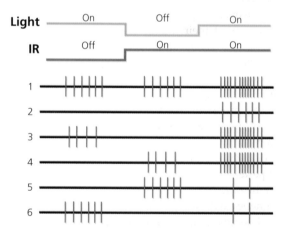

Figure 4.3 Potential responses of cells in the optic tectum of infrared (IR) sensitive snakes in the presence of light, IR, or both. Cell 1 responds to light or IR ('OR' cell), cell 2 responds to IR and light together ('AND' cell), cell 3 responds to light alone but more strongly to both light and IR (IR-enhanced visual cell), cell 4 responds to IR alone but more strongly to both IR and light (vision-enhanced IR cell), cell 5 responds to IR but is inhibited by light (light-depressed IR cell), and cell 6 responds to light but is inhibited by IR (IR-depressed visual cell). The different cell types act as feature detectors for objects in the environment. Adapted from Newman and Hartline (1982).

detectors (see Chapter 3), because they seem to respond to a small, warm, moving object (like a rodent; Newman and Hartline 1982). Note that other work has shown that snakes do not just respond to warm moving objects, but also to cool objects against warm backgrounds (e.g. Van Dyke and Grace 2010), such that thermal contrast against the background rather than the warmth of the object itself may be important in detecting prey and predators (as would be expected given that snakes prey upon both ecto- and endothermic animals). IR-depressed visual cells may be used for detecting other object classes, such as wind-blown vegetation, and so they would be important in categorization of stimuli. Vision-depressed IR cells might respond to inanimate objects warmed by the sun (like a rock), which are not prey animals, again helping in object classification.

4.4.3 Orientation and Migration

Most animals have to orientate in their environment as they move through their home range, and this should favour fast and efficient routes, even on a local scale. In addition, many species travel or migrate over long distances at certain times of year or periods of their life. Migratory movements in birds and turtles, for example, can occur over thousands of miles. To achieve this, they also need to use information about direction and/or location to orientate and navigate successfully. Migration has long been an important area of biology to understand the use of multiple sensory modalities in guiding behaviour.

For compass directions, animals often use celestial patterns (such as polarization patterns of light and the position of the sun or stars) and the earth's magnetic field. Many animals integrate cues from multiple modalities in order to calibrate the other system and gain reliable compass information (Gould 1998; see below), and the sun compass, for example, requires an internal clock to interpret the sun's position. Cross-calibration is important because, for example, the sun, polarized light patterns, and stars move across the sky depending upon latitude, time of day, and season. The earth's magnetic field also drifts in time and can be affected by local anomalies, although most of these changes

would be too slow and weak to affect short-lived species. To be able to navigate, animals need to first determine where their goal is in relation to their current position (a map), and then calculate a how to orientate to their goal based on available cues (a compass) (Gould 1998; Kramer 1953; Lohmann *et al.* 2008a). Maps could simply be a memory of a route taken away from the goal, a list of instructions, or a bearing and distance, rather than the two-dimensional type map humans tend to use. Many ants, for example, are well known to learn landmarks in foraging and to use an approach whereby they match and compare a memorized 'snapshot' of a landmark to a current retinal image in order to orientate (e.g. Lent *et al.* 2010; Müller and Wehner 2010; Reid *et al.* 2011).

In birds, cues from a wide range of modalities have been shown or suggested to play an important role in orientation and migration, including magnetic, olfactory, polarized sunlight, starlight, visual landmark cues, and more (Gould 1998; Lohmann *et al.* 2008a; Wiltschko and Wiltschko 2009). The relative use of different cues, and how they are integrated, seems to vary with ontogeny and phases of migration across species (Wiltschko and Wiltschko 2009; Wiltschko *et al.* 1998). For example, celestial cues may provide initial information to move in a specific direction and this is then modified by magnetic information once the migration begins (Prinz and Wiltschko 1992; Weindler *et al.* 1996). In contrast, some migrating birds, such as savannah sparrows (*Passerculus sandwichensis*) have been shown to repeatedly calibrate their magnetic compass perhaps based on celestial cues during their lives (Able and Able 1995b). Individuals migrate between North and South America and seem to undertake magnetic compass recalibrations, possibly based on polarized light patterns in the sky at sunset and sunrise (Able and Able 1995a; Muheim *et al.* 2007). Conversely, this species' celestial compass does not appear to be calibrated based on magnetic information (Able and Able 1997). This discrepancy between studies may arise because in some experiments birds are only able to see a proportion of the sky, and so information may be limited. When birds have the whole sky to view, they might still use this to calibrate the magnetic compass, as suggested by Muheim *et al.* (2006). However, the situation is complex because other studies and authors have not found similar results and dispute the interpretation of work apparently showing that the magnetic compass is recalibrated based on polarized light patterns (e.g. Wiltschko *et al.* 2008a,b). More work is needed in order to resolve this issue and to learn more about how and when different compass components are recalibrated.

Sea turtles demonstrate the range of cues that can be used by animals in navigation, moving across wide expanses of ocean as well as at more local scales to feeding and nesting grounds (Lohmann *et al.* 2008a). They, like presumably many other animals, use information from different cues sequentially at different stages of navigation. Newly hatched turtles head in the direction of the water based on differences in the amount of light that is reflected from the surface of the sea and the land (water reflects more moon and starlight than land), and also down elevation gradients from the beach into the water (Lohmann and Lohmann 1996a).

Following this, turtles head into the open ocean based on wave movements (Figure 4.4). Waves head directly into the beach and so provide a reliable indication of the direction to the open ocean. Turtles of various species swim directly into the waves, and continue to orientate directly into them both in real settings and in wave tank experiments, based on characteristic rolling movements of wave patterns (Avens *et al.* 2003; Lohmann *et al.* 1990, 1995; Wang *et al.* 1998; Wyneken *et al.* 1990). Once into the ocean, turtles use magnetic information to migrate large distances to feeding grounds and to return to nesting beaches (Fuxjager *et al.* 2011; Lohmann 1991; Lohmann and Lohmann 1996b). The final stages of navigation back to an end goal (such as a nesting beach) may be based on both magnetic (Lohmann *et al.* 2004) and non-magnetic cues, such as odour cues and visual landmarks, although work is needed to determine the importance of these (Lohmann *et al.* 2008b).

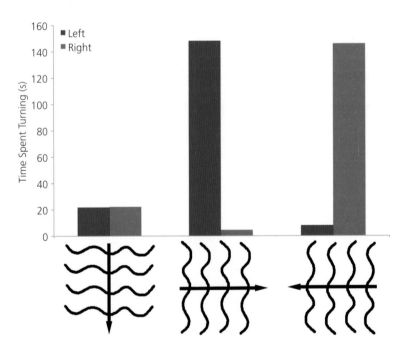

Figure 4.4 When turtle hatchlings are presented with simulations of wave movements in different directions, they orientate to swim head-first into the waves. When the waves head directly to the turtles, they turn left and right equally often but generally do not turn much. When the waves are coming towards their left side, however, they turn left, whereas they turn right when the waves come towards their right side. Data from Lohmann and Lohmann (1996a).

4.5 Future Directions

In future, it would be valuable to understand in more detail how energetic costs constrain the evolution of sensory systems and potentially limit their performance, especially outside of vision. Likewise, although trade-offs are well known to exist in sensory systems, our understanding of some key aspects needs improving. For example, it would be important to know how common trade-offs in processing different streams of information within a modality are, and whether such trade-offs occur across modalities. More generally, an important area of future work is in understanding whether sensory integration is widespread outside of vision and hearing, and what advantages it brings. This includes further testing of how and when sensory integration helps animals to correctly identify stimuli and the neural processes that facilitate this. Finally, more work is needed to determine how common it is for sensory systems to be optimized for a specific task as opposed to being generalists for many tasks, and what selection pressures favour specialization. Linked to this, it is still unclear whether cases of apparent tuning are really instances where the stimulus has evolved towards features of the sensory system, rather than changes in the sensory system towards the stimulus, or where both have co-evolved together.

4.6 Summary

Sensory systems are crucial to the survival and reproduction of animals, yet they incur significant energetic costs. In addition, most sensory systems are used by animals in numerous tasks, each of which may require different features of sensory processing. This can create trade-offs in the way that sensory systems are constructed in order to most effectively function in a wide range of tasks. There are some good examples where sensory systems are optimized for one task of high importance, but more often they seem broadly evolved to work in many tasks. Finally, it is important to remember that sensory systems do not work in isolation. Instead, animals integrate information from several modalities, and combine this during processing. This decreases uncertainty about a stimulus, facilitating detection, localization, identification, and allows for more effective performance in tasks such as navigation.

4.7 Further Reading

Readers are encouraged to consult the literature on sensory integration and multimodal enhancement, and a recent review by Munoz and Blumstein (2012) covering the psychological approach to understanding outcomes from sensory integration ('multisensory perception'), and how this can help animals to overcome environmental uncertainty. Readers may also consult Chapter 6, which discusses multimodal communication. There has been a range of reviews on migration and navigation (see above), and a recent book by Gould and Gould (2012) on this subject. Readers should also consult recent reviews on the energetic costs of sensory systems (e.g. Niven and Laughlin 2008).

PART 3

Communication

CHAPTER 5

Signalling and Communication

Box 5.1 Key Terms and Definitions

Communication: A process involving signalling between a sender and receiver, resulting in a perceptual response in the receiver, which extracts information from the signal, potentially influencing the receiver's behaviour.

Cue: An incidental source of information that may influence the behaviour of a receiver, despite not having evolved under selection for that purpose.

Deceptive Communication: A process that exploits a pre-existing normally reliable communication system and that benefits the signaller but not the receiver.

Eavesdropper: An unintended receiver that is able to detect and use the signals of other individuals (hetero- or conspecific, such as a rival mate or a predator) for its own benefit.

Handicap: Signals that are reliable because they are too costly to produce by cheats.

Indices: Signals that reliably convey information about quality because they physically cannot be produced by inferior individuals.

Signal: Any act or structure that influences the behaviour of other organisms (receivers), and which evolved specifically because of that effect.

Signal Efficacy: The form or structure of a signal, influencing the effectiveness of the signal to transmit its strategic element to the receiver.

Signal Reliability: The information that a receiver can acquire from a signal must, on average (over evolution), be reliable and useful, otherwise the receiver should stop responding and the system would break down. Note that deceptive communication can still arise to exploit such systems.

Strategic Element: The information or 'message' that can be extracted from the signal by the receiver.

Animals obtain a great deal of important information from inspecting the world around them; for example, in navigation or finding prey. However, many organisms also regularly communicate with individuals from the same and other species (Figure 5.1). Communication is fundamental to many aspects of behaviour, yet the terms 'signal' and 'communication', while often intuitive, have been fraught with definitional and semantic disagreement. Here, we discuss some of the key theory and ideas associated with signalling and communication, using a range of examples. The following chapters in this section of the book also cover key aspects of communication and build on many of the concepts introduced here.

5.1 Signals and Cues

Organisms communicate with each other via signals. Here, one organism, the *signaller* or *sender*, emits the signal, and another, the *receiver*, picks it up (Figure 5.2). Signals can also be intercepted by *eavesdroppers*. These are unintended receivers: animals that a signal is not aimed at, but who exploit the signals of others to their own advantage. For example, predators may

Sensory Ecology, Behaviour, and Evolution. First Edition. Martin Stevens.
© Martin Stevens 2013. Published 2013 by Oxford University Press.

Figure 5.1 Many animals have bright coloration in order to attract mates or to signal to rivals, such as this rainbow lorikeet (*Trichoglossus haematodi*) and lantern bug (*Pyrops whiteheadi*).

detect the mating calls of prey, or potential mates may assess something useful about a rival. We discuss eavesdropping and its implications in Chapter 7.

Signals are different from *cues*, where an animal can acquire information from the environment through incidental sources (Marler 1961, 1967). For example, some birds of prey hunt voles based on ultraviolet reflectance from the voles' urine that they excrete along the trails where they move (Viitala *et al.* 1995). Many predators also locate their prey by virtue of the rustling sounds that the prey animal makes when moving on the substrate. Here, the prey animal is not signalling to

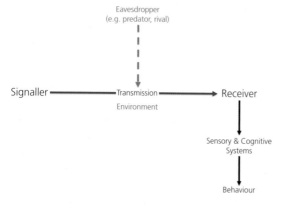

Figure 5.2 The various components of signalling. Signals have to be sent by a signaller, transmitted through the environment, and then picked up by the sensory system of the receiver, who may then change their behaviour accordingly. During transmission, signals are also vulnerable to interception by eavesdroppers, such as rivals or predators.

the predator (or any other animal), but the sound cues generated by its movement are important nonetheless. Likewise, when we hear leaves rustling in a tree, it tells us that it is a windy day and, although this may be useful information, it is not a signal. Note that cues and signals are not always easy to distinguish (Schaefer 2010; see below), and it has been suggested that if cues are beneficial to the sender then they may evolve into signals (Schaefer and Ruxton 2011). In contrast, if cues are costly to the sender, then selection may act to minimize them. For example, many prey animals will walk quietly in order to prevent sound cues from giving their presence away to a predator.

5.1.1 What is a Signal?

So far, we have not actually defined precisely what a signal is. There has been much confusion regarding the evolution and maintenance of signals, often reflecting the different approaches and interests researchers have. On one level, signals are regularly thought of as transmitting information between signallers and receivers. However, there is more to signalling than this; signals may be selected not for transferring information *per se*, but rather to elicit favourable responses in the receiver. We will deal with these issues more thoroughly below.

Maynard Smith and Harper (2003) define a signal as: 'any act or structure which alters the behaviour of other organisms, which evolved because of that effect, and which is effective because the receiver's response has also evolved.' The crucial part of this definition is 'evolved because of that effect'. As Maynard Smith and Harper (2003) point out, this distinguishes a signal from a cue. Signals evolve specifically under selection to influence a receiver. We can contrast this with the example of body size, which is often a cue. Some male animals may decide whether or not to engage in a contest with a rival depending on their relative body sizes (and hence potential strength). However, body size will often be simply a function of factors such as nutrition, energy requirements, and so on, and larger sizes have not evolved specifically to cause a response in a receiver.

A potential problem with the definition, however, regards the final statement about the receiver's response also having evolved. This is logical but presents two problems. First, it does not address the initial evolution of a signal—should the receiver's response or the signal evolve first? Second, it implies a level of coevolution between signaller and receiver, which may often not be the case. As we shall see in Chapter 8, many animals utilize signals that receivers respond to because the signal has been modified from another context. This includes pre-existing biases, where the receiver has a perceptual bias towards certain stimuli that exists before the evolution of the signal. For example, some male sexual signals have evolved features that exploit pre-existing preferences that exist in the female sensory system, such as towards certain visual stimuli or frequencies of sound (see Chapter 8). A more widely applicable definition has been presented by Ryan and Cummings (2005): 'Signals evolved to communicate information and manipulate [influence] receivers to the signallers benefit'. We can insert the word 'influence' here because manipulate implies a negative outcome for the receiver, which is often not the case. Therefore, we define a signal here as:

Any act or structure that influences the behaviour of other organisms (receivers), and which evolved specifically because of that effect.

An important aspect of signals is that the information they allow the receiver to acquire must (on average over evolution) be honest or reliable and of value to the receiver, otherwise the receiver would stop responding and the system would break down. Both the sender and receiver should, on average, benefit in a signalling system. The addition of 'on average' is important, because although both parties should benefit over time for signalling to persist, the encounter may not be beneficial on every occasion. However, as we shall see below, the concept of signal honesty is not always clear. For one thing (as we discuss on various occasions during this book) signallers often present false information and are deceptive. For example, many harmless prey exploit the conspicuous warning signals of toxic species (the model) in order to 'trick' a predator into avoiding them (Batesian mimicry). How such deceptive systems can be maintained is a source of debate, but on average, provided that deceptive signals are relatively low in frequency compared to reliable signals, then the receiver should still respond (Hasson 1994). In addition, receivers may undergo selection to discriminate between honest and deceptive signals better (which may in turn lead to further improvements in mimicry by the deceptive party).

5.1.2 Some Examples of Cues

Unlike signals, cues have not evolved specifically to modify the behaviour of a receiver. Cues usually benefit the receiver, but often do not benefit the sender (although they can). Likewise, while signals should have undergone selection to improve efficacy, there need not be any such selection pressure on cues as they are by-products of other processes. In fact, the organism emitting the cues may be under selection to minimize them if they are costly to the sender.

Elegant experiments by Harland and Jackson (2000, 2002) have shown how the jumping spider *Portia fimbriata* adopts different types of stalking behaviour when attacking different prey types based on the presence of specific prey cues. When hunting, *P. fimbriata* stalks its prey, but adopts a specific 'cryptic stalking' approach (involving retracted palps, slow walking, and freezing when the prey orientates towards it) when hunting other jumping spiders in order to remain undetected. Harland and Jackson presented spiders with different lures made from dead spider species and flies, with the presence, size, and shape of different body parts such as legs, abdomen, and eyes altered. They found that *P. fimbriata* only adopted cryptic stalking when the lures were made with large anterior-median eyes, characteristic of jumping spiders; otherwise, ordinary stalking was adopted. Subsequent experiments, presenting *P. fimbriata* with computer-generated three-dimensional lures based on its main jumping spider prey, revealed that it is not only the presence and absence of eyes that is important, but also the size, shape, and position of the eyes. For example, small eyes and square-shaped eyes were less likely to induce cryptic stalking than large or circular eyes.

Because cues have not evolved specifically under selection to influence a receiver, they have not undergone selection to be distinctive from general features of the environment. This presents a problem for animals relying on cues because failure to detect a cue against environmental noise could result in a lost meal or even death (if needing to detect a predator), whereas repeated erroneous detections could lead to lost energy or reduced foraging opportunities. As expected, some animals can distinguish the subtle differences between important and unimportant cues. For example, red-eyed treefrog (*Agalychnis callidryas*) embryos will hatch early and drop into water below when a predator, such as a snake or wasp, is detected in order to escape. Caldwell *et al.* (2010) showed that they can distinguish between the vibration characteristics of rainfall and those of an approaching predator, which differ in some aspects of frequency composition and temporal characteristics, in order to avoid hatching early simply due to the presence of rainfall. Hatching too early would lead to an increased exposure to predators in the pond below.

In other organisms, the level of refinement in using cues can vary. For example, different species of brood (social) parasitic cuckoo bumblebees exploit other bumblebee species to bring up their offspring (Chapter 9). Cuckoo species that specialize on just one host species need to utilize a complex set of chemical compounds to identify their specific host correctly, whereas generalist species that exploit a range of hosts only require selected compounds common to all their hosts (Kreuter *et al.* 2010).

5.1.3 Distinguishing Between Signals and Cues is Not Always Easy

Individuals of some animal species are capable of distinguishing related from unrelated individuals that are genetically dissimilar (to avoid inbreeding and homozygosity, or for nepotism); for example, mate preferences based on urine odour in mice (Isles *et al.* 2001). In some species such preferences are not due to individuals learning the smell of their parents and then selecting partners with different smells, but rather are a product of a genetic mechanism that influences the olfactory system (Isles *et al.* 2001). Such mechanisms may aid individuals to avoid inbreeding, or even to select partners with a genotype that would result in offspring with an advantageous genetic make-up. These effects can sometimes be complex, however, with individuals potentially choosing mates both on the basis of genetic dissimilarity and inherent genetic quality ('good genes'; e.g. Roberts and Gosling 2003). This provides an example of how it can be difficult to distinguish signals from cues. In this example, if the scent has evolved specifically to influence the response of the receiver and benefits both parties (on average), then it fits our definition of a signal. However, if the individual's smell is simply a by-product of its physiology and is incidentally related to its genetic make-up then it is a cue, even if it is used by the receiver to make a decision. It is possible that individual odours began as a cue and have since evolved into a signal. For example, behaviours that cause urine to be deposited in prominent places where it will more readily be detected by others might be considered a signal.

5.2 Signal Components

Following the work of Guilford and Dawkins (1991) and Endler (1993), we can think of signals as having two aspects. First, the *strategic* element refers to the actual content or 'message' of the signal; i.e. the information that can be extracted from it (Figure 5.3). This determines how and why, from a functional perspective, a receiver should respond to a signal. Second, *efficacy* refers to the form ('design') or structure of the signal; i.e. how it is constructed over evolution to influence the response of the

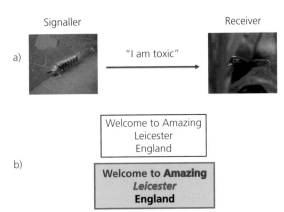

Signaller   Receiver

a)   "I am toxic"

Welcome to Amazing
Leicester
England

b)

Welcome to Amazing
Leicester
England

Figure 5.3 a) Illustrates the idea that signals have a strategic element, being the information or message that a receiver can extract from the signal. b) Illustrates the idea of signal efficacy; here, both signs say exactly the same thing (i.e. they have the same strategic aspect) but the signal form varies greatly in order to convey that message.

receiver effectively. This determines the likelihood that a signal will reach the receiver, be detected, and influence a response at all. Of these two aspects, efficacy is often most relevant to sensory ecologists because we want to know how the signal actually transmits through the environment and stimulates the sensory system of the receiver. As Guilford and Dawkins (1991) emphasized, the strategic component of a signal is far from sufficient to explain the diversity of signals in nature. The role of the environment in signalling (and sensory systems) is considered in depth in Chapters 10 and 11.

5.2.1 Signal Efficacy

Guilford and Dawkins (1991) stated that three features of a signal's efficacy are particularly important (of which many similar ideas were discussed by Marler; e.g. 1957, 1967): (1) how easily a signal can be distinguished from the background, (2) how easily a signal is discriminated from other signals in the environment, and (3) how memorable a signal is (and also how easily a given signal form is learnt). It is important to remember that different receivers may have different sensory and cognitive abilities, and may detect, discriminate, and remember different aspects of signal form (Chapter 1). In general, however, more intense and contrasting (with the

background environment) signals should be better at eliciting responses, although costs can constrain signal form (see Chapter 7). Signal redundancy may also be widely found, whereby signals are composed of the same repetitive elements, as opposed to having a wide range of different components. Redundancy may enhance detection and reliability, but may be more limited in the amount of information available to the receiver (but see below). A related idea is that of *alerting components*, which are aspects of a signal with no strategic element, but which simply attract the receiver's attention before the main signal is given. Similarly, signal *amplifiers* (Bradbury and Vehrencamp 2011), enhance the efficacy of the main signal, increasing information acquisition by the receiver. More direct empirical work needs to be done on signal amplifiers, but we will return to alerting components in Chapter 10.

Stuart-Fox *et al.* (2007) have shown in a comparative analysis of 21 populations of different species of dwarf chameleon (*Bradypodion*), that habitat characteristics strongly predict features of the chameleon display coloration (quantified in terms of the chameleon visual system). As predicted, signal form can be substantially modified over evolution in order to maximize efficacy given prevailing environmental conditions (see Chapter 10). Similar results have been found in comparative studies of manakin birds, where plumage contrast against the background in males is greater when higher levels of sexual dimorphism occur, implying selection for signal efficacy (Doucet *et al.* 2007). Further work modelling a wide range of bird colours from 111 species, in terms of bird vision, also shows evidence that the diversity of avian colours used in communication has diverged from other colours found in the environment, such as vegetation (Stoddard and Prum 2012). More generally, artificial neural network modelling indicates that specific signal forms, including symmetry, signal exaggeration, and repeated signal elements can result as by-products of features of sensory processing or, for example, when there is a chance that the signal will be partially obstructed in the environment (e.g. Enquist and Arak 1993; Kenward *et al.* 2004; see Chapter 8).

Signal efficacy is not just important for animals; many flowers not only attract pollinators with bright colours, but also have specific markings that

Figure 5.4 The image of the flower on the left shows its appearance in human visible light. The image on the right contains information in the ultraviolet, shortwave, and mediumwave parts of the spectrum, broadly corresponding to the information available to bee vision. The lines leading into the centre of the plant may act as 'nectar guides' to pollinators, and are visible as patterns in ultraviolet light.

act as 'nectar guides', thought to direct pollinators to regions of the flower where there is a nectar reward (Figure 5.4). These often consist of lines and spots, frequently visible only in ultraviolet light (which many pollinators can see). Recent work, manipulating the appearance of specific flower markings has shown that the form of the signal does not change the likelihood of pollinators approaching the flowers, but does significantly influence flower fertilization through proboscis insertion from pollinators (Hansen *et al.* 2012). Thus, nectar guides seem to function predominantly in close-range orientation rather than long distance detection. The latter may be mediated more by the overall flower colour and its contrast with the general background.

Many bats are revered for their ability to detect fine-scale features of insect prey that they hunt by echolocation. However, some echolocating bats are not insectivorous but are nectar feeding and important pollinators. They may detect the species that they pollinate by virtue of characteristics of the returning echoes from the plants. Sometimes the relationship can go beyond this, however, with plants evolving structures that are conspicuous to echolocation calls. von Helversen and von Helversen (1999) have suggested that some plants have the equivalent of an acoustic nectar guide, directing pollinating bats to their flowers. They found that the vine *Mucuna holtonii* from Central America has a small concave structure (the vexillum) that acts like an acoustic mirror, reflecting much of the energy of an echolocation call directly back at the bat. They broadcast 1 ms calls

similar to that of the main pollinator *Glossophaga commissarisi* through speakers over a range of angles and measured the returning echo from flowers with the vexillum intact, with the vexillum closed, or with the vexillum filled with cotton wool. Only flowers with the unmodified open vexillum reflected the echo effectively, and most sound energy was reflected back at angles in the direction of the speaker, even when the vexillum did not directly face the sound source (Figure 5.5). In field trials over a two-hour period, bats successfully visited 88% of flowers with the vexillum intact, compared to just 21% of flowers with the vexillum removed. Similar findings resulted when instead of removing the vexillum of some flow-

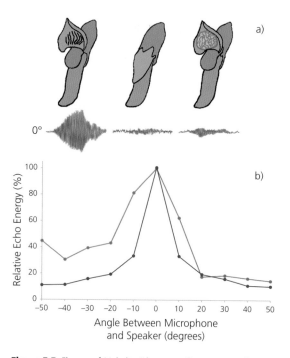

Figure 5.5 Flowers of *M. holtonii* have a vexillum structure that reflects the echolocation calls of bats strongly back towards the sender. a) Playback experiments show that when the vexillum is present much more sound energy is reflected from the flower than when it is removed or filled with cotton wool. Redrawn from von Helversen and von Helversen (1999). b) The sound reflected back from the vexillum is strongest in the direction of the sound source (0° angle between speaker and micrphone), regardless of whether the flower directly faces the sound source (at 0°, blue line) or is rotated away from the sound source by 30° degrees in the horizontal plane (red line). Data from von Helversen and von Helversen (1999). Reprinted by permission from Macmillan Publishers Ltd: Nature, von Helversen and von Helversen, Vol 398:757–760, copyright 1999.

ers they filled it with cotton wool instead. Finally, the authors note that flowers that are pollinated by non-echolocating bats seem to lack a vexillum, as predicted if the structure acts as signal to attract echolocating pollinators.

A later experiment by Simon *et al.* (2011) shows that another vine (*Marcgravia evenia*) from Cuba also has a leaf structure that acts as an acoustic beacon to bats. This vine has one or two dish-shaped leaves near its flowers, where the stalks are twisted to position the leaf blade upright and with the concave side facing any pollinators. The authors found that echoes from these leaves cover a wider range of angles at greater amplitude than normal leaves and remain more consistent in structure over various angles. When presented with a task to find a feeder in a foliage background, bats (*Glossophaga soricina*) have a significantly shorter search time when presented with a dish-shaped leaf than when the feeder was presented either alone or with a normal leaf. The dish leaves increased foraging efficiency in the bats by approximately 50% (Simon *et al.* 2011). The structures on these vines may not be strictly speaking analogous to the visual nectar guides found on some flowers that direct pollinators to appropriate parts of the flower, but are more analogous to the bright colours used by some flowers to increase detection by visually guided pollinators.

5.3 Strategic and Efficacy Costs of Signals

Maynard Smith and Harper (2003) describe how signals can have both efficacy costs and strategic costs. The former involves costs associated with transmitting a signal effectively when faced with other selection pressures, and is discussed in depth in Chapter 7. Strategic costs, however, are relevant to understanding how signal reliability may be maintained. Many (but not all) signals are costly to produce; for example, large conspicuous structures or loud calls may involve considerable energetic investment and may increase the risk of predation. Such factors can present considerable efficacy costs, affecting the evolution of signal form, but these are necessary in making signal transmission and detection effective. Signals can also have strategic costs, whereby an organism directly develops an elaborate

signal *in spite of the additional costs associated with elaboration*. As Maynard Smith and Harper (2003) point out, strategic costs are costs that can be associated with a signal to keep it honest, and their costs should be greater than the efficacy cost alone. Or, more specifically, it is the increase in strategic costs associated with higher quality individuals that can keep signals reliable; if the cost of a signal does not vary according to individual quality, then the signal is costly but not honest (Számadó 2011). The widespread existence of efficacy costs means that simply demonstrating that a signal is costly does not mean that it is a handicap or honest (see below) and carries a strategic cost. Overall, many costs associated with signals have little to do with the strategic component.

5.3.1 Are Signals Honest and What Maintains Signal Reliability?

Zahavi (1975, and subsequently) has suggested that *all* signals are an honest reflection of an individual's quality. Because it should not pay receivers to attend to dishonest signals, these would become ignored and only honest signals would prevail. Zahavi argued in his '*handicap principle*' that honest signals are maintained because they are costly (in terms of strategic costs), such that only a genuinely 'good' individual can afford to make the signal. Consider tail length in birds and mate choice. Females may prefer males with longer tails, and only the fittest males can produce and maintain such tails because only they have the quality to cope with the costs that this incurs (increased risk of predation, loss of flight efficiency, and so on). Because the signal is costly, only the fittest individual can produce and maintain such a signal while coping with the costs, and tail length becomes an honest signal of quality (Grafen 1990; Zahavi 1975). The handicap principle gained widespread acceptance, especially in sexual selection theory. However, subsequently it has been shown to be just one of numerous factors that can lead to signal reliability, and handicaps probably only exist in some situations; not all signals are costly, and not all costs are handicaps (Ryan and Cummings 2005).

First, it is clear that many signals are not honest (they are manipulative, deceptive, or exploitative;

Hasson 1994; Chapter 8), including our example of Batesian mimicry above. The fact that many signals are clearly dishonest means that the key questions to ask are not whether signals are always honest but rather what keeps signals generally reliable, what keeps individuals from constantly cheating, and how is signal reliability maintained?

Second, most discussion of the strategic components of signals has been dominated by studies concerned with sexual selection, where a prevalent assumption has been that an individual, such as a male trying to attract a mate, produces a costly signal because it conveys information about his quality. These and other signals, however, can be honest without being handicaps as Számadó (2011) discusses. Importantly, he emphasizes that what actually stops dishonest signals is if they impose an additional cost on the signaller (relative to the costs of honest signalling); that is, what really matters is not the additional cost wasted by honest signallers, but the potential costs of cheating. Handicaps offer one way to achieve honesty, and with handicaps even honest signallers pay the additional cost of the handicap. However, Számadó discusses a number of other biological scenarios that could lead to additional costs for cheaters (and hence honest signalling) but which do not require that honest signals have strategic costs. In many cases, if the cost of cheating is high, it can maintain reliability even if the current cost paid by the honest signallers is low. There are various examples of honest signals that are not handicaps. For example, some male animals may be more attractive to females when they have calls of lower frequency. Producing low-frequency calls may not be more energetically demanding than high-frequency calls, but only larger individuals can physically produce such calls owing to their morphology. If large size is a valuable attribute for females, then call frequency can be an honest signal of quality that cannot be cheated without being a handicap. This type of signals is called an *index* (Maynard Smith and Harper 2003). Signals can also be reliable if both parties benefit from the encounter, such as when they share a common interest (e.g. not getting hurt in a contest), or when they repeatedly encounter and interact with each other (whereby cheats may be punished). For example, many primates live in groups of closely related individuals

and give alarm calls when faced with a predator. Here, the receivers benefit from reduced predation risk, and the signaller benefits by helping its relatives, who share many of the same genes and help with group defence, vigilance, and food acquisition. In this case, the signals should be reliable and of low cost and high mutual benefit. In general, such *minimal-cost signals*, as they were termed by Maynard Smith and Harper (2003), are signals whose reliability does not depend on their cost (i.e. they are not handicaps), and which can be made by most members of the population (i.e. they are not indices).

In general, one of the problems that has arisen in much of the signalling literature is that far too little attention has been paid to efficacy costs, even though these are a crucial factor shaping signal form (Ryan and Cummings 2005; Schaefer 2010). Costs imposed by the need to transmit the signal through the environment are a major selection pressure on signal generation. Distinguishing between strategy and efficacy costs is not easy (Számadó 2011), and undoubtedly, many costs that are traditionally assumed to be strategic costs presented as handicaps are really efficacy costs, and simply exist because the sender has to invest resources to ensure the signal is successfully transmitted and received. The use of economic models of behaviour, such as game theory, has been a central component of communication theory to understand what maintains, for instance, signal reliability. However, Ryan and Cummings (2005) rightly point out that this approach is insufficient to explain much of what we know about signal form and communication, including things like supernormal stimuli (see Chapter 8) and categorization behaviour. This does not mean that economic models are not useful, but we need to consider evolutionary contingency, sensory systems, and cognition more fully too. In general, work on animal communication has been dominated by issues of signal cost and reliability, even though strategic models cannot explain much of the diversity in signal form that exists, whereas much can be explained by understanding perceptual processes (Hurd *et al.* 1995; Rowe and Skelhorn 2004). However, we are now rapidly acquiring a much better understanding of signal efficacy, as discussed in subsequent chapters.

5.3.1.1 Selection for Strategic Components of Signals: Dominance as an Example

Dominant individuals may have greater access to resources, such as food, shelter, and mates. Consequently, signals of dominance status can be beneficial for individuals in allowing them to assess the potential likelihood of winning a contest and to avoid engaging in a fight with a stronger competitor that could be costly in terms of potential injury or death. In some species, signal forms are directly linked to body size, and body size in turn is directly related to dominance. Furthermore, in at least some cases the signal is honest in that individuals are simply not physically capable of producing a signal associated with higher dominance than their status; i.e. they are indices. If so, we would predict that such signals should give reliable information about quality (e.g. fighting potential) and weaker individuals should be unable to fake the signals of stronger individuals.

A study by Davies and Halliday (1978) on the toad *Bufo bufo* tested the role of calling in competition between males for access to females. Males often wrestle for possession of a female during the spawning season. Davies and Halliday showed that large males could sometimes displace smaller individuals from females, but not *vice versa*. Attacks to displace males in possession of a female predominantly occurred when the attacking male was larger in size than the possessing male. In addition, there was a strong negative relationship between the fundamental frequency of male calls and male body size, such that male calls give reliable information about strength. This is because bigger males have a larger larynx and therefore only they can produce deeper calls. During staged contests with different sized males, when accompanied by calls differing in frequency, males were more likely to attack an individual accompanied by a call of higher frequency than individuals presented with low-frequency calls.

Similar situations seem to exist in various other species and a range of sensory modalities. In red deer (*Cervus elaphus*), males will engage in acoustic 'roaring' contests when one individual challenges another, and characteristics of roaring (e.g. minimum frequencies) seem to relate to body size, fighting ability, and hence reproductive success (Clutton-Brock and Albon 1979; Reby and McComb 2003). In addition,

experiments also show that oestrous females also prefer the calls of males with frequencies that indicate larger body size (Charlton *et al.* 2007). Likewise, in some jumping spiders, such as *Phidippus clarus*, the rate of substrate-borne vibrations is positively correlated with male size, and signalling rate predicts male mating success with females (Sivalinghem *et al.* 2010). It has also been suggested that short-term changes in electric organ discharge (EOD) can signal dominance status in electric fish (e.g. Hagedorn and Zelick 1989). Recent work in the gymnotiform electric fish *Sternarchorhynchus* spp., by Fugère and Krahe (2010) found that EOD frequency was positively correlated with body size. In experiments, individuals with EODs of higher frequency were dominant over those with lower frequency EODs in gaining access to a rock refuge. In further playback tests, Fugère and Krahe found that fish were more likely to approach and attack the stimulus electrodes when they played back EODs of lower frequency. As with the red deer and jumping spiders, when electric signal form reflects male body size, it may also be used in mate choice. Curtis and Stoddard (2003) performed an experiment with pintail knifefish (*Brachyhypopomus pinnicaudatus*) whereby they allowed females to choose between fish in a two-way choice experiment in fish tanks. Females preferred males with EODs of longer duration and higher amplitude, and these features correlate positively with male body size (Figure 5.6). Interestingly, in another species (*B. gauderio*) male EOD signals become more reliable indicators of body size when under greater competition with other males (Gavassa *et al.* 2012). When competition is higher, individuals that are further away from the predicted EOD values (based on body size) exaggerate their signals more, strengthening the relationship between signal amplitude and body length. Therefore, under situations of greater competition, the male signals became more honest, whereas under low competition, males reduced their signalling effort potentially to reduce costs (Gavassa *et al.* 2012).

5.3.2 Distinguishing Between Signal Strategy and Efficacy can be Difficult

It is not always easy to determine what types of cost are associated with signals, and to distinguish

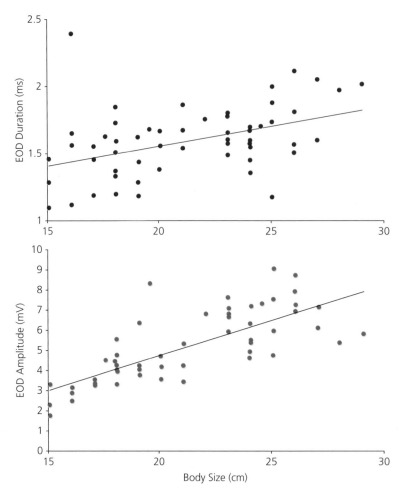

Figure 5.6 Males of the pintail knifefish, *B. pinnicaudatus*, show a positive relationship between body size and both the amplitude and the duration of their electric organ discharges (EODs). Females prefer individuals with EODs characteristic of larger body size. Data from Curtis and Stoddard (2003).

between efficacy and strategic aspects. For example, Gomez *et al.* (2009) showed that female tree frogs (*Hyla arborea*) choose males in mating based on visual signals, even under nocturnal conditions. They presented females with video playbacks of calling males (with identical calls) and modified the emission properties of the display in order to present stimuli controlled with respect to the visual system of the frogs. They then adjusted the coloration of the vocal sac of males, either pale or colourful, and also reduced a prominent flank stripe on some males by colouring it the same as the average coloration of the body. They then presented females with videos of the males in pairs (pale and colourful vocal sacs,

and flank stripe or no flank stripe). They found that females preferred males with a colourful vocal sac and with a flank stripe, and suggest that these signals enhance the conspicuousness of males to females, and hence detection and localization (i.e. selection for signal efficacy). In addition, they suggest that because vocal sac coloration is based on carotenoid pigments, which are limiting and obtained from the diet of most species, that vocal sac coloration may also indicate male quality (i.e. a strategic aspect).

A series of studies by Hare and colleagues has investigated the role of different components of acoustic alarm signals in Richardson's ground

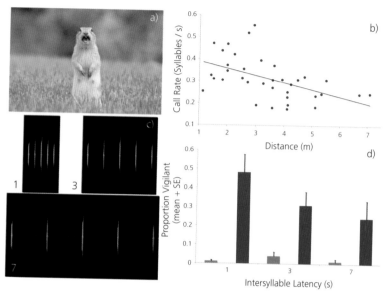

Figure 5.7 a) Richardson's ground squirrels make alarm calls when a potential threat such as a predator appears. b) The rate at which they call (syllables per second) is inversely proportional to the distance from the threat. c) Playback experiments with modified calls with the same number of syllables but different inter-syllable latencies (1, 3, or 7 seconds) show that d) ground squirrels are more likely to respond with vigilance behaviour when they hear calls of a higher call rate (blue bars). The red bars show baseline vigilance behaviour occurring before any playback. Image a) and c) playback calls reproduced with permission from James Hare. Data for b) and d) from Warkentin *et al.* (2001).

squirrels (*Spermophilus richardsonii*) in Canada. Experiments presenting ground squirrels with a model predator at varying distances show that individuals produce repeated call syllables at a higher rate when the risk is closer to them (Warkentin *et al.* 2001). Furthermore, when the authors used playbacks of calls to squirrel colonies, manipulated to produce three stimuli with different call syllable rates, a greater proportion of squirrels responded with alert vigilance behaviour to calls presented at higher rates (Figure 5.7). Thus, call rate indicates distance to a predator and potential threat, rather than simply increasing the likelihood of signal detection by incorporating signal redundancy, as is often assumed in such systems.

In addition, the structure of the repetition of call syllables also influences receiver responses. Monotonous call structures with syllables at constant intervals are more likely to elicit vigilance behaviour than call durations of the same length but where the calls have variation in intersyllable latency (Sloan and Hare 2004). Given that the intersyllable call rate can allow receivers to extract information about current risk, monotonous calls should be more reliable than variable ones, where information would be lost. Furthermore, the presence of call components with different structures, such as 'chucks' and 'chirps', can also increase vigilance behaviour and even facilitate the orientation of the receiver to the caller (Sloan *et al.* 2005). Specific syllable types may also act as alerting components, increasing the receiver response to the information that can be extracted from subsequent call characteristics (Swan and Hare 2008). Finally, alarm calls seem also to be sensitive to the type of threat presented; for example, from terrestrial or airborne predators (Thompson and Hare 2010). Overall, alarm calls in Richardson's ground squirrels nicely illustrate how different features of a signal can act both to increase the response and detection likelihood of the receiver, but also convey specific information about the level and type of predator threat.

5.4 What is Communication?

5.4.1 Defining Communication: Both Information and Influence are Important

In a chapter dealing with communication, we have so far avoided explicitly defining it. This is because communication is a concept that seems intuitively simple but has been difficult for biologists to define formally, and consequently constitutes a source of disagreement (see Ruxton and Schaefer 2011). As Greenfield (2002) has said, there are probably as many definitions of 'communication' and 'signals' as there are books about the subject. Although it often helps not to get too bogged down in semantics, the concept of communication is essential and thus needs an appropriate conceptual discussion.

Probably beginning with Marler (e.g. 1961), many (although by no means all) behavioural ecologists have argued that communication involves the transfer of information between a sender, via a signal, to a receiver, following which the receiver responds (Bradbury and Vehrencamp 2011; Hailman 1977; Schaefer 2010). Most authors argue that in 'true communication', information transfer is not accidental but occurs because it benefits both the sender and receiver, on average (e.g. Bradbury and Vehrencamp 2011; Greenfield 2002).

However, a problem is that defining information has always been problematic and the emphasis on information in communication has long been controversial (Bradbury and Vehrencamp 2011; Hasson 1994). If we return to our discussion of information from Chapter 1, recall that information can be thought of as something that reduces uncertainty about an aspect of the world (e.g. the environment or another individual), and in communication this is with respect to alternative states or responses of a sender (Laughlin 2011). Thus, we might consider communication to be a process that reduces uncertainty about the state or motivation of another organism sending that information. The problem is that there is simply no common measure of information in biology, and so while it is a useful concept, it lacks practical value in many cases. Information could be defined in various forms and stages of interaction and receiver processing; for example, some measure of signal structure or specific nerve cell responses in the receiver's sensory system. In addition, as discussed by Schaefer and Ruxton (2011), not all information is relevant or important to the receiver, and even relevant information may only be so in some contexts. For example, that a male has elaborate plumage signifying high quality would be irrelevant if the mating sea-

son has just ended. As Marler (1961) highlighted, any information content of signals alone cannot be discussed independently of the response of the receiver.

If we take information on its more technical term, as a reduction in uncertainty, there are still issues that arise. First, we need to somehow specify the behavioural options available to receivers, and how much the receiver already knew about the sender or environment to reduce uncertainty by a certain level (Font and Carazo 2010). Second, in an evolutionary sense it is not specifically the information that matters in communication, but rather the response of the receiver to a signal that ultimately determines each individual's fitness (Schaefer and Ruxton 2011). In many cases, measuring individual responses matters more than knowing the precise information that was transmitted. Likewise, it can be very hard to know when a trait used in communication actually reduces uncertainty in the receiver (Schaefer 2010). On the other hand, the information that a receiver obtains will often be crucial in influencing their response, and researchers often need to know broadly what the content of the signal was in order to be able to interpret this response correctly.

Other authors have been more critical of the information concept in communication. Rendall *et al.* (2009) and Owren *et al.* (2010) criticize information-based definitions of signalling and communication because, they argue, the terminology involved often implies anthropomorphic or metaphorical terms somewhat akin to human language containing information as an entity in its own right. They are especially critical of primate studies of 'language-like' communication and the metaphorical way that information and communication are discussed. Their criticism is a good one in the sense that we should be very careful not to imply conscious intent in organisms communicating, or some sort of theory of mind, but remember that individuals should respond optimally owing to selection pressure over evolution. However, the criticism of the use of information on these grounds seems excessive, as most studies do not imply that information transfer has any kind of conscious meaning (Seyfarth *et al.* 2010), even if their terminology is not as precise as it could be. The key problem is in determining how we measure and define information, and a lack of consideration given to signal efficacy by many researchers (which Rendall *et al.* 2009 also point out).

Rendall *et al.* (2009) and Owren *et al.* (2010) argue in favour of basing the idea of communication on 'influence' by the sender on the receiver (akin to Schaefer and Ruxton's 2011 argument about the receiver response being what affects fitness, not information content). Owren *et al.* make the valid point that studies of communication in mate choice have been dominated by arguments of signals providing information about health, good genes, and other aspects of fitness, and tend to ignore more fundamental issues to signalling, such as simply whether or not the signal is detectable and localizable, and receiver-perceptual systems. For example, sensory exploitation, where signals exploit pre-existing features of the receiver's sensory system (see Chapter 8), has little to do with encoding signaller quality but rather seek to influence the receiver's behaviour (Owren *et al.* 2010). In short, Rendall *et al.* (2009) and Owren *et al.* (2010) argue that communication is fundamentally an attempt to 'influence the current or future behaviour of another individual'.

Owren *et al.* (2010) also exclude the common idea that the receiver must benefit in communication. Again, this is on the grounds of concepts like sensory exploitation, whereby the receiver does not gain information about mate quality. However, these arguments have been criticized on the grounds that sensory exploitation may not be widespread. In addition, the problem here is that even when sensory exploitation is present, the receiver still benefits in the sense that they have detected the presence of, for example, a potential mate. Part of their stance is the (valid) argument that signals do not have to convey honest information about mate quality, as is evidenced by the fact that various signals can evolve by exploiting pre-existing biases in the sensory system of the receiver, for example male mating signals (Chapter 8). However, it would be wrong to argue that information does not apply here—these signals, although not conveying any honest information about quality, are still telling the receiver that a potential mate is present. The response of the female is thus a combination of the fact that she has detected a potential mate (information component that has resulted in a reduction in uncertainty about the presence or location of a potential mate), and that the signal is highly effective at stimulating her sensory system (no information needed). Thus, information is not the sole factor determining a receiver's response, but it remains an important consideration. In defining communication, we should ask 'would a communication system arise or persist were no information at all transferred'? Here, the answer is likely to be 'no', because neither party would gain anything from such a system. Seyfarth *et al.* (2010) argue that information is central to communication studies. They rightly argue that if the receiver does not extract any information then it is unclear why they should respond at all.

Font and Carazo (2010) make the interesting point that Rendall/Owren *et al.*'s approach is incomplete without incorporating the concept of trading information. They argue that senders benefit by influencing the receivers' behaviour to their advantage (not because they 'want' to share information with the receiver), whereas the receiver responds to the signal because they benefit from the information that they obtain. Thus, senders do not have to send a specific message, but receivers should benefit from information acquisition. Font and Carazo (2010) argue that communication involves trading signals, not information, and that information is not transferred or transmitted (as is regularly stated in studies of communication), but rather *extracted* by receivers (see also Ruxton and Schaefer 2011). They argue that terms like 'information transmission' or 'information sharing' are at fault in generating confusion. The argument that information is not transmitted in a signal but rather extracted by and a property of the receiver (Font and Carazo 2010; Ruxton and Schaefer 2011) can fit with our definition of information as providing a reduction in uncertainty in the receiver (Chapter 1). In short, influence and information are compatible and both have value.

In summary, it is certainly true that many researchers of animal communication use the term 'information' loosely and do not consider signal efficacy and sensory systems. However, understanding signals or cues in terms of information can be very helpful and compatible with understanding efficacy. Neither the manipulation definition of Rendall/Owren or concepts of information transfer/extraction alone are sufficient to explain all of communication and signalling. There is no reason to adopt a purely 'response' or purely 'information' based stance of communication. We can reconcile these different opinions by considering communication as involving signals that have evolved under selection to influence the behaviour of the receiver, and where the receiver extracts at least some information from the signal. Therefore, the definition of communication we will follow here is that:

Communication is a process involving signalling between a sender and receiver, resulting in a perceptual response in the receiver, which extracts information from the signal, potentially influencing the receiver's behaviour.

5.4.2 Examples of What is and is Not Communication

Given our above definition, we can think about some examples of things that are or are not considered as instances of communication. Schaefer and Ruxton (2011) define communication as occurring 'between one individual (the sender) and another individual (the receiver) if trait values of the sender stimulate the sensory systems of the receiver in such a way as to cause a change in the behaviour of the receiver (compared to a situation where the trait values of the sender were different)'. This definition does not directly invoke arguments relating to information, but does clearly state the importance of the receiver's sensory system and the concept of influence. However, it does not specifically rule out cues; it merely states that changes in the trait value in one individual cause a change in behavioural response in another individual. In fact, Schaefer and Ruxton (2011) include cues (which they refer to as 'inadvertent communication') as a type of communication. As with Owren *et al.* (2010) and many other authors, here we do not consider cues as communication, as they are not signals that have evolved under selection specifically to influence the receiver (Marler 1967).

Schaefer and Ruxton (2011) also argue that deceptive traits, such as Batesian mimicry, are not signals because the receiver does not benefit. While this is true, the traits are exploiting an already present communication system, which on average does benefit the receiver (e.g. warning signals), and so such cases

are still best thought of as deceptive signals. Likewise, they argue that crypsis (a type of camouflage that prevents the detection of an object) is an example of communication, albeit deceptive communication. This does not fit most definitions, whereby camouflage is a process that hides information that would otherwise give away an organism's presence. Crypsis is about not influencing the behaviour of the receiver because the predator does not detect anything in the environment of interest. This does not fall under most definitions of communication because no signalling is involved. It is also not in keeping with a general assertion that in communication both the signaller and receiver should benefit (Greenfield 2002).

We have said that crypsis is not a type of communication, but what about masquerade (a type of camouflage that prevents recognition by resembling an unimportant object in the environment like a leaf or stone)? Here, as with crypsis, we adopt the stance that masquerade is neither a signal nor a communication system. This stance is taken because communication systems involve benefits to both parties, and once these have evolved, other organisms may exploit these systems with deceptive signals (such as orchids attracting pollinators with sexual mimicry). However, masquerade does not exploit an already present communication system; it exploits the fact that predators should pay no attention to things like twigs and leaves, neither of which are communicating with the predator. Batesian mimicry therefore, is an example of (dishonest) signalling and communication, whereas masquerade is not.

5.5 There is More to Communication than Just Sensory Systems

Clearly, there are many instances where animals show an 'innate' or 'hard-wired' response to signals that requires no cognitive input (e.g. stopping and looking up when an alarm call is heard). Here, the level of sensory stimulation alone can be crucial in eliciting a response. However, in many species, especially those that are long lived or live in complex societies, cognition can play a key role in influencing the response to a signal, and in fact the value that a signal may have. We will illustrate this with an example.

Signals in some animals may act as signs of fertility or receptivity. For example, female rhesus macaques (*Macaca mulatta*) show marked changes in facial lightness during their reproductive cycle. This variation is related to the timing of the female fertile period; females have darker faces when receptive (Higham *et al.* 2010). Higham *et al.* (2011) took photographs of female rhesus macaques at different stages of their fertility cycle. They then printed pictures of the females that were perceptually, to rhesus colour and luminance vision, indistinguishable from the coloration exhibited by each female when originally photographed. They presented these images to males in a choice test (a preferential looking procedure) and analysed the propensity of males to look at either image in a pair, where one image corresponded to the female at ovulation and the other image was when she was not fertile. They presented the stimuli to two sets of males; one set where the males were familiar with the female (coming from the same group), and another set of males that were not familiar with the female. The rationale was that highly social animals such as primates have access not only to potential signals like face colour, but also are familiar with group members. Higham *et al.* found that males distinguished ovulatory from pre-ovulatory faces, looking more at the former and showing that they do respond to this signal. However, a significant proportion did this only among males with prior experience of a female (i.e. individuals from the same group). This latter finding makes sense; while females tend to have darker faces when fertile, the absolute darkness and level of variation changes from one female to the next. The appearance of some females during their fertile phase is still lighter than the faces of other females that are not ovulating (Figure 5.8). Thus, prior experience of a given female may help a male extract the correct information about her reproductive state.

Figure 5.8 Female rhesus macaques have darker facial coloration around the time of ovulation (a) than when not in the fertile phase (b). However, there is great variation across females. For example, female (c) has a lighter face than female (d), yet it is female (c) that is ovulating. In fact, female (c) never has a face coloration that is as dark as female (d). Therefore, a simple rule of thumb, 'prefer females with darker faces' would not be appropriate for males to judge when females are receptive. Instead, prior familiarity with a female and her coloration patterns may help. Images from Higham *et al.* (2011) reproduced with permission.

5.6 Future Directions

Despite its long history, signalling and communication remain vibrant areas of research today, with many unresolved questions. Perhaps some of the most important issues require a better understanding of where different signal forms reflect selection for both signal efficacy and strategic elements. Linked to this, we need a better understanding of the relative importance of different components of signal structure in communication, and how different components allow receivers to extract different pieces of information. In addition, more work should be conducted investigating how different aspects of the same signal could achieve different features of signal efficacy, such as detection and localization of a sender. Finally, all of the above would be strengthened by a more comprehensive understanding of how different signal forms stimulate specific neurophysiological processes found in receiver sensory systems.

5.7 Summary

In this chapter we have introduced key concepts relating to signals, cues, and communication. Signals and communication are essential in many areas of animals' lives, from finding mates, to alerting individuals to danger, to fending off rivals. Signals have evolved to elicit a response in the receiver, who can extract relevant information from signals in communication. Cues on the other hand, are incidental sources of information, which have not been directly selected in order to influence the behaviour of another organism. Signals have two key aspects: their strategic aspect, which describes the information that can be extracted by a receiver, and their efficacy, which is how the signal has evolved into a particular form in order to effectively influence the response of the receiver. In the past, most research in communication has focused on strategic aspects and ideas relating to signal honesty, yet it is clear that to understand the diversity of signals and communication systems in nature fully we also need to consider signal efficacy carefully.

5.8 Further Reading

Bradbury and Vehrencamp's (2011) book presents an in-depth look at communication, including some of the more technical aspects associated with this field. Maynard-Smith and Harper's (2003) book on signalling is a highly readable account of the subject, and although somewhat biased towards strategic aspects of signals and economic models of signal honesty, it also has some nice discussion on signal efficacy too. Ruxton and Schaefer (2011) present an interesting discussion about influence and information in communication. Finally, Schaefer and Ruxton (2011) have published a very clear and comprehensive account of plant–animal communication, dealing with a wide range of examples and key concepts, and being an excellent companion to the more animal-centric accounts found in other books (including this one!).

CHAPTER 6

Multimodal Signals and Communication

Box 6.1 Key Terms and Definitions

Emergent Response: A response in the receiver to a multimodal signal that does not occur when either signal component is presented alone.
Multimodal Signal: A signal comprising components that are received by more than one sensory channel.
Non-redundant Signal Components: Two or more signal components allow the receiver to extract different information.
Signal Dominance: Where one signal component takes precedence over another component in influencing the receiver.
Signal Enhancement: Where the presence of one signal component produces an enhanced response to another component.
Signal Redundancy: Where two or more signal components each allow the receiver to extract the same information.

In the last chapter, we introduced key concepts regarding signalling and communication. Most studies of signalling and communication deal with just one modality at a time, often categorizing signalling interactions by the main sensory channel used. However, in many cases, signals are transmitted using several modalities; that is, many instances of communication are 'multimodal' (Partan and Marler 1999, 2005; Rowe 1999b). In the last 10–15 years, a growing number of researchers have investigated multimodal signals, providing some illuminating findings about signalling and communication and how multisensory information is processed by animals (see also Chapter 4).

6.1 Multimodal, Multicomponent, and Complex Signals

Multimodal signals comprise components that are presented to the same receiver across two or more sensory modalities. In addition to multimodal signals, there has been a trend to call some signals 'multicomponent'. This encompasses signals with multiple aspects that have different effects even when expressed in the same modality (Hebets 2011). For example, work in spiders has shown how seismic signals comprise multiple components produced in different ways, and these have different influences on the receiver (Elias *et al.* 2006). Often the term 'complex signalling' is used to encompass both multimodal and multicomponent signals, regardless of the number of modalities used.

In this chapter we will deal only with multimodal signals. This is partly because in nature many signals are multicomponent. Most visual signals, for example, comprise attributes of colour, pattern, size, and brightness, and most acoustic calls have elements of different frequency and amplitude. Although the idea of multicomponent signals can be helpful, a discussion of this at present could either encompass almost all signals in nature, or be restricted to only the few cases where we can be confident that different signal components are each under selection to provide some additional aspect of signal efficacy or strategy, rather than simply covarying or being a by-product of the production of other signal features. In addition, the terms 'complex signalling' and 'multicomponent signals' are not always helpful because these ideas can be somewhat subjective. A signal that is complex to us need not be so to the real receiver. However, readers should note that in our discussion below, many (but not all) of the suggested advantages of multimodal signals may also apply to genuine multicomponent signals.

Before beginning our discussion of multimodal communication, we should address two notes of caution. First, researchers in complex signalling sometimes discuss work where they or others have isolated different modalities (or aspects) of a display and talk about having isolated different 'cues'. However, as we discussed in the last chapter, the term 'cue' has a very specific meaning different from signals and communication (in fact, a cue is specifically not a signal). Here, we will therefore refer to different 'components' of multimodal signals (as is also common in discussions of unimodal multicomponent signals). Second, Hebets (2011) makes the point that we should be careful with defining components based on a specific modality, because although receivers may respond to signals of different physical properties (e.g. sound waves, photons of light), the distinction between modalities may not always be clear. For example, chemical signals could produce sensations of smell or taste depending on the specific receptor types that they target, and chemoreceptors can be very different in form, response, and the regions of the nervous system they innervate. Likewise, sound waves may be transmitted through the air to hearing organs, but can also be transmitted through the substrate as vibratory/seismic signals. Greater appreciation in multimodal studies of the specific receptor types involved will help in clarifying such issues, and perhaps an ultimate goal of defining a multimodal signal is whether they target different receptor classes and neuronal coding mechanisms (see Chapters 2 and 3), rather than the physical properties of the

signals themselves. For the time being, however, in this chapter we consider multimodal signals as simply stimulating more than one sensory modality.

6.2 What Advantages do Multimodal Signals Provide?

Given that multimodal signals seem to be widespread (see examples below), we need to ask what advantages multimodal communication brings. There is a wide range of, not mutually exclusive, explanations, which invoke both strategic (content-based) and efficacy arguments (see Chapter 5).

6.2.1 Strategic Benefits

Multimodal signals may provide additional information over unimodal signals in the form of 'multiple messages'. For example, different signal components could provide varied information about things like individual or species identity or individual quality (Hebets and Papaj 2005). This may provide greater benefits if the receiver is consequently more informed and acts in a more appropriate way. Multimodal signals may therefore reduce the likelihood of things like mating with the wrong species and increase the likelihood of identifying a suitable mate or genuine rival.

Krakauer and Johnstone (1995) make some interesting suggestions about how multimodal signalling may affect aspects of honesty in communication (see Chapter 5). They suggest that when the number of possible signal dimensions is large (e.g. including several modalities), then signallers of different quality may evolve more divergent displays if it allows receivers to more easily distinguish between them and potential cheats. They also suggest that cheats may find it more difficult to 'chance upon' an effective display when the possible dimensions of display components are large, meaning that multimodal signals may be more reliable in encoding individual quality. The output from a neural network model that they present is consistent with this prediction, but the generality of these ideas remain to be more fully explored. Overall though, it seems reasonable that if a signal comprises multiple components it may be harder to fake.

There are other mechanisms that might make a signal comprising multiple diverse components harder to fake. For example, many displays are physiologically or energetically costly (see Chapter 7). Multiple signal components in the same modality may be more able to share costs (e.g. of construction of signalling structures) than components in different modalities. For example, once a peacock has invested in a large tail, then this provides a structure that facilities other visual traits (coloration in the tail, behaviours that draw attention to the animal), but does not offer similar benefits to signal traits involving, for instance, vocalization. Signals comprising several multimodal components (e.g. visual movement and generating vibrations or vocalizations) may, therefore, be especially demanding and only possible for the highest quality individuals. Thus, multimodal signals may arise under selection pressure to maintain signal reliability. However, costs on multimodal signals may be greater than for unimodal signals if they increase the risk of predation or parasitism more than unimodal displays (Partan and Marler 2005; Roberts *et al.* 2007; see Chapter 7). Exploring the costs of different signal components within and between sensory modalities would be very welcome.

6.2.2 Efficacy Benefits

As Rowe (1999b) has outlined (following Guilford and Dawkins 1991), multimodal signals may have a number of key efficacy benefits. Probably the most prevalent idea is that multimodal signalling incorporates redundancy (different components that convey the same information), and this increases the likelihood that at least one signal component will be detected and identified by the receiver's sensory and cognitive systems, and that reaction times may be shorter (see Chapter 4). Multimodal signals may also produce a stronger response in the receiver if different components of the signal stimulate the various sensory systems of the receiver at about the same time, allowing for greater discrimination and faster learning (Hebets and Uetz 1999; Partan and Marler 1999).

Some authors place redundancy under content-based (strategic) explanations. However, with redundancy the information available to extract by the receiver is the same, whereas the chance of the receiver detecting, localizing, and/or discriminating the signal may be increased (see Chapter 4; Sensory Integration). Therefore, a key benefit of redundancy is in getting the message across more effectively (efficacy), rather than altering the message itself. As with the multiple messages theory, this could also decrease the likelihood that the receiver will mistake the identity or quality of the signaller because each modality provides an independent assessment of a stimulus (Munoz and Blumstein 2012). Recall from Chapters 1 and 5 that we can define information as a reduction in uncertainty in the receiver about potential outcomes, events, or states. If the different components each allow the receiver to extract information, although imperfectly (such as with noise or some degradation of the signal), but the different components are consistent in the information that they provide, then this may reduce uncertainly in the receiver (in effect, increase the information acquired). As such, redundant signal components act as back-ups for each other. Each additional component from another modalities is consistent in providing the same 'message' for the receiver to extract, and so there is no change in the strategic aspect of the signal, although the fidelity of the information obtained may be increased. Back-up signals may be especially useful when individuals communicate on different substrates or in different environments, each with varying transmission properties for the different components (e.g. Gordon and Uetz 2011; see Chapter 10).

Different modalities also have relative strengths and weaknesses for communication (see Chapters 2 and 4; Marler 1967), including at various stages of a signalling interaction (e.g. signal transmission, reception). For example, visual signals travel quickly but are easily obstructed by obstacles, whereas chemical signals travel slowly, but move around and past objects. Combining multiple modalities may allow beneficial aspects of several modalities to be combined, and the limitations of each modality to be overcome. Related to this, different components of multimodal signals may be effective in different environments. For example, electric or acoustic signals may be more beneficial than visual signals in cluttered habitats or when visibility is low (e.g. in turbid water). Multimodal signals may therefore allow effective signalling under varied environments or when there is a lot of environmental uncertainty (Munoz and Blumstein 2012). Different components of a signal may act as backups if they are differentially affected by environmental noise or degraded by specific features (temporal or spatial) of a given habitat (Hebets and Papaj 2005). Specific modalities of a signal may also be important at successive stages of an interaction. For example, a woodland plant may utilize olfactory signals to lure insect pollinators from a distance, but the final orientation of the pollinators to the flowers may rely more on visual aspects of the floral signal. In such cases, it is debatable whether the plant's signal is best interpreted as multimodal or as a sequence of unimodal signals (see section 6.5 for further discussion).

With regards to signal reception, Hebets and Papaj (2005) argue that when speed of signalling is important, this may select for multimodality because a greater amount of information can be obtained in a given time if several different sensory systems are utilized simultaneously. The existence of multisensory neurons that pool information from multiple sensory modalities (Chapter 4) may further increase the saliency of signal detection specifically when signals are multimodal. It seems reasonable to ask what came first, the multimodal signal or the multisensory neural architecture. We might suggest that the detection architecture may have evolved to detect multimodal cues about the environment and have then been exploited by appropriate signals.

Although it is common to assume that each signal component presented in isolation will produce an effect, this need not be the case. Instead, some components of a multimodal signal may have no strategic content (convey no information for the receiver to extract). Instead, they may alert the receiver to another signal component, from which the receiver can extract information, or amplify the efficacy of another component (Hebets and Papaj 2005; Rowe 1999a; Chapters 5 and 10). There is a wealth of empirical evidence that multimodal signals can provide better recognition, faster discrimination learning, and multidimensional generalization (see Hebets and Papaj 2005; Rowe 1999a). What remains to be explored is how much these benefits are enhanced (or costs reduced) if different components are spread across different sensory modalities.

6.3 A Framework for Multimodal Signals Based on Influence?

Rather than presenting an historical account of discussions of terminology and definitions regarding multimodal communication, here we discuss a framework that can be used to investigate multimodal signals. Because there are difficulties in describing communication based on information provision alone without considering influence (see Chapter 5), and because not all multimodal signal components need to have a strategic component (no message for the receiver to extract), we describe and extend a framework by Partan and Marler (1999, 2005) for multimodal signals based on their influence on the receiver (outlined in Figure 6.1). In general, most work on multimodal signals involves presenting receivers with the different components in isolation and then together to determine their influence on receiver response. Note below and in Figure 6.1 that we are using the term 'response' in a broad sense, which could encompass responses such as detection or inspection behaviour, or some effect on learning.

Partan and Marler (1999, 2005) argue that redundant signal components should provide the same information as other parts of the signal and each should produce the same influence on a receiver if presented independently, whereas non-redundant components may have different effects. When presented together, two or more redundant components of a signal may either lead to the same effect (*equivalence*), or may increase the receiver's response owing to a reduction in uncertainty in the receiver (*enhancement*).

For non-redundant components, one component may have *dominance* when presented together and take precedence over the other in producing a specific outcome. Alternatively, both may have *independence* and each lead to specific changes in behaviour (for example, two different behavioural outcomes occurring). The presence of one component may also *modulate* the response of the receiver. This could include enhancement, but also reduction in response. Finally, the combination of two

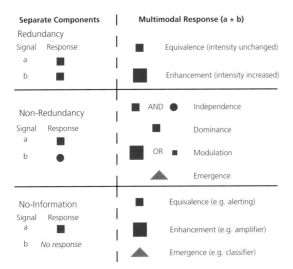

Separate Components | **Multimodal Response (a + b)**

Redundancy
Signal Response

Equivalence (intensity unchanged)

Enhancement (intensity increased)

Non-Redundancy
Signal Response

AND Independence

Dominance

OR Modulation

Emergence

No-Information
Signal Response

Equivalence (e.g. alerting)

Enhancement (e.g. amplifier)

No response

Emergence (e.g. classifier)

Figure 6.1 Classification of multimodal signals based on influence on the receiver (revised and extended from Partan and Marler 2005). The left column depicts the response of the receiver (as a symbol) when presented with a multimodal signal component in isolation, when the multimodal signal is one of three types, where different components are either: 1) redundant, 2) non-redundant, or 3) where one component presents no information for the receiver to extract at all. The response to the complete multimodal signal is on the right, depicting when effects can be seen as *equivalence* (response is the same as to either component alone), *enhancement* (the response is greater than to either component alone), *independence* (both responses are observed), *dominance* (one component dominates in its influence over the other), *modulation* (one signal component affects the response to the other), or *emergence* (an entirely new response is observed). When one component carries no information, this may act as an alerting component, drawing the receiver to respond to another component, act as a signal amplifier, increasing the response of the receiver to the other component, or act as a classifier, allowing the receiver to correctly interpret or classify another component.

components may produce *emergence*; a new response not observed to either component alone.

Where one component of a signal has no strategic component, presenting this in isolation may lead to no response by the receiver at all. When presented with another component from which the receiver can extract information, this may lead to equivalence (i.e. same response as to the message-containing component alone) if the non-strategic component acts as an alerting component, or enhancement if the non-strategic component acts as a signal amplifier. Finally, if a component does not by itself contain any useful information, but provides context to another component of the signal, then we

may consider it as a classifier. Such a component may allow the receiver to correctly interpret another component of the display.

6.4 Examples of Multimodal Communication

6.4.1 Multimodality in Warning Signals (Aposematism)

Animals that are toxic, distasteful, or generally unprofitable often signal this to predators with conspicuous signals (aposematism; Chapter 9). Although most work has tended to consider visual components of aposematism (such as bright colours), warning signals can also be chemical, vibratory, or acoustic, and often seem to combine several such components (see review by Rowe and Guilford 1999a). Two key functions of warning signals are to prevent initial attack, and to promote, enhance, or maintain learnt avoidance over repeated encounters. One of the earliest experiments to investigate multimodal warning signals was conducted by Marples *et al.* (1994) with seven-spot ladybirds (*Coccinella septempunctata*; Figure 6.2) and Japanese quail (*Coturnix coturnix japonicus*).

Ladybirds are often toxic by virtue of alkaloids, and in addition to their distinctive coloration they can also 'reflex bleed', where fluids are exuded through pores in the knee joints of their legs. The fluid also often strongly smells of pyrazines. Marples *et al.* presented ladybirds to quails either whole,

Figure 6.2 Many ladybirds, such as the seven-spot ladybird (*Coccinella septempunctata*) have bright warning signals coupled with unpleasant tastes, smells, and/or toxins that deter predators. Image reproduced with permission from Lina Arenas.

or with components of their display added to a palatable beetle species. For example, in some treatments they either added ladybird elytra to beetles (only presenting the ladybird appearance), or presented beetles to quail with the smell or taste of a ladybird. They presented the different components singly or in combinations and found that initial avoidance was relatively effective for colour alone and especially for the combination of taste plus colour, compared to whole ladybirds. Smell alone was rarely used by the birds as a signal of toxicity. However, although when other cues were paired with taste avoidance persisted for longer than for any component alone (including just taste), avoidance was only properly maintained for whole ladybirds. That no single or pair of components was as effective as whole ladybirds in maintaining avoidance, illustrates that a combination of factors makes ladybird aposematism effective. Colour, taste, and smell together was almost as effective in preventing initial rejection, but avoidance is not maintained unless coupled with the toxicity of the real ladybird (perhaps because the birds learn that the signal is a fake). This work shows that different components can have different value in promoting both initial and maintained avoidance, and that combinations may produce enhanced effects, or be essential to maintain avoidance. It suggests that combined components in aposematic signals may increase the degree of initial avoidance, and the speed and persistence of avoidance learning, although more work is needed. Finally, Marples *et al.* (1994) suggest that the multimodal nature of the defence may provide protection at different stages (e.g. detection, capture, subjugation).

Presentation of novel odours often associated with warning signals, such as pyrazines, can enhance avoidance of warning colours (e.g. Rowe and Guilford 1999b). In some cases, odours can produce emergent avoidance responses to conspicuous warning colours, when neither component is aversive when presented alone (Lindström *et al.* 2001; Rowe and Guilford 1996). A range of novel odours can produce this effect, suggesting that aposematic prey are exploiting a general feature of predator psychology rather than predators having an evolved response to avoid aposematic prey with specific odours (Jetz *et al.* 2001). Emergent responses make

sense here, because for aposematism to work the predator needs to make an association between a conspicuous signal and some element of unprofitability, which in the case of toxic chemicals may also have characteristic smells. On the other hand, if this is true it is strange that there is no response to smell alone. It may be that instead the smell allows the correct interpretation of the colour, given that there are other red objects that a predator may encounter, including edible foods. In this case, the component may act as a signal classifier. Interestingly, sound can also produce unlearnt avoidance behaviour in birds to warning colours and enhance discrimination learning (Rowe and Guilford 1999a; Rowe 2002), though the reason for this is unclear.

6.4.2 Multimodality in Plant–Pollinator Interactions

Many plants advertise flowers to pollinators with signals in a range of sensory modalities. Kulahci *et al.* (2008) trained bumblebees (*Bombus impatiens*) to discriminate between artificial flowers that were rewarding or non-rewarding (with a sucrose solution or just water, respectively), that differed either in their visual appearance (shape: cross versus circle) or odour (peppermint or clove), or in both appearance and odour. Bees learnt about rewarding flowers faster when they differed from non-rewarding flowers in both modalities than when flowers differed only in shape. Bees also continued to visit flowers that they had been trained on for longer when the flowers were multimodal than when they were unimodal. In general, attending to signals in both modalities allowed bees to increase their accuracy in visiting rewarding flowers (73% correct choices when trained multimodally, compared to 64% correct choice for shape or odour alone). Here, both odour and visual components present the same information to the receiver (rewarding or unrewarding), and so are redundant. The effect on receiver response (learning) seems to be enhanced by the presence of both. Subsequent work also shows that the presence of one type of signal (e.g. olfactory information) can reduce uncertainty about another signal component (e.g. colour) and the information it provides about whether a flower is rewarding or not (Leonard *et al.* 2011).

In Chapter 5 we introduced the idea that some plants can have flowers with nectar (floral) guides that lead pollinators to the location of nectar rewards. Goyret (2010) showed that nectar guides can be more effective when multimodal. When presented with artificial flowers, hawkmoth (*Manduca sexta*) proboscis placement was influenced by visual information whereas tactile guides influenced proboscis placement. Here then, the two signal components are non-redundant and provide different information, and seem to produce independent receiver responses. We might expect this relationship to be common in plant-pollinator interactions if, for example, flower colour allows initial detection by the pollinator and odour presents information about nectar reward (quality).

6.4.3 Begging in Avian Chicks

Many begging displays in young birds are multimodal, including bright gape colours and loud calls. Kilner *et al.* (1999) showed that when feeding young, reed warbler parents monitor visual and acoustic aspects of begging displays. Parents adjust their feeding rate based on the visual area of the gape displays of the entire brood. This correlates with brood size and the age of the chicks, and their need over time. In contrast, the vocal displays of the brood allow the parents to adjust their feeding with respect to current hunger levels. As such, both visual and vocal components provide independent information (they are non-redundant) about required provisioning levels. However, it is unclear how much of the display components' features are important in signal efficacy.

6.4.4 Courtship Behaviour in Spiders

6.4.4.1 Wolf Spiders

Spiders can detect cues (e.g. from prey or predators) or signals (e.g. from conspecifics) using a variety of sense organs, to extract acoustic, vibratory, tactile, and visual information (see Uetz and Roberts 2002 for a review with reference to communication). A range of work on spider mate choice and courtship has investigated multimodal communication, primarily using wolf spiders from the genus *Schizocosa*.

Here, males of all species use vibratory signals by virtue of a stridulatory organ found on the male palp. In addition, some species also use visual signals, including dark pigmentation and hair tufts on their forelegs.

Hebets (2008) showed that in *S. stridulans* the seismic component of courtship signals seems to dominate over visual components (black foreleg pigmentation and brushes of hair). Hebets tested copulation frequency when females were presented with males with both seismic and visual signal components (light room and filter paper substrate), just the seismic component (dark room), just the visual component (granite rock substrate), or neither seismic nor visual components (dark room and granite substrate). There was no effect of presence or absence of visual signal on courtship success, whereas absence of seismic components significantly decreased copulations (Figure 6.3).

Hebets then presented females with video playbacks of males with modified forelegs with either enlarged brushes, normal brushes, or no ornamentation, all in the presence of seismic components from a visually isolated courting male. She found no effect of visual stimulus on female receptivity. This contrasts with previous work, where playbacks of males with enlarged brushes increased females' receptivity (Hebets and Uetz 2000). However, in the earlier study, females were presented with playbacks of males in the absence of seismic components (Figure 6.4). These results indicate that in *S. stridulans* the seismic signal component dominates the visual component, and that the signal components are not redundant because the visual signal is not necessary for copulation (Hebets 2008). This raises the question of why the visual component exists. Hebets suggests that in environments where vibratory signals do not travel far (e.g. leaf litter where the spiders live) that the visual components may increase detection probability, especially given that other work has shown that females are more responsive to males of greater ornamentation (Hebets and Uetz 1999, 2000). When in close proximity, the seismic component may then become dominant.

Interestingly, the situation seems to vary even across closely related species. For example, *S. uetzi* females are more receptive to playbacks of ornamented males when they are presented at the same

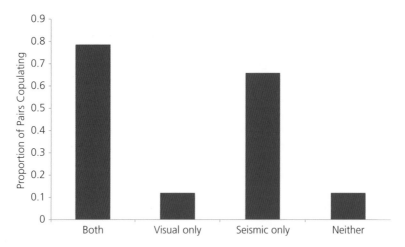

Figure 6.3 Proportion of copulations between male and female *Schizocosa stridulans* when either visual or seismic signal components were presented alone, both components (multimodal), or neither component. The results show that the seismic but not the visual components were key in copulation frequency. Data from Hebets (2008).

time as a seismic signal component (Hebets 2005). This indicates the two components can have an enhancement effect on receiver response. Thus, in one species, presence of seismic signal components causes visual components to have more influence, whereas in another species, seismic components dominate visual components. In *S. uetzi*, seismic components may increase female response towards visual displays or ornamentation; that is, they could work as signal amplifiers or alerting components.

Uetz *et al.* (2009) investigated female responses to unimodal and multimodal signals in two species of wolf spider, *S. ocreata* and *S. rovneri*. Females of both species detected and orientated towards male signals more quickly when the signals were multimodal (seismic and visual) than when they were unimodal

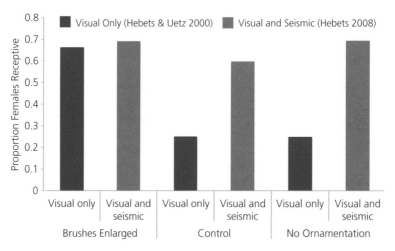

Figure 6.4 Female *Schizocosa stridulans* receptivity when presented with video playbacks of males with leg ornamentation that was unmodified, with enlarged brushes, or no ornamentation, when in the presence of seismic signal components or not. The results show that females only prefer males with enlarged brushes in the absence of seismic display components. Data from Hebets and Uetz (2000) and Hebets (2008).

(Figure 6.5). Thus, multimodal signals in both species increase detection probabilities of the signaller by the receiver, which may indicate enhancement. However, there were differences between the species once orientation had occurred. Once detected, female *S. ocreata* approached males displaying either seismic or visual components at the same level, but approached males with multimodal signals more often and showed greater receptivity towards multimodal signals. This indicates that in *S. ocreata* seismic and visual signals are redundant, producing the same response and level of response in females but together produce an enhanced effect. *S. rovneri* females, on the other hand, did not differ in their approach towards multimodal or unimodal signals, but responded with receptive behaviour more to either multimodal or seismic components alone than to visual signals alone. Here, male courtship components are non-redundant, with seismic signal components dominating over visual components. Differences between these two species could be due to their respective habitats, although more work is needed in this respect.

Finally, work has started to investigate the relative importance of signal content (strategy) and signal efficacy in multimodal signals, by rearing individuals of *S. floridana* on high- or low-quality diets and presenting them to females and investi-

gating mating success (Rundus *et al.* 2010, 2011). Work so far indicates that both the male visual ornaments and their seismic signals are condition-dependent (i.e. a strategic component); males raised on a high-quality diet have greater leg pigmentation and higher rates of seismic signalling to females (Rundus *et al.* 2011), and therefore may be redundant components. In addition, females chose males based on their seismic signal components (preferring those from high quality diets), but only when visual signals were absent, indicating that if the signals are redundant then they could allow signalling under different environmental conditions (i.e. selection for signal efficacy). Finally, when female mate preferences reflected male condition (in the dark using seismic signals), this translated into more offspring. These experiments show that the way multimodal signal components are used reflects selection for both strategic and efficacy components of signals, just as with unimodal signals. However, another study using a different species, *S. retrorsa* (Rundus *et al.* 2010) failed to find an association between male copulation success and either signalling environment or high- versus low-quality diet. Gordon and Uetz (2011) demonstrate that males of *S. orcreata* have the behavioural flexibility to compensate for environmental constraints, for example

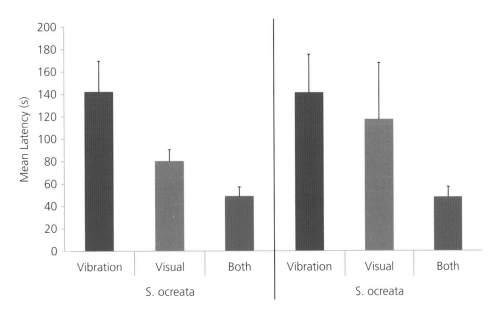

Figure 6.5 Mean (plus SE) time that female wolf spiders of two species take to orientate towards a male when presented with either seismic or visual signal components or both together. Females orientate more quickly when the signals are multimodal. Data from Uetz *et al.* (2009).

increasing visual signalling on substrates where seismic communication is less effective. Overall, *Schizocosa* wolf spiders have provided an excellent model system for multimodal research, with sometimes marked differences between species. It will be interesting in the future to determine why some species differ so much in terms of construction and influence of their multimodal signal components.

6.4.4.2 The Peacock Spider

The male Australian peacock spider (*Maratus volans*) has a remarkable elaborate courtship dance that it uses in attracting females. This includes raising brightly coloured flaps that bear resemblance to the peacock's tail, and which when not courting are kept lowered down around the abdomen (Figure 6.6). Girard *et al.* (2011) have shown that, like the wolf spiders above, the peacock spider's display involves not only colour signals and movement, but also vibratory signals. The entire display comprises a range of signal components and behaviours, including moving the abdomen or opisthosoma rapidly up and down, waving the pair of third legs that have black and white markings, performing a fan dance involving moving the colourful fan back and forth, and various other related movements. The males also use some of these components at different stages of the interaction with a female, along with vibration components throughout, especially when the female is further away or not orientated towards the male. It is possible, therefore, that the vibration components partly act as alert components, although work is yet to be done analysing female responses to the displays.

6.4.5 Alarm Signals in Squirrels

There has been a growing use of robotic animals to study animal communication. These are useful because they can resemble real species, but behaviour and signal form can be controlled in a way rarely possible when using live animals. Robots can then be presented to real animals and behavioural responses measured. Partan and colleagues (2009) recently used robots resembling eastern grey squirrels (*Sciurus carolinensis*) in the USA to determine responses to unimodal or multimodal alarm signals (Figure 6.7).

Displays used by the squirrels to communicate in their groups include tail flagging and barking calls. The robots had a mechanical tail and hidden speakers, and by adjusting the signal components, Partan *et al.* (2009) showed that squirrels presented with either component alone responded with only moderate alarm responses, whereas when presented together alarm behaviour increased greatly. Therefore, the two signal components are redundant, because both produce the same response alone, but have an enhancement effect when presented together. Interestingly, in a similar subsequent study of squirrels in urban and rural areas, Partan *et al.* (2010) also found that squirrels responded more to multimodal than unimodal signals, but that urban squirrels responded more to tail-flagging than those in rural areas. They suggest that in noisy conditions, the receivers switch to rely more on the visual signals. This idea of plasticity in multimodal signals and the relative importance of different components will be valuable to explore more in the future.

6.5 True Multimodality or Multiple Receivers?

One of the difficulties in studying multimodal communication is in distinguishing between cases where all components of a signal are aimed at a single receiver (and hence truly multimodal) and when different components are relevant to different receivers. For example, many tiger moths have bright conspicuous colours combined with toxins that deter visually hunting avian predators. However, as we shall see in Chapter 9, some tiger moths also present acoustic signals to bats, which learn to avoid toxic species. Ratcliffe and Nydam (2008) showed that tiger moth species that are more active in the spring rely more on visual aposematic signals rather than sound, perhaps because birds present a greater predation threat than bats at this time. Conversely, species active later in the season seem to be more visually cryptic and likely to produce ultrasonic clicks, which may function as acoustic aposematism to bats (see Chapter 9). Thus, different selective pressures from predators with different sensory capabilities could lead to the evolution of apparent multimodal displays, both within and across species. This means that the conventional

Figure 6.6 Males of the peacock spider are elaborately coloured, including abdominal flaps that are raised during courtship displays to females (top images). This is accompanied by other movements including of the legs, and vibratory signals in sequences as the male approaches a female (middle image). The bottom panels outline the different movements used during a fan dance display. All images reproduced with permission from Madeline Girard, bottom panels modified from and © Girard *et al.* (2011).

Figure 6.7 A robotic squirrel that can produce visual and acoustic alarm-call components, as used by Partan and colleagues. Image reproduced with permission from Hampshire College.

assumption that different components of multimodal displays may have interactive effects may not be true if instead the components are directed at different types of receiver.

Another example where different components of multimodal signals influence the behaviour of multiple receivers comes from butterfly anti-predator signals. The peacock butterfly (*Inachis io*) has four large conspicuous wingspots ('eyespots') that it flashes to predators when under attack by opening and closing its wings in a startle display (Figure 6.8; Chapter 9).

In addition, the wing flicking is accompanied by a 'hissing' sound made by rubbing the wing veins

Figure 6.8 A peacock butterfly (*Inachis io*) with the four prominent wingspots that are used during startle displays to deter predators and halt attacks.

together and ultrasonic clicks. Vallin *et al.* (2005) presented peacock butterflies to birds in aviaries and showed that birds would not attack the butterflies when the eyespots were intact. In addition, the authors also cut parts of the wings to prevent sound production. When presented along with eyespots, there was no noticeable effect of sound on predator startle response. This could simply be that the visual display (spots) dominate the acoustic display, because only one of 34 butterflies with eyespots intact were killed in the 30-minute trials. However, when presented without the eyespots, there was no difference in startle effect for butterflies with or without sound. Therefore, although Vallin *et al.* (2005) and Olofsson *et al.* (2012) suggest the signal is multimodal, this does not seem to be the case. There are some limited data to suggest that the sound may startle a different receiver, namely bats (Møhl and Miller 1976), but this requires further work and how often these diurnal butterflies come into contact with bats is unclear (although they may roost/over-winter in the same locations). More promising, however, is the idea that the sound display is used to deter rodent predators like mice and rats (which, like bats, can also detect ultrasonic calls). This is supported by the fact that peacock butterflies hibernate in colonies over the winter in locations where mice are common, as well as data showing that mice represent a major predation risk during this time, and that peacock butterflies perform wing flicking when mice approach (Olofsson *et al.* 2011; Wiklund *et al.* 2008). Recently, Olofsson *et al.* (2012) presented peacock butterflies to mice in laboratory experiments, with or without the ability of the butterflies to produce sound. They showed that mice were more likely to flee when faced with a butterfly that could produce sound than towards butterflies with manipulated wings that could not produce sound. Therefore, it seems that the different signal components are actually different signals presented to different receivers. It would still be useful in future work to test if the mice are startled by any control sound, or specifically the sound of the peacock display and what aspect of this makes it effective.

However, distinguishing between selection from different receivers and 'true' multimodality may not always be straightforward, and the distinction between multiple unimodal and multimodal signals may become more blurred when there are multiple receivers of the same species towards which different components are aimed. For example, multiple receivers of the same species could also be presented with different information from different modalities. Higham *et al.* (2009) suggest that female baboons (*Papio hamadryas*) may have mating signals in some modalities that are only available to consorting males that are in close proximity to the female (e.g. odour and taste components), whereas other components, such as sexual swelling size and colour, may be available more widely. The authors suggest that if the information receivers gain from these signal types differs (owing to the 'active space' that signals/components in different modalities cover; i.e. how far they are transmitted), then females could control which individuals have access to which information, allowing females simultaneously to confuse and assure different males about paternity. This general idea, that the active space of signal components in different modalities could allow signalling different information to different individuals could be an important selection pressure on multimodal signals, and merits more research.

6.5.1 Summation and Integration of Multimodal Signal Components

We have discussed above that apparent multimodal signals may not be so if multiple receivers are involved. We might also ask when a signal is multimodal if the components are temporally or spatially separated. For example, if an animal releases a chemical mating pheromone to attract a mate, and then displays with a visual signal when the potential mate is within viewing distance, is this a multimodal signal or two unimodal signals? Here, the latter is more likely and multimodal signals should occur approximately at the same time and location to be genuinely multimodal (although some signals, like odours, may persist in the environment for a long time). Some minor temporal and spatial separation may often occur, but interactions separated over a longer period of time or space are better thought of as two separate signals because the effect on the receiver's sensory systems will be separated (although effects on cognitive systems may be more

prolonged). Admittedly, this distinction is ambiguous but it is difficult to be more specific generally because different distances and timings will be relevant to different species.

At present, little work has been done in this area. However, an exception is work by Narins *et al.* (2003), who have shown that in territorial disputes between males of the poison frog *Epipedobates femoralis*, both an advertisement call and the visual presentation of vocal-sac pulsations are needed to induce aggressive behaviour. Following this, Narins *et al.* (2005) presented these two signal components either overlapping temporally, or separated in time. They found that when the signal components were overlapping, aggressive behaviour was consistently invoked, but this was not the case when there was temporal separation. Spatial separation of the two components, however, did not seem to reduce aggressive responses. This work indicates that temporal disparity in signal components reduces the capacity of the receiver to integrate different pieces of sensory stimulation (Rowe 1999b; see Chapter 4). Taylor *et al.* (2011) have also shown in amphibians that temporal disparity between acoustic and visual components can reduce female responses to multimodal displays. This is an effective approach that could be used to tease apart the possible presence of two unimodal signals versus a true multimodal signal, and how components are integrated.

6.6 Future Directions

Studies of multimodal signalling have progressed quickly in the last few years, although the field still lacks a clear and consistent theoretical framework with which to interpret the results of the various studies. Work has generally focused on determining whether multimodal signals are redundant or non-redundant, and related aspects of this. However, as yet, little work has investigated why differences in multimodal signals exist across species; not just why some species use multimodal signals and others do not, but also why some species show, for example, dominance of one signal component whereas others show signal redundancy. What is it about the ecology of these species that leads to such differences in signalling form, and are the selection pressures favouring multimodal signals different across closely related species and why? More work is also needed to test when multimodal signals have arisen owing to strategic or efficacy benefits. Work has so far rarely aimed to distinguish between these, and so little is known about the relative importance of these. In addition, we need to know more about the neural integration of multimodal signals in receivers in order to be able to properly understand differences in response, especially outside of vision and hearing (see Chapter 4). This includes how components of multimodal signals are integrated over time. Finally, more work is needed to determine if multimodal signals do allow the receiver to extract greater information about the signaller or environment (perhaps by increasing attention, having amplifiers, or improving fine scale discrimination by having two informative components), and whether multimodal signals promote signal reliability. It will also be interesting to test further whether there are costs to multimodal signals, both in terms of eavesdropping by rivals or predators/parasites, and in terms of energetics to produce and receive such signals.

6.7 Summary

Multimodal signals are seemingly widespread in nature and involve the use of signal components across two or more sensory modalities. Explanations for the existence and benefits of multimodal signals are largely centred on whether signal components are redundant and provide the same information, or whether signal components are non-redundant with

each providing a different 'message', and the influence this has on the receiver. The former may allow effective signalling under noisy or variable signalling environments, whereas the latter may allow receivers to extract several different aspects of information from the signal. Other components may convey no information at all, but may alert the receiver to a different component, or amplify the effect of another component. Although still early days, evidence for many of the main hypotheses is gathering, including non-redundancy of signal components (e.g. flower signals and chick begging displays), signal dominance (e.g. wolf spiders), emergent responses (e.g. warning signals), increased detection or attention (wolf spiders), and signal redundancy (e.g. wolf spiders, squirrels). Finally, researchers should be careful not to assume a display or behaviour is multimodal just because it contains multiple modalities; sometimes these may be different signals aimed at different receivers.

6.8 Further Reading

An excellent place to start is reading the reviews by Hebets and Papaj (2005) and by Partan and Marler (2005). Although work has progressed since these were published, and that theoretical frameworks vary between researchers and over time, these both present a range of important ideas regarding multimodal signals. Readers should also consult Rowe (1999b) for more discussion with regards to the psychological/cognitive literature, and Chapter 4 for more discussion on sensory integration.

CHAPTER 7

Trade-Offs and Costs in Signalling

Box 7.1 Key Terms and Definitions

Eavesdropper: An unintended receiver that detects and uses the signals of other individuals (hetero- or conspecific) for its own benefit, such as a rival mate or a predator.

Efficacy Cost(s) of a Signal: The cost(s) associated with producing a signal that is effectively transmitted through the environment and picked up by the receiver, influencing the receiver's behaviour. These can be energetic/physiological or arise due to increased risk of eavesdropping.

Private/Hidden Communication Channel: A communication system where the signal form involves a structure that an eavesdropper is insensitive to because its sensory system is not tuned to the specific signal form.

Strategic Cost of a Signal: A cost of a signal over and above the efficacy costs that is important in maintaining reliability of the signals.

In the last two chapters we discussed theory and concepts regarding signalling and communication. Here we discuss how signal form can be influenced by costs and trade-offs. Signalling involves significant energetic investment and can incur a range of physiological costs. In addition, many signals, such as mating calls, are often vulnerable to eavesdroppers, including predators and parasites. The same signal form may also be used in multiple tasks; for example, animal colours may be used for both camouflage from predators and also communication with conspecifics. These factors mean that signals often reflect the outcome of trade-offs between potentially conflicting selection pressures. Here, we will focus predominantly on costs stemming from eavesdroppers and on trade-offs in signal form, because although physiological costs of signals are important, they often do not directly relate to sensory systems. In Chapter 5 we discussed how signals can have strategic costs and efficacy costs (Maynard Smith and Harper 2003). In the former, an

individual may, for example, indicate that it is especially fit to a mate with an honest signal of quality that is costly to produce and maintain. Here, we focus more on efficacy costs, which are costs needed for a signal to function and transmit information effectively, and relate to the specific signal form that evolves. Such costs involve generating the signal, transmitting it, and influencing the receiver (potentially against competing selection pressures), rather than introducing additional costs to indicate quality.

7.1 Physiological Costs

The actual production of signals involves a range of costs, including energetic and physiological investment (see Stoddard and Salazar 2011). For example, male drumming wolf spiders (*Hygrolycosa rubrofasciata*) drum their abdomen against dry leaves to attract females, producing substrate vibrations and airborne signals. Females prefer males with a higher

Sensory Ecology, Behaviour, and Evolution. First Edition. Martin Stevens.
© Martin Stevens 2013. Published 2013 by Oxford University Press.

drumming rate. Mappes *et al.* (1996) restricted food availability to males (low, intermediate, and high levels), and found that only males on the high food treatment maintained a high drumming rate. In addition, males induced to increase their drumming rates (by being presented with females close by) lost more weight and had higher mortality than males kept alone. This increase is mortality relates to the fact that the metabolic rate of males during drumming is 22 times higher than when at rest, and four times higher than when males are moving (Kotiaho *et al.* 1998a). Energetic costs of signalling have also been investigated with respect to acoustic signals across various species (see Gillooly and Ophir 2010; Ophir *et al.* 2010). For example, the ultrasonic signals used by some male moths to attract females incur a high metabolic rate (Reinhold *et al.* 1998), and birds singing at higher intensities may have substantially increased daily energy expenditure (Hasselquist and Bensch 2008). However, overall, although acoustic signals are costly to produce, the exact metabolic cost will vary across species/taxa (Ophir *et al.* 2010).

Electric signals are also energetically costly, especially under limiting environmental conditions, such as low oxygen (Reardon *et al.* 2011; Salazar and Stoddard 2008; Stoddard and Salazar 2011). The wave-type gymnotiform electric fish *Sternopygus macrurus* from South America produces electric organ discharges (EODs) that vary in amplitude based on the time of day and social circumstances. Markham *et al.* (2009) showed that individuals boost their electric signals by 40% at night and during social encounters by allowing more sodium ions to cross the electrocyte membrane, producing larger action potentials. This is achieved by rapidly (within minutes) moving voltage-gated sodium channels into the excitable membranes of the electrocytes, under hormonal control. These extra ion channels are removed from the membrane when there is reduced need for EOD signalling, thus reducing energy expenditure.

The energetic costs of visual signals have been less well quantified, but it is clear that displays like rapid movements often undertaken by birds and reptiles when presenting visual signals will lead to significant investment. Other types of visual signal, including some bird plumage patches or chick begging displays often use pigments like carotenoids (that produce red and yellow colours). These are both derived from the diet and also seem to play a role in body functions, such as antioxidants (see McGraw 2006). Therefore, devoting pigments like carotenoids to visual signals may compromise other requirements, for example immune function.

7.2 Eavesdropping by Predators and Parasitoids

Competitors such as rival males aiming to attract a mate may eavesdrop on the courtship displays of each other. In doing so, they may gain information about their rival and adjust their own display accordingly. This seems to happen in wolf spiders (*Schizocosa ocreata*), for example (Clark *et al.* 2012). Although this may be common, most work investigating eavesdropping and its cost to the signaller is based on examples of predators or parasitoids locating prey/hosts. Many predators detect, for example, their prey using cues, such as prey rustling sounds when walking (e.g. Goerlitz and Siemers 2007; see Chapter 5). However, communication signals may also be exploited by predators and parasitoids to find their prey/hosts (reviewed by Zuk and Kolluru 1998). For example, the begging calls of nestling birds can attract predators to the nest, especially when the calls are at an increased level (Haff and Magrath 2011). Some female fireflies (*Photuris*) eat other fireflies and locate their prey (males of other species) by eavesdropping on the bioluminescent signals that those males make to females (Lloyd and Wing 1983; see Chapter 4). Interestingly, in addition to the metabolic costs of signalling that drumming wolf spiders (*Hygrolycosa rubrofasciata*) incur (see above), the spiders also are at heightened risk of predation from lizards and birds when signalling (Kotiaho *et al.* 1998b). Additionally, experiments have shown that spiders themselves can exploit the vibratory communication systems of their leafhopper (*Aphrodes*) prey (Virant-Doberlet *et al.* 2011). In Chapter 6 we discussed how signals can be multimodal. Work by Roberts *et al.* (2007) shows that wolf spiders (*S. ocreata*) may be more vulnerable to jumping spider predators when signalling to mates comprises more than one modality (seismic and visual as opposed to visual alone).

These kinds of costs may be common because mating signals are often under selection to be highly conspicuous. Therefore, we expect signals to often reflect a trade-off in attracting the intended receiver's attention, while minimizing the risk of eavesdropping from unintended receivers. One of the most well-studied examples of predators that exploit the mating signals of their prey comes from the fringe-lipped bat (*Trachops cirrhosus*), which locates its amphibian prey by the males' vocalizations (see Box 7.2). Eavesdropping is also found in a range of other modalities and taxa, from visual to electric senses, and we will discuss various examples throughout this chapter.

It is not just predators that target prey based on their mating calls, but also parasitoids and pests like blood-sucking flies. For example, túngara frog (*Engystomops pustulosus*, formerly *Physalaemus pustulosus*) males are not only targeted by bat predators, but also by blood-sucking flies of the genus *Corethrella* which are also attracted by male calls (Bernal *et al.* 2006, 2007). A range of work has investigated how insects like crickets are targeted by parasitoid flies, which also locate hosts by their mating calls (e.g. Cade 1975; Robert *et al.* 1992). Cade (1975) found that the parasitoid fly, *Euphasiopteryx ochracea*, was attracted to speakers playing callbacks of its field cricket host, *Gryllus integer*, but not to speakers playing control sounds. Lehman *et al.* (2001) showed that in bush crickets (*Poecilimon thessalicus*), calling males are targeted by parasitoid flies (*Therobia leonidei*), and males that are most attractive to females are also the most vulnerable to parasitism.

Many parasitoids locate hosts by chemical signals, such as by their sex pheromones or by the chemical 'alarm signals' given off by plants being attacked by herbivorous insects (see reviews by

---

Box 7.2 Signals are Exploited by Predators: Túngara Frog Mating Calls

One of the most extensively studied systems showing predator eavesdropping on mating calls involves male túngara frogs (*Engystomops pustulosus*) and an amphibian predator, the fringe-lipped bat (*Trachops cirrhosus*). A series of experiments by Ryan and colleagues has investigated the dynamics of this system in Panama. Túngara frog males call in aggregations from ponds to attract females. Females arrive at the pond, and freely choose a male with which to mate. Males have a specific call type, called a 'whine-chuck' that consists of two components: a frequency modulated sweep, the 'whine', and short suffix called a 'chuck' (Figure Box 7.1). Calls can be either simple, comprising just the chuck, or complex, comprising the whine plus one to seven consecutive chucks (Gridi-Papp *et al.* 2006).

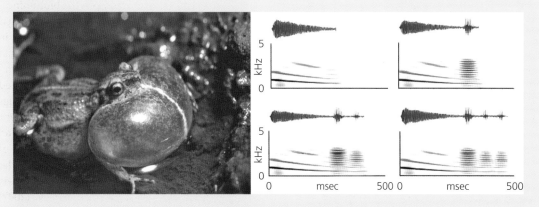

Figure Box 7.1 Left: a male túngara frog calling to attract a female. Right: the calls of males with no chuck components (a simple call), or one, two, and three chucks (complex calls). Images and graphs reproduced with permission from Mike Ryan.

continued

Box 7.2 *Continued*

Females prefer complex male calls with chucks to simple calls at a probability of about 0.86 (Gridi-Papp *et al.* 2006; Ryan *et al.* 1982). However, the system is also exploited by the fringe-lipped bat, which specializes in finding amphibian prey by the frog mating calls. The bats circle the pond and capture frogs floating on the surface.

Do Bats Find Frogs Based on Their Calls?

Tuttle and Ryan (1981) showed that the success rate of hunting bats depends on detecting male frog calls. They monitored 14 breeding sites over 35 nights and categorized the level of frog chorus into 1) a full chorus, 2) a partial chorus, 3) just a few frogs calling, and 4) no frogs calling. They then

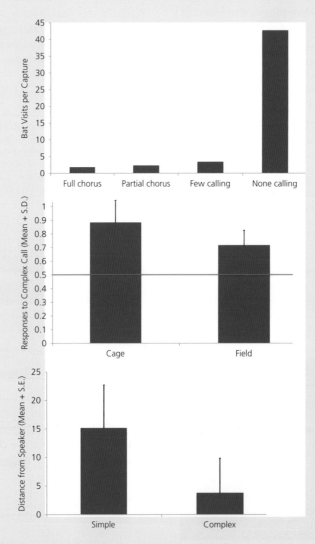

Figure Box 7.2 Top: Bat-capture success for frog choruses involving different levels of calling. Data from Tuttle and Ryan (1981). Middle: Proportion of bat responses to complex over simple frog calls in flight cages and the field. The red line is what would be expected if the bats were selecting calls at random. Data from Ryan *et al.* (1982). Bottom: Distance bats landed from a speaker playing continuous complex or simple frog calls in the presence of noise. Data from Page and Ryan (2008).

produced a measure of bat hunting success, being the number of visits made by bats per frog capture. This is an inverse measure of success, because less successful hunting periods require more visits to a pond to capture a single prey item. Tuttle and Ryan found that bats caught fewer frogs when the chorus was silent. However, only a few frogs needed to be calling for capture success to be greatly improved (Figure Box 7.2).

Do Bats Prefer Complex Calls?

Male túngara frogs only use simple calls when no other males are calling, but will produce complex calls with chucks in response to the presence of other males' calls. This implies that complex calls may be more costly than simple ones, and a trade-off between being more attractive to females and heightened predation risk may exist. Ryan *et al.* (1982) tested whether the complex calls of túngara frogs are preferred by bats. In flight cage experiments, bats were presented with two speakers in the corners of a cage an equal distance from the bat (4 m). One speaker played simple calls with no chuck, and the other speaker played complex calls with three chuck components. Bats strongly preferred to fly to the speakers playing complex calls (at almost 90%). In field experiments, results were much the same, with bats making significantly more passes at speakers playing complex rather than simple calls (Figure Box 7.2).

Why do Bats Prefer Complex Calls?

There is evidence that fringe-lipped bats prefer complex male túngara calls over simple ones because complex calls facilitate localization of the caller. Marler (1955), when discussing bird alarm calls, argued that complex calls with lots of components will provide more information for binaural comparisons for the receiver, making them easier to localize. In particular, localizable calls should have lots of components of short duration, fast rise and fall times, and a broad frequency range. The male túngara frog calls fit these characteristics, being short in duration, with several components, and a broadband structure. A series of flight-cage experiments by Page and Ryan (2008) tested if complex túngara frog calls are easier for bats to locate than simple calls. They found that bats flew more quickly and landed closer to speakers playing complex calls than to those playing simple calls. The bats' ability to land close to the speakers declined in treatments when the calls stopped once the bat started to fly, and in cases where bats were presented with background noise or obstacles. Bats flew more closely to speakers playing complex calls than simple calls when they were presented with background noise (with continuous frog calls; Figure Box 7.2), and when bats were presented with obstacles (with calls that stopped).

Fatouros *et al.* 2008; Stowe *et al.* 1995). Many males have evolved methods that inhibit or reduce the likelihood that females will re-mate with another male, including mate guarding, copulatory plugs, and male accessory gland substances (Andersson 1994). In addition, a range of male insects, including for example beetles (Happ 1969), butterflies (Andersson *et al.* 2000; Gilbert 1976), and bees (Kukuk 1985) transfer 'anti-aphrodisiac' pheromones to females during mating. These repel other males from the female and reduce male harassment. Recently, Fatouros *et al.* (2005) showed that a tiny (0.5 mm) parasitoid wasp *Trichogramma brassicae* exploits the anti-aphrodisiac (benzyl cyanide) of its host the large cabbage white butterfly (*Pieris brassicae*). The authors showed that the parasite is attracted to mated butterfly females more than virgin females, but that virgin females could be made to be attractive to the wasps by applying benzyl cyanide. The wasps then hitchhike on the butterflies (phoresy) until they reach the host plant where the butterfly eggs are laid. One would predict that in this system males should be under constraining selection pressure to limit the use of anti-aphrodisiacs. Egg parasitoids of stink bugs can also show refined responses to host calls, being more responsive to female vibratory signals than those of males or non-host species, as this would increase the chances of finding eggs (Laumann *et al.* 2011).

7.2.1 Predators and Parasitoids Have Sensory Systems Tuned to the Signals of Their Prey/Hosts

The fringe-lipped bat (*Trachops cirrhosus*), which specializes on finding amphibian prey by their mating calls, is like many bats, sensitive to ultrasonic frequencies used in echolocation (between about 50 and 100 kHz) (Barclay *et al.* 1981). However, unusually for bats, it also has an additional sensitivity peak at just below 5 kHz (Ryan *et al.* 1983), and adaptations in its cochlea to detect low-frequency sounds (Bruns *et al.* 1989). This sensitivity corresponds to the frequency of the túngara frog calls.

Sensory adaptations to detect prey and hosts based on their mating signals seem common. For example, female parasitoid flies of the genus *Ormia* lay maggots on their cricket hosts. The maggots burrow into the host, feed inside, and kill it within about 10 days, at which point they emerge. Male crickets call to females in mating, and parasitoid flies locate males with these calls. Many parasitoids like *Ormia* have hearing organs that are specially adapted in both morphology and sensitivity to detect the calls of hosts (Robert *et al.* 1992). Insects like crickets, grasshoppers and cicadas produce sounds of high frequency and intensity (above 3 kHz) over distances greater than 10 m. They have a tympanal hearing organ comprising a specialized external cuticular membrane (tympanum), and a receptor organ consisting of sensory cells sensitive to changes in air pressure from distant sound sources. Robert *et al.* (1992) made neural response recordings of the auditory cells of the parasitoid fly *Ormia ochracea* when presented with acoustic stimuli. They found that the sensitivity of the hearing organ is closely tuned to the peak song energy (4–5 kHz) of its host the field cricket *Gryllus rubens*. The organ is most sensitive to sounds of 4–6 kHz at sound pressure levels of 20 dB, which is consistent with long range detection of the host mating signals (Robert *et al.* 1992). The structure of the hearing organ also shows strong convergent evolution with the hearing organs of crickets—the parasitoid hearing organ is more like a tympanal organ than the hearing organ of non-parasitoid flies. Anatomically, it is made of transparent membranous enlargements on the prosternum, which form a wall to an air-filled chamber in the thorax housing sensory organs. Thus, the parasitoid hearing organ is convergent not just in terms of sensitivity, but also in terms of anatomy. Robert *et al.* (1992) also showed experimentally that intact or sham-operated female flies would successfully locate a speaker playing cricket mating calls, whereas flies with their prosternal tympanal membranes punctured could not detect the speaker.

7.3 How do Animals Balance the Risks of Eavesdropping with Communication?

Clearly, the potential risks from predators and parasitoids during communication are substantial for many species, yet communication is often essential in finding a mate or fending off rivals. We therefore expect selection to modify signals to reduce the risks ('victim adaptations'; Zuk and Kolluru 1998). In theory, this could be achieved in a number of ways, including:

1. Changes in behaviour.
2. Reducing the detectability of a signal.
3. Signalling in a different sensory modality.
4. Utilizing a hidden or private channel of communication.

We will deal with each of these in turn.

7.3.1 Changes in Behaviour

Perhaps the most obvious way to reduce the risk of eavesdropping is with modifications to the signaller's behaviour. Tuttle *et al.* (1982) presented model bats to túngara frog choruses at ponds and found that frogs stopped vocalizing when they detected models of a predatory species. The shutdown in the chorus can be very rapid (less than 1 second). Frogs further away from the bats stopped calling but remained floating on the water surface with their vocal sac still inflated, ready to resume calling again. However, if a bat was nearer, frogs deflated the sac and dived underwater. Thus, the frogs have a graded behavioural response to perceived risk, enabling them to reduce the risk of predation while lowering the costs of interruptions to their calling. Other frogs, such as males of the treefrog, *Smilisca sila*, produce calls of varying complexity to attract a mate. Like the túngara frog, they are also eaten by the fringe-lipped bat. Male frogs seem able to synchronize their calls with other individuals, and bats respond less to synchronous than asynchronous calls, potentially if localization of a sound is more difficult when receiving multiple calls simultaneously (Tuttle and Ryan 1982).

Some crickets have reduced signalling behaviour and conspicuousness at times when parasitism risk is greater. Male field crickets *Teleogryllus oceanicus* from the Hawaiian islands are parasitized by the parasitoid fly *Ormia ochracea*, with parasitism rates varying between islands. Lewkiewicz and Zuk (2004) took males from populations originating from three different islands and measured the latency for males to resume calling following a disturbance (puffs of air, which may indicate an approaching fly). Male crickets from the island with the highest parasitoid prevalence (25% of males infested) had the longest latency to resume calling, whereas crickets from the island with the lowest parasitoid levels (8% infestation) resumed calling after the shortest time. In some crickets, aggregations comprise both calling males and non-calling 'satellite' males, which attempt to mate with females attracted to the calling males. Cade (1975) found that in the field cricket, *Gryllus integer*, calling males had very high infestation rates, whereas just one out of 17 satellite males was parasitized. A silent strategy also occurs in populations of *T. oceanicus* on Hawaiian islands. Here, Zuk *et al.* (2006) have shown that in less than 20 generations more than 90% of males have moved to a new wing morphology which does not involve sound production.

Other animals, such as nestling birds, stop calling and become silent when they detect a predator or when detecting an alarm call (e.g. Kleindorfer *et al.* 1996.; Madden *et al.* 2005; Magrath *et al.* 2007), a strategy termed 'acoustical avoidance' by Curio (1976). In addition, some nestling species stop begging calls when they assess danger independently from the parental calls. A study by Haff and Magrath (2010) shows that nestling white-browed scrub wrens (*Sericornis frontalis*) in Australia will cease calling most frequently when they hear cues of a predator, a pied currawong (*Strepera graculina*) walking on leaf litter, but less often to general broadband or novel sounds. Thus, the nestling responses are tuned to specific predator cues.

Some electroreceptive predators are capable of detecting their prey based on the prey's electric communication signals (e.g. Hanika and Kramer 1999). In response, some electric fish will adopt behavioural patterns of inactivity and hiding in order to evade detection by predators (Machnik and Kramer 2010), and show plasticity in their electric organ discharge patterns by adopting discharges of shorter duration when not engaging in courtship (Hanika and Kramer 2008). Many mammals scent mark in order to advertise territory ownership or to signal status to rivals in mating, and often overmark discovered scent marks of their rivals. Roberts *et al.* (2001) have shown that male mice (*Mus musculus*) will scent mark at different rates. When the authors simulated the presence of a potential predator by using the urine of ferrets (*Mustela putorius furo*) they found that some mice would reduce their overmarking behaviour of other individuals scent. Further tests with neutral odours showed that this response was not simply due to the presence of a novel odour.

Finally, in visual communication, some species of lizard will vary their signalling behaviour to be less conspicuous after a predatory attack (e.g. brown anole lizards, *Anolis sagrei*, will reduce movement-based displays; Simon 2007). Male guppies (*Poecilia reticulata*) tend to display more at the beginning and end of the day, when light intensity is lowest and the males' colour signals should be less detectable to predators but still visible to females (Archard *et al.* 2009; Endler 1991).

7.3.2 Changes in Signal Detectability or Localization

One way to reduce the risk of attracting unwanted attention is to reduce signal detectability. Experiments by Endler have shown that male guppies evolve more constrained mating signals when predation risk is greater. Guppies are small freshwater fish found in streams in places like Trinidad and Venezuela. Males are extremely colourful, with a range of bright spots and blotches of different colours and sizes over their body (Figure 7.1). There are large differences between individuals, underpinned by high genetic variation (Endler 1978). Females prefer males with more colourful spots and greater numbers of spots.

In initial field studies, Endler (1978) sampled guppies at 112 locations (53 streams) in Trinidad and Venezuela, and scored each location for the level of predation pressure from various predatory species. When predation risk was higher, male

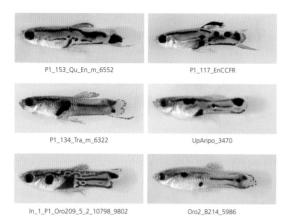

P1_153_Qu_En_m_6552

P1_117_EnCCFR

P1_134_Tra_m_6322

UpAripo_3470

In_1_P1_Oro209_5_2_10798_9802

Oro2_8214_5986

Figure 7.1 Variation in male guppy patterns and colours. Images reproduced with permission from Christine Dreyer.

guppies had fewer and smaller spots. Following this, Endler (1980) established a population of guppies in his laboratory. He divided the guppies into ten artificial streams mimicking those found in the field. Into each stream he established a founding population of 200 individuals, and let these settle and evolve for six months. In the field, guppies are eaten by visually hunting predatory fish, including the highly dangerous pike cichlid *Crenicichla alta*, and less often by *Rivulus hartii*, which presents a much less significant threat than *C. alta*. After six months, Endler added a single dangerous *C. alta* to four streams, whereas in another four streams he added six benign *R. hartii*. These approximately correspond to the densities of each fish species that would be found in equivalent stream areas in the wild. Two control streams had no predators. In addition, Endler subdivided the streams into two types: those with substrate gravel of fine size, and those with coarse gravel. Endler allowed the guppy populations in each stream evolve for a further 14 months, and took a census at approximately four and ten generations of male guppy colours (Figure 7.2). In both control streams and streams with the innocuous *R. hartii*, guppies evolved brighter spots and more spots under female choice. However, in streams with a dangerous predator, *C. alta*, guppies had clear declines in the conspicuousness of their ornamentation to enhance camouflage. This was primarily driven by a loss of the blue and iridescent spots, presumably because these would be most

visible against the largely black, white, and green substrate. In addition, spot sizes on the fish generally evolved to resemble the size of the gravel substrate in the *C. alta* streams to facilitate camouflage, whereas in the other streams spots were mismatched to background gravel size to make the guppies more conspicuous against the substrate to potential mates.

Endler (1980) also performed a field experiment in Trinidad. Here, he relocated guppies from a stream with lots of predators (where guppies had relatively dull patterns) to a guppy- and predator-free tributary stream at another location. After two years, released from predation pressure, male guppies had rapidly evolved brighter and more spots under sexual selection. Later experiments have also shown that another predatory fish from Trinidad, blue acara cichlids (*Aequidens pulcher*) preferentially attack more brightly coloured guppies when presented with two sized-matched males (Godin and McDonough 2003), showing directly effects of predation on brighter males.

Marler (1955) suggested that the calls of some small songbirds may have properties that make the emitter difficult to localize by a predator (reviewed by Ruxton 2009). Such 'seet' alarm calls, made by birds when they detect an aerial predator, are relatively pure-toned and of high-frequency (6–9 kHz). Alternatively, Klump *et al.* (1986) suggest that such calls may have a structure that the predator is less sensitive to, such as high-frequencies, making them difficult to detect. In line with this, Klump *et al.* (1986) found that great tits (*Parus major*) were more sensitive to pure tone sounds of around 8 kHz than were predatory sparrow hawks (*Accipiter nisus*). A study by Wood *et al.* (2000) tested if two species of songbird, the New Holland honeyeater (*Philidonyris novaehollandiae*) and the noisy miner (*Manorina melanocephala*), were more responsive to some classes of sounds than a predator, the brown falcon (*Falco berigora*). They found that all birds located broadband distress (mobbing type) calls easily, but had trouble locating the alarm calls. The performance of the songbirds and the falcons was similar, indicating that in these species at least, the alarm calls of passerine birds likely prevent localization by predators, as suggested by Marler (1955), rather than prevent detection. Jones and Hill (2001) tested

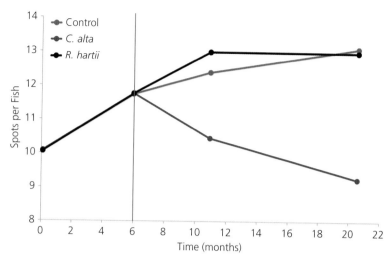

Figure 7.2 Results of Endler's (1980) laboratory experiments, analysing guppy spots over time. The green line shows the results of the control streams without predators, the black line those with the benign *R. hartii* predators, and the blue line those streams with a dangerous *C. alta*. The vertical red line shows the point when the predatory fish were added to the streams. After this point, the number of spots on the male guppies declines rapidly when faced with strong predation pressure but continues to rise in the other two treatments under selection by female mate choice. Data from Endler (1980).

four species of owl and four species of hawk in their ability to respond to mobbing calls and seet alarm calls. The birds found the seet calls more difficult to detect than the mobbing calls, and they were not always able to localize the source of the seet calls. Thus, there is evidence for both the detection and localization hypotheses. Most likely the situation varies not just based on the call structure, but also based on the sensory system of the receiver and the acoustic environment (Ruxton 2009).

Predation risk can also lead to a reduction in signal detectability in chemical communication. For example, the pine engraver bark beetle (*Ips pini*) found in the Great Lakes region of North America, spends the majority of its life within the tissues of its host plant (pine trees) and relies mainly on chemical communication for most signalling tasks, including mating (see Raffa *et al.* 2007). The beetles are attacked by a relatively narrow group of predators, including *Thanasimus dubius* and *Platysoma cylindrica*, which are attracted to *I. pini*'s pheromones. However, interestingly, *T. dubius* and *P. cylindrica* are most attracted to two different components of the pheromone, whereas *I. pini* itself prefers an intermediate mixture of these components. Furthermore, an additional component that is inactive by itself

increases the attraction of the beetles to the pheromone components, but has no effect on the predators and is largely used in communication when predators are most abundant (Raffa *et al.* 2007). Thus, the beetle pheromone components seem to have resulted from selection pressure to reduce detectability to predators as a form of chemical crypsis (see Ruxton 2009).

7.3.3 Utilize a Different Sensory Modality

In the case of many of the above examples, the risk of signal detection may directly constrain signal form. However, there are potential ways that animals could simultaneously reduce the risk of eavesdropping, while maintaining the ability to signal effectively. One strategy could involve switching to use a different sensory modality that the predator does not use. Just as with crickets and frogs, neotropical katydids (also called bush-crickets) are also preyed upon by bats that detect katydids by virtue of the calls that males make to attract females. A study by Belwood and Morris (1987) investigated a range of katydid species in Panama. They found that in habitats comprising mature forests with various species of foliage-gleaning bats (that take prey from the surface of vegetation) katydids spend significantly less time singing (shorter song-duty cycle) compared to species occurring in secondary growth forest lacking foliage-gleaning bats. Interestingly, in those species that reduce their song-duty cycle, individuals compensate by adopting species-specific vibratory (tremulation) signals that are transmitted though the plants and are inaudible to the bats. The one species of katydid that occurs in primary forest that retains a call relatively easy to localize, sings from bromeliads that are covered with long sharp spines that would endanger a bat, and is not taken as prey (Belwood and Morris 1987).

7.3.4 Private Channels of Communication

Another way that animals may reduce the risk of eavesdropping is to use private or hidden channels of communication to which predators are insensitive. This does not involve signalling in a different modality, but instead exploits differences in sensitivity between the eavesdropper and signaller to information within the same modality. Whereas the strategy of reducing detectability generally involves reduced signal amplitude, private communication channels involve shifting the signal form to properties that the predator's sensory system is insensitive to. We discuss various examples of these below.

7.3.4.1 Ultraviolet Signals and Vision

Many animals have the ability to detect and utilize ultraviolet (UV) light (Tovée 1995). Modelling work by Håstad et al. (2005) indicates that the plumage coloration of some songbirds may be more conspicuous to conspecifics (e.g. potential mates) than to birds of prey (but see Gomez and Théry 2007; Stevens and Cuthill 2007). This is because birds of prey seem to be less sensitive to ultraviolet light than many songbirds (although they can still see UV; see Chapter 4). In fish, Siebeck et al. (2010) showed that male ambon damselfish (Pomacentrus amboinensis) use UV face patterns to discriminate between rival males and individuals of a sympatric species, the lemon damselfish (P. moluccensis). To humans, the two species look similar, yet they have elaborate UV face markings (Figure 7.3). The yellow coloration of the damselfish may provide camouflage against the reef background, with the main predators of the fish

probably insensitive to UV. In addition, in underwater environments, UV wavelengths scatter quickly, restricting the distance that UV could be used in detection.

An earlier study in the northern swordtail fish (Xiphophorus nigrensis) and its major predator the Mexican tetra (Astyanax mexicanus) by Cummings et al. (2003) also shows evidence of private communication with UV signals, and tested the response of the predators to the signals. Males of X. nigrensis have significant levels of ultraviolet reflectance used in mate choice, whereas females have less UV. Female X. nigrensis spend longer associating with males in choice experiments in tanks when UV is present than when UV is blocked from the ambient light. However, Mexican tetras show no such difference. Visual modelling indicates that the UV signals of the male swordfish are three times more conspicuous against the background to females than to the predators.

7.3.4.2 Red Bioluminescence and Visual Sensitivity in Deep-Sea Dragonfish

Due to wavelength-dependent light attenuation, most light in the deep sea is lacking in longer wavelengths as these are removed more quickly than shorter wavelengths as depth increases (see Chapter 10). In addition, almost all bioluminescence is also of shorter wavelengths ('blue-green') (for recent reviews of bioluminescence in the ocean see Haddock et al. 2010; Widder 2010). Consequently, very few deep-sea animals have a visual system sensitive to longwave light. A significant exception is three genera of deep-sea dragonfish (Malacosteus, Aristostomias, Pachystomias). These species have blue-green photophores like many other deep-sea organisms, but also suborbital photophores that emit far-red light, with peak emissions around 700 nm (Bowmaker et al. 1988; Denton et al. 1970, 1985; Herring and Cope 2005). In addition, they also have a LW pigment with sensitivity remarkably similar to the emission spectra of the red bioluminescence. Denton et al. (1970) suggested that the light may be helpful in hunting red or brown animals in the deep sea. Because LW sensitivity is so rare in deep-sea environments, these species could both communicate with each other unnoticed by predators, and even detect and covertly sneak upon

Figure 7.3 Images of ambon male damselfish in human visible light and ultraviolet, showing the detailed facial markings in ultraviolet. Images reproduced with permission from Ulrike Siebeck.

their prey. Some dragonfish species, such as *Aristo-stomias tittmanni*, eat other fish like lantern fish. These prey have a single-layered retina with rods of peak sensitivity around 490 nm. Partridge and Douglas (1995) calculate that the prey would be able to detect *A. tittmanni*'s blue bioluminescence at about 36 m away, yet would only be able to detect the red bioluminescence at about 1 cm, due to the lack of sensitivity to LW light in the prey's visual system.

In *A. tittmanni*, sensitivity seems to be based on an opsin gene allele that confers LW sensitivity. In con-trast, *Malacosteus niger* lacks this LW sensitive pig-ment but instead has a photostable pigment complex located in the outer segments of the photoreceptor (Bowmaker *et al.* 1988). This absorbs light with a peak of 667 nm, very close to the emission spectra of the red bioluminescence (Douglas *et al.* 1998; see Chapter 4). Douglas and colleagues (1998, 1999) found that this complex seems to act as a photosen-sitizer, transferring the energy that the complex derives to the visual pigment. This pigment complex seems to contain a chlorophyll derivative obtained from the diet of *M. niger* (Douglas *et al.* 2000).

7.3.4.3 Polarization Vision

Some animals, including various birds, reptiles, amphibians, and fish, but particularly invertebrates from cephalopods to insects, are capable of perceiving polarized light (see Chapter 2). In addition, some species, such as mantis shrimp (Marshall *et al.* 1999), beetles (Pye 2010), and cuttlefish (Shashar *et al.* 1996), present polarization patterns on their bodies. Because many species (most vertebrates) seem unable to detect polarization, species that can use this to signal may have a private communication channel (Mäthger *et al.* 2009; Shashar *et al.* 1996; Warrant 2010), although direct evidence is generally lacking. Some species may also be able to discriminate between different types of polarized light (such as linear or circularly polarized), and this may enable them to signal in a way that other animals cannot detect, potentially for example in some mantis shrimp (Chiou *et al.* 2008) and scarab beetles (Brady and Cummings 2010).

7.3.4.4 Spectral Shifting and Signal Cloaking in Electric Fish

The ability to produce weak electric signals, in addition to evolving tuberous electroreceptors (see Chapter 2) that can detect EODs, opened up a large potential for weakly electric fish to communicate in a modality that was either entirely unavailable to predators (because they cannot detect electric signals or are absent from some geographical regions), or with signals to which predators were insensitive to (having only ampulary electroreceptors and lacking sensitivity to high frequencies). This ability to communicate relatively unnoticed seems to have enabled rapid divergence in some groups of electric fish (Arnegard *et al.* 2010a; Carlson and Arnegard 2011; Carlson *et al.* 2011; see Chapter 11). Some weakly electric fish are, however, still vulnerable to eavesdropping by electroreceptive predators, but they can compensate for this by having spectrally shifted signals that occur outside the sensory range of the eavesdropper. This can involve having EODs of higher frequency (reviewed by Stoddard and Markham 2008; Chapter 11). For example, gymnotiform and mormyrid electric fish are preyed upon by some electroreceptive catfish. All groups have ampul-

lary electroreceptors sensitive to low-frequencies (0–60 Hz). However, in addition, gymnotiform and mormyrid fish also have a second type of 'tuberous' receptors that are specifically adapted to detect EOD characteristics and sensitive to these higher frequencies (Chapter 2). At present, it seems like most catfish do not have sensitivity to these higher frequencies (Hanika and Kramer 2000). In addition, Stoddard and Markham (2008) discuss how fish may sometimes have a 'signal cloaking' mechanism, where low-frequency parts of their signal cancel each other out at a distance of a few centimetres. Some fish, such as *Brachyhypopomus* spp. have EODs that seem to restrict the low-frequency components of their signals to within five cm of their bodies, while allowing higher frequencies to spread further (Stoddard 1999, 2002). The head of the fish produces a single phase EOD, whereas the tail end produces a biphasic EOD; these two fields sum over a distance to produce a waveform with little low-frequency energy corresponding to the ampullary receptor sensitivity. Without this characteristic, the overall EOD signal would be detectable to predators at a distance of one metre (Stoddard and Markham 2008).

7.3.4.5 Ultrasonic 'Whispering' in Moths and Calling by Ground Squirrels

Recent work in the Asian corn borer moth (*Ostrinia furnacalis*) found in Japan by Nakano *et al.* (2006, 2008, 2009a, 2009b) shows good evidence for private acoustic communication (Figure 7.4).

Many moths have hearing organs sensitive to ultrasonic frequencies, which likely evolved to allow moths to detect bats and take evasive action (see Chapter 9). Subsequently, many moth species have evolved the ability to use loud ultrasonic calls (25–100 kHz) to attract mates. However, these can also be detected by predators and parasitoids. The Asian corn borer moth also uses ultrasonic mating calls, yet these are very quiet (around 46 dB sound pressure level at 1 cm, compared to 76–125 dB at 1 cm in many other species) (Nakano *et al.* 2008, 2009b). During courtship, males produce ultrasound, comprising a series of groups of pulses ('chirps') (Figure 7.5; Nakano *et al.* 2006). These are produced by rubbing together specially adapted

Figure 7.4 Female (left) and male (right) Asian corn borer moths. Images reproduced with permission from Ryo Nakano.

scales on the wings and thorax that only males have (Nakano *et al.* 2008).

These quiet ultrasonic calls are important for males to gain copulations, and the male song works by suppressing escape behaviour in females (Nakano *et al.* 2008). Initial work by Nakano *et al.*

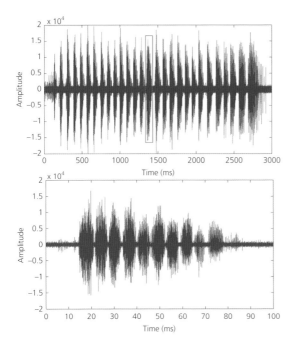

Figure 7.5 Song characteristics of male Asian corn borer moths, showing a train of sound emission lasting 3000 ms (top) and a chirp lasting 100 ms (red box top and below). Data courtesy of Ryo Nakano.

(2006) showed that deafened females rejected males more often than sham operated females. Later Nakano *et al.* (2008) showed that 90% of female moths would accept males that had intact scales or were sham operated, compared to 65% when males were muted. In addition, males gain significantly more copulation attempts with females when they produce either their natural song or are accompanied by a playback of songs, compared to males that are silenced or accompanied by playbacks of noise. These experiments show that muted males accompanied by playbacks will gain significantly more copulation attempts, provided that the sound level of the song reaches 46 dB SPL or above. Nakano *et al.* (2008) made electrophysiological recordings from female auditory neurons (tympanal nerves) when presented with sound stimuli at different sound pressure levels changing in frequency in steps of 5–10 kHz over a range between 5 and 120 kHz. This enabled them to calculate an auditory threshold curve showing the sensitivity of hearing to sounds of different frequencies and loudness (Figure 7.6).

The work shows a close match between the frequencies of the male song and the sensitivity of the female hearing. Based on the hearing sensitivity to different sound pressure levels, and the loudness of the male song, Nakano *et al.* (2008) calculate that females can only detect (and therefore respond to) song when the male is within 3 cm. Beyond this, the call is too quiet. This means that rival males or females only a few centimetres away will not be able to hear another male's courtship song.

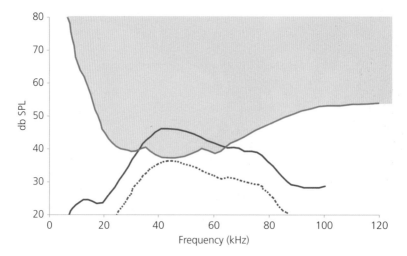

Figure 7.6 Female hearing sensitivity and male song characteristics in the Asian corn borer moth. The red line and shaded region shows the sensitivity curve of females to sounds of different frequencies and pressure levels. The solid and broken blue lines are the frequency and pressure levels of the male song from 1 cm and 3 cm away respectively, showing that the male call is not detectable to the female beyond 3 cm distance. Data from Nakano *et al.* (2009a).

In addition, the authors also calculate that bats with an auditory threshold of 20 dB SPL would not hear calls of moths from more than 20 cm away. Recent work has indicated that quiet courtship songs in moths may be more widespread than previously thought. A cross-species analysis of tympanate moths in four families by Nakano *et al.* (2009b) showed that nine out of 13 species produced low intensity (43–76 dB SPL at 1 cm) songs, which vary greatly in form across species. The implication is that quiet songs in moths may be a widespread adaptation to reduce the risk of eavesdropping by rival males, predators, and parasitoids. However, on a semantic level, whether this example is a clear case of a private communication channel is debatable because the system may correspond more closely to the way that signals can evolve reduced detectability by eavesdroppers, analogous to the duller coloration of guppies (see above).

Ultrasonic signals have also been found in ground squirrels. Wilson and Hare (2004) found that Richardson's ground squirrels (*Spermophilus richardsonii*) have an ultrasonic component of around 50 kHz to their alarm signals, as well as the human-audible 8 kHz component. They presented playbacks of calls to receivers and found that the ultrasonic frequencies would elicit high levels of vigilance behaviour. Such ultrasonic calls may work both by

decreasing detection, because ultrasonic frequencies are highly directional and attenuate rapidly, but also could be a hidden signal if predators are insensitive to ultrasonic frequencies. Because ultrasonic calls function over a short distance, they may be especially important in alerting juveniles to danger (Wilson and Hare 2006). Ultrasonic calls in tarsiers have also been suggested to function as private communication channels (Ramsier *et al.* 2012).

7.4 Multiple Functions and Trade-Offs in Signal Form

Signals not only face trade-offs from costs associated with factors such as avoiding predation or parasitoids, but also in combining multiple functions, which may require opposing signal forms. Here, we discuss some examples regarding visual signals. A key cost of signalling is that it can compromise camouflage from predators because communication displays tend to be intrinsically conspicuous. Although the principles of camouflage are often opposing to those of signalling, there are potential ways that animals might combine the two within the same signal form (Hailman 1977; Stevens 2007). The most widely discussed possibility relates to distance-dependent effects. For example, because the ability to see fine pattern declines with increasing viewing distance, patterns with high internal contrast or conspicuous colours could blend into one uniform camouflaged colour patch when viewed from sufficient distance, because the different colours 'blur' to match that of the background, especially if faced by a predator with a low spatial acuity. This has most frequently been discussed with respect to warning signals, where animals often have bright contrasting colours to warn potential predators that they are unprofitable (aposematism; Chapter 9; e.g. Edmunds 1974; Hailman 1977; Rothschild 1975). However, the concept should also apply to any communication strategy, especially for signals that comprise bands or stripes, because at a given distance an observer may be unable to resolve the different stripes.

Currently, firm evidence for distance-dependent effects of combining camouflage and signalling is limited, but the most stringent work comes from a study by Marshall (2000) of reef fish. Marshall took reflectance spectra from the blue and yellow coloured angelfish (*Pygoplites diacanthus*) of the Great Barrier Reef. He used a model of fish vision to analyse the conspicuousness of the patterns against a range of reef backgrounds, both in terms of the individual colour components, as well as the additive combination of yellow and blue (as viewed from afar). The results showed that the blue colour was well camouflaged against the open blue water, but conspicuous against the colour of the coral reef. In contrast, the yellow coloration was conspicuous against the water but well matched to the reef background. However, the average colour of the blue and the yellow was well matched to both backgrounds, meaning that the fish could be effectively concealed to a predator beyond about 5 m. In general, blue and yellow colours are common in reef fishes and so there is great potential for distance-dependent camouflage and communication in reef fish (Marshall 2000). Similarly, some weevils (e.g. *Entimus imperialis*) have a body colouration of different brightly coloured scales. Analysis and modelling of the weevil coloration indicates that the coloured scales could be seen by a conspecific (e.g. for communication) up to 6 cm away, yet at a distance the overall coloration of the body should be camouflaged against the foliage background (Wilts *et al.* 2012).

Evidence for distance-dependent colour strategies is weaker for anti-predator coloration. One experimental test of distance-dependent effects of aposematism and camouflage was conducted by Tullberg *et al.* (2005). They presented human 'predators' with images of (unpalatable) swallowtail butterfly (*Papilio machaon*) larvae against two background types, photographed at different distances. They used imaging software to either make the larvae match the background better or to be more conspicuous, and measured the time for subjects to detect the caterpillars. Larvae manipulated to be more camouflaged took longer to find at close distance than the normal or more conspicuous items, and there was some evidence that the larvae made to be more conspicuous were more detectable than the natural coloration. Tullberg *et al.* (2005) concluded that the natural larvae colours combine warning coloration at close range with camouflage at a distance. Although the results are suggestive, humans are not the natural receivers of the larvae and have a different visual system to the real predators. This idea is therefore promising but needs to be tested with appropriate predators or models of predator vision.

Instead of using the same body region for different signals, animals may adopt appearances in different areas of their body directed at different receivers. For example, some crabs display stronger signals in body regions that are more likely to be detected by conspecifics but are less visible to avian predators from above (Cummings *et al.* 2008a). In addition, the arrangement of colour patches also seems to be important in some spiders, where viewed from above the individuals' coloration matches the background leaf litter (enhance camouflage against aerial predators), but viewed from the side contrasts with the background, to enhance intraspecific communication (Clark *et al.* 2011).

Another strategy involves adopting different appearances at different stages of a species' life-history. In some insect larvae, changes in coloration occur over with ontogeny and growth (such as in successive instars). A range of lepidopteran larvae are initially camouflaged in appearance in the early instars, but when they grow larger they become more conspicuous and apparently aposematic (e.g. caterpillars of the panic moth *Saucrobotys futilalis*;

Grant 2007). This is possibly because bright warning signals would be ineffective when the animal is very small and hard to detect in the first place, whereas camouflage should be an effective defence (but see Mänd *et al.* 2007). Alternatively, polymorphisms across individuals in a species may reflect a trade-off between predator avoidance and sexual selection. For example, male wood tiger moths (*Parasemia plantaginis*) are unpalatable and can have either white or yellow hindwings. Recent aviary and field experiments in Finland have shown that predators (birds) avoid yellow morphs more than white ones. In contrast, white morphs have higher mating success, and so the trade-off between avoiding predation and attracting a mate can cause intraspecific variation (Nokelainen *et al.* 2012).

Finally, some animals, such as cuttlefish or chameleons, can quickly change their coloration to adopt different patterns over time. In this way, the animals can either signal to conspecifics during tasks such as mate choice, or adopt camouflage pat-

terns when under higher risk of predation (Langridge *et al.* 2007; Stuart-Fox and Moussalli 2008; Stuart-Fox *et al.* 2008; Stuart-Fox and Moussalli 2009; Zylinksi *et al.* 2011; Figure 7.7).

Although there are clear instances where we may expect multiple functions with the same signal form in visual stimuli, it is less clear how often this should occur in other sensory modalities. In acoustic, electric and vibratory communication, many signals are quickly produced and transient, meaning that an animal could quickly move from one call function to another with little trade-off (somewhat similar to rapid changes coloration in cuttlefish or chameleons). In olfaction, signal production is likely to be slower to change, meaning that compromises may arise. Overall, more work in these areas would be useful.

7.5 Future Directions

There is now a wealth of studies showing that parasitoids and predators will utilize the mating calls of hosts/prey to find them. In some systems, males with more attractive signals to females also appear more vulnerable to eavesdroppers. This may be that these individuals are easier to locate, but more work is needed to test other ideas, such as prey quality. There is now good evidence for private communication channels, but there remains great potential to investigate this, especially in non-visual systems. In addition, most work on private communication channels has been based on modelling approaches, and few studies have measured predator response to different signals behaviourally. Most work investigating how signal form may reflect multiple communication tasks has been done in visual signals, and work on other modalities is needed. Again, experimental work would be particularly valuable.

7.6 Summary

In this chapter we have discussed how signals and communication carry significant costs. Signals are not free to produce, but rather often require substantial energetic and physiological investment. In addition, signals in a range of modalities are exploited by eavesdroppers, including predators, prey, and competitors. Consequently, there is often a trade-off in

Figure 7.7 Some animals such as chameleons can change their coloration to be used in either camouflage or communication. The left image shows camouflage in the dwarf chameleon *Bradypodion taeniabronchum*, and top-right shows a dwarf chameleon (*B. atromontanum*) camouflaged against predatory a shrike (*Lanius collaris*). The bottom image shows a contest between males of the dwarf chameleon *B. damaranum* with the winner on the left and the loser on the right. Images reproduced with permission from Devi Stuart-Fox and Adnan Moussalli.

making signals conspicuous to receivers, while not attracting unwanted attention, such as that from predators. To reduce such risks, animals can evolve less conspicuous or localizable signals, they can change their signalling behaviour, or they can signal in a modality that the eavesdropper cannot detect or with a signal form that the predator is insensitive to. Finally, visual signals often combine multiple functions with the same signal form, such as enhancing camouflage while still communicating with potential mates or using warning signals. Alternatively, organisms may change their signal form over time, either rapidly or during development.

7.7 Further Reading

Bradbury and Vehrencamp (2011) discuss the different issues regarding the optimization of different signal types across different modalities, including more about the physics involved. Maynard-Smith and Harper (2003) discuss in detail economic approaches to communication and signals with more focus on strategic costs of signals and signal honesty. Zuk and Kolluru (1998) review various issues surrounding eavesdropping on sexual signals by parasites and parasitoids.

CHAPTER 8

Deception, Mimicry, and Sensory Exploitation

Box 8.1 Key Terms and Definitions

Aggressive Mimicry: When a predator or parasite mimics a harmless or beneficial stimulus to lure, capture or otherwise exploit its prey/host.

Batesian Mimicry: When a harmless organism (the mimic) resembles a toxic, dangerous or otherwise unprofitable species (the model).

Masquerade: A type of camouflage involving an organism resembling an uninteresting or unimportant object preventing recognition by the receiver.

Pre-existing (Sensory) Bias: An inherent bias (or 'preference') towards certain stimuli in the sensory and/ or cognitive system of the receiver that evolves before any trait that exploits this. The most frequently discussed type is sensory bias, which exists at relatively early stages in sensory processing.

Sensory Exploitation: The evolutionary modification of traits to elicit a stronger response in the receiver's sensory system, potentially due to a pre-existing bias in the receiver. Perceptual exploitation involves stimuli exploiting sensory, neuronal, or cognitive processes.

Sensory Trap: Traits similar to another stimulus that produce an out of context response in the receiver.

Supernormal Stimuli: Exaggerated versions of an existing stimulus that elicit a stronger response in the receiver than the normal stimulus.

We discussed in Chapter 7 how signals are vulnerable to eavesdroppers, such as parasites or predators. However, animals can also exploit the sensory systems of others to their own advantage: sensory exploitation. Here, the term 'exploit' does not need to mean that the trait elicits costly responses by the receiver. They may just evolve to best utilize how the receiver's sensory apparatus work. However, in many instances, animals do present deceptive and costly signals, including mimicking other organisms. In this chapter we discuss the related issues of mimicry, deception, and sensory exploitation, and how these different concepts link together, both conceptually and in evolutionary terms. We will deal with a wide range of examples, outlining how exploitation and mimicry are used by animals and plants to obtain prey, pollinators, and mates, using stimuli from a variety of modalities.

8.1 Exploitation and Biases

Unlike selection for strategic information aspect of a signal (see Chapter 5), sensory and perceptual exploitation theories argue that a major force shaping the form of signals is the sensory system and subsequent (e.g. cognitive) processing by the

Sensory Ecology, Behaviour, and Evolution. First Edition. Martin Stevens.
© Martin Stevens 2013. Published 2013 by Oxford University Press.

receiver. In short, signals that exploit pre-existing features of perceptual systems will be favoured (Christy 1995; Ryan 1990). Many different terms and definitions have been used to describe the different types of exploitation that may exist, and these are often closely related and describe the same or similar phenomena (see Box 8.1). Almost exclusively, these have been developed with regards to sexually selected traits, such as male ornaments (e.g. Endler and Basolo 1998), and in Box 8.2, we discuss how sensory bias and exploitation models relate to 'traditional' theories of sexual selection. However, concepts such as sensory bias and exploitation should also readily apply to other signalling systems, such as for brood parasites (Ryan 1990) or warning signals (Enquist and Arak 1993). In this chapter, we will consider a range of potential examples of sensory exploitation both with respect to and outside of mate choice. However, before doing so, it is important to discuss the different types of model that exist more closely, and their relationship with each other.

Some signals evolve as a result of a *pre-existing bias* in the receiver's sensory or cognitive system. Here, the bias in the perceptual system of the receiver occurs before the evolution of the trait. Most work and discussion has focused on *sensory biases*, which involve a bias towards certain stimuli or features occurring relatively early in sensory processing. Two further models are often discussed. *Sensory exploitation* describes how traits that are most successful in stimulating features of the sensory system of the receiver are favoured (Ryan *et al.* 1990b). *Sensory traps* describe circumstances where an out-of-context response is given to a trait similar to that used in a different situation (Christy 1995). For example, male movement patterns during courtship may mimic some prey features that females detect during foraging. The key idea is that sensory trap models require an inappropriate response and place more emphasis on cognitive rather than sensory processes (Endler and Basolo 1998; Schaefer and Ruxton 2009) because the receiver misclassifies a stimulus as something else (see below).

In reality, many of the above terms amount to overlapping phenomena (Endler and Basolo 1998), and it is not always clear how to determine if a trait better fits one definition or another. Sensory bias and exploi-

tation in particular have often been considered interchangeable (e.g. Fuller *et al.* 2005). Furthermore, constraining the use of sensory exploitation to occur only when specific pre-existing biases are present seems overly restrictive because many traits may evolve and exploit general features of a sensory system. Therefore, in this chapter we will use the term 'sensory exploitation' in a broader sense to encompass all examples where a trait exploits either specific biases or broad properties of the receiver's sensory system (therefore, encompassing both sensory biases and sensory traps). One other term recently introduced by Schaefer and Ruxton (2009) is *exploitation of perceptual biases*, to cover traits that evolve by exploiting features of (or a combination of) sensory, neuronal, or cognitive processes of the receiver. This is useful because we should remember that cognitive processes also have an important part to play in many of the examples discussed here, particularly processes such as categorization and generalization.

Finally, Endler (1992) and Endler and Basolo (1998) have described sensory exploitation as contained within the idea of sensory drive, which describes how environmental conditions, different behaviours, and aspects of sensory systems and signals interact to produce different communication systems and behavioural (e.g. mate) preferences. However, sensory drive is perhaps more effective as a model that specifically focuses on the role of the environment in signal and sensory system divergence or tuning. Sensory exploitation need not relate to environmental features and so is best separated from sensory drive. We will discuss sensory drive in Chapter 11 with respect to divergence and speciation.

8.1.1 Sensory Exploitation and Mate Choice

Models of sensory exploitation and traps were originally developed in the context of mate choice, and have most frequently been applied to this area. A vast body of work has been devoted to more 'traditional' models of mate choice (see Box 8.2), but sensory biases may explain many mate preferences and traits. To show that pre-existing biases have resulted in the evolution of a male trait, studies need to show female choice towards the trait, that the absence of the trait in males is the ancestral state

---

Box 8.2 The Relationship Between Receiver Bias and 'Traditional' Female Choice Models

Traditional Female Choice Models

Darwin (1871) did not propose a mechanism to explain why females choose, but rather simply suggested that females have a sense of aesthetics. To this day, explaining the evolution of female preferences remains controversial in sexual selection theory. *Fisherian runaway selection* (Fisher 1930) is the self-reinforcing evolution of female preferences and male traits (Andersson 1994). Fisher argued that if most females in a population prefer to mate with males with a specific trait, such as a long tail, it will pay an individual to prefer this too because her sons will have the long tail trait and be more attractive to females (Maynard Smith and Harper 2003). This could lead to a 'runaway' process involving an elaboration of the preferred trait and the preference. The preference evolves as a correlated response to selection on the male trait. This runaway process could end when survival costs (such as increased predation risk or physiological costs) of the trait outweigh any mating advantages it confers (Andersson 1994; Kokko *et al.* 2003). The *good-genes* theory of sexual selection posits that females are more likely to mate with fitter males, passing their good genes to her offspring, which in turn are more likely to survive and reproduce themselves (Andersson and Simmons 2006; Kokko *et al.* 2006; Maynard Smith and Harper 2003). Here, the male trait may act as an *index* and be causally related to some aspect of fitness, such as resistance to parasites (Hamilton and Zuk, 1982). An alternative mechanism to indices in good genes models is Zahavi's (1975) *handicap principle*, where a male display is an honest signal of quality that poor males cannot produce and maintain. Maynard Smith and Harper (2003) argue that indicator traits should be relatively cost free and correlate with a trait that contributes to fitness outside of mating. In contrast, handicaps should be costly to produce and reduce fitness in contexts outside of mating. Finally, female preference for a male trait may also evolve because it confers *direct benefits* on female or offspring fitness, including resources such as protection from predators and parental care.

Traditional Models and Sensory Exploitation

Ryan (1990), and other researchers before, emphasized the importance of considering sensory mechanisms in sexual selection, yet recent discussions of mate choice still give little coverage to signal efficacy and sensory systems, and focus most on theories devoted to explain signal content. As Endler *et al.* (2005) state, insufficient attention has been paid to how sexual signals are processed by sensory systems and the evolutionary implications of this. This is a shame, because an understanding of the evolution of mate choice will remain incomplete until more researchers integrate signal efficacy into 'traditional' models. The sensory exploitation hypothesis for mate choice predicts that signal efficacy is more important than strategy (content), and that females will prefer traits in males that elicit the greatest sensory stimulation (Ryan 1990; Ryan and Keddy-Hector 1992). Sensory exploitation mechanisms are often complementary and can be accommodated with good genes and runaway selection models, and studies of mate choice would benefit from a greater appreciation of sensory mechanisms (Endler and Basolo 1998; Ryan 1990; Ryan and Keddy-Hector 1992).

As Christy *et al.* (2003b) point out, 'traditional' sexual selection models argue that processes involved in mate choice lead to variation in female or offspring fitness, and to specific female mating preferences. However, these models do not consider how preferences arise, but only make predictions about the content (strategic component) of the signal. They rarely predict why or which specific signal forms evolve (e.g. particular plumage colours, song frequencies, and so on), though they may explain why such traits become exaggerated (Endler and Basolo 1998). The unique prediction of the sensory exploitation hypothesis is that the preference will exist before the evolution of the trait (Ryan and Keddy-Hector 1992). As Ryan (1990) and Ryan and Keddy-Hector (1992) point out, a range of studies across different taxa indicate that the total amount of sensory stimulation can be crucial in determining directional mate trait preference. In addition, there seems to be a preference for complexity or novelty, perhaps because this reduces the likelihood of habituation and enables signals to better stand out from background noise or competitor's signals. Thus, sensory bias models can explain why a specific trait may evolve in the first instance. For example, trichromatic primates may have a pre-existing bias for red colours derived under selection for foraging effectively for ripe fruit, and this preference may lead to the evolution of red skin and pelage coloration in sexual signalling (Fernandez and

Morris 2007). Once such red mating signals have evolved they may be exaggerated under further sensory exploitation, and/or processes like Fisherian runaway selection. Fisherian runaway selection may favour the elaboration of traits and preferences, and good gene models logically will often predict brighter, louder, and more intense displays that also elicit greater sensory stimulation (Ryan and Keddy-Hector 1992). Another intriguing link between runaway selection and sensory exploitation could arise with sensory traps. Garcia and Lemus (2012) have shown that some female fish have evolved resistance to male sensory trap ornaments that resemble food, perhaps because such traps induce females to respond more often to males than is optimal for them, taking time away from other tasks such as foraging. The authors suggest that this could lead to exaggeration of male traits in order to remain effective and greater female resistance, in a runaway type fashion.

The point at which selection for exaggerated traits stops may be when the physiological (e.g. energetic) or ecological costs (e.g. predation risk) of producing the signal become too high. Alternatively, as Ryan (1990) discusses, directional selection for the exaggeration of traits may stop when sensory processing limits are reached. For example, some male butterflies prefer females with faster wing beat frequencies than normal (8–10 Hz), up to the point when the temporal acuity (flicker-fusion rate) of their visual system has been reached (140 Hz), and males would no longer be able to see the wing beats (Magnus 1958, cited in Ryan 1990). In this instance, exaggeration of female wing beat frequencies above 10 Hz is presumably constrained by other factors (flight performance, energetics). Were these costs not present, counter-selection on trait exaggeration might not come from predation as Fisher imagined, but rather by sensory processing limits of the receiver (Ryan 1990).

(and that the trait has not been simply gained and subsequently lost), and importantly that in species without the trait females still have a preference for it (Basolo 1990).

8.1.1.1 Sensory Bias

In Chapter 7 we discussed how male túngara frogs (*Engystomops pustulosus*) produce calls to attract females, composed of a whine only (a simple call) or a whine plus one to seven chucks (a complex call), with females preferring complex over simple calls (Gridi-Papp *et al.* 2006). In addition, females also prefer chucks of males that have a lower fundamental frequency. Work by Ryan and colleagues was among the first to provide good evidence for sensory exploitation, demonstrated in the male túngara frog calls. Ryan *et al.* (1990b) used a model to calculate how male chucks would stimulate the female auditory system, and a range of modified chucks with either higher or lower frequencies than real calls. They showed that lower frequency calls would elicit greater neural stimulation (Ryan 1990). Phonotaxis experiments with females presented with synthesized male calls showed that

increased call energy or duration did not explain female preference, but the level of stimulation of the auditory system did (Ryan and Rand 1990). Ryan *et al.* (1990b) also showed that a closely related species of frog, *Physalaemus coloradorum*, also has tuning of the female basilar papillar to the same frequencies as the túngara females, yet males of this species do not produce chuck-like calls. Female *P. coloradorum* also prefer the calls of males with three túngara chuck components added over the normal call of their species (Ryan and Rand 1993). Thus, the sensory bias in the auditory system evolved before the evolution of the chuck component in túngara frogs.

Work by Basolo has shown similar findings in a visual communication system. Swordtail fish, *Xiphophorus helleri*, are small freshwater fish from Central America. Females prefer males with longer swords (an elongation of the lower portion of the caudal fin). In addition, the platyfish, *X. maculatus,* lacks the sword in males but preference tests show that females still have a preference for platyfish males with an artificial sword added (Basolo 1990, 1995b). This is despite evidence that the common ancestor of both the platyfish and swordtails lacked the sword (Basolo 1990). Subsequent work by Basolo (1995a) showed that in a species from a sister genus, *Priapella olmecae*, where males also lack swords, females also preferred males with an artificial sword added. Females also had a stronger preference for tails of greater length. Thus, it does seem that the preference for swordtails evolved before the evolution of the tails in these fish groups. Interestingly, other work has shown that predators of swordtails also have a preference for males with swords, and that this preference is also likely to have arisen before the predators came into contact with the swordtails (Rosenthal *et al.* 2001). Studies with mosquitofish (*Gambusia*) have also shown evidence for pre-existing biases in females towards exaggerated male traits that do not naturally occur, even though species in this genus do not seem to show female choice in the wild (Gould *et al.* 1999).

In the last chapter we discussed how the quiet ultrasonic male courtship calls of the Asian corn borer moth (*Ostrinia furnacalis*) are an example of a private communication channel, hidden from rivals and predatory bats (Nakano *et al.* 2006, 2008). Ultrasonic hearing probably initially evolved in moths to detect bats (see Chapter 9), and so ultrasonic male calls may also be an example of sensory exploitation. Interestingly, male calls apparently work to increase copulations by inhibiting female moths from fleeing. This seems to be a type of behavioural exploitation, because the freeze response in females seemingly evolved as a response to detecting bats, whereby many moths become motionless or drop to the ground. Recently, Nakano *et al.* (2010) tested the idea that male ultrasonic calls represent an example of sensory exploitation in the common cutworm moth (*Spodoptera litura*) of which males produce ultrasonic courtship calls with a tymbal organ that the females lack. Nakano *et al.* found that only 40% of silenced males (with their tymbal structures muted) were accepted by females, compared to more than 95% female acceptance of naturally calling males. Female courtship acceptance of silenced males was increased back to *ca.* 100% when accompanied by playbacks of male song. The really exciting finding, however, was that female courtship acceptance of muted males was also increased

almost to the same extent (*ca.* 95%) by presenting playbacks of simulated bat calls!

Although there are still a limited number of studies investigating sensory exploitation and mate choice, there is now good evidence that sensory bias is an important process in female choice and associated male traits. Further studies are needed to investigate how widespread this is, especially as most work has been done in acoustic and visual senses. Recently, Machnik and Kramer (2008) suggested that longer electric organ discharges (EODs) in the electric fish *Marcusenius pongolensis* that are more attractive to females, may have evolved by sensory exploitation. They argue that, like all electric fish, *M. pongolensis* has ampullary receptors (see Chapter 2) tuned to low frequencies and used for detecting live objects such as food. These receptors also weakly respond to the low-frequency components of EODs (primarily picked up by the tuberous receptors found in electrogenic fish), especially when these components are long in duration. Ampullary receptors likely evolved before tuberous ones, and males with longer EODs would stimulate these receptors more, increasing the level of female preference. Interestingly, sensory exploitation has also recently been invoked to explain the evolution of iconic representations (imagery and art) in humans (see Verpooten and Nelissen 2010) and red pelage and skin in trichromatic primates (Fernandez and Morris 2007).

8.1.1.2 Sensory Traps

A range of work on sensory traps has been done on fiddler crabs, *Uca spp*, where male crabs build mud pillars next to their burrows to attract females. Christy *et al.* (2002, 2003a, 2003b) tested the sensory trap idea in the fiddler crab *Uca musica*. In this species 20–60% of males build hoods on their burrows. These increase male attractiveness because females preferentially approach males with hoods, although the hoods do not influence female mating decisions after reaching the male. Females are more likely to approach burrows because these seem to exploit the behaviour of crabs as they move through the environment via landmarks that offer temporary cover from predators. Males of a related species, *U. stenodactylus*, do not build burrows, yet females are more likely to go to empty burrows with hoods than those without hoods when they are threatened by a predator (Christy *et al.* 2003a). In addition, some fiddler crabs will spontaneously approach a range of naturally occurring objects in the environment without males or burrows, both when reproductively active and when non-receptive (Christy *et al.* 2003b), demonstrating the latent preference that exists.

Proctor (1991, 1992) has shown that in some water mites, *Neumania papillator*, which are sit-and-wait predators that locate their copepod prey by detecting water vibrations, males will mimic prey vibrations to attract females ('courtship trembling'). Females respond to the males in a similar way to their prey, by clutching males by their forelegs. In addition, females are more likely to mate when they have not eaten compared to females that have been well fed (Proctor 1991). Mapping of predatory and courtship behaviour of mites onto a cladogram indicates that hunting using vibrations may have evolved before the male use of the vibrations to attract females and is a sensory trap (Proctor 1992).

Although there is some good evidence for sensory traps, this still comes from a limited number of species and studies. Much more work is needed to establish how widespread sensory traps are, their evolutionary significance, and relationship with other aspects of mimicry and exploitation (see below). Male ornaments acting as sensory traps may also take time and energy away from females for other tasks (especially foraging) if females are induced to inspect males more often than they would otherwise do so. This could be costly for females and lead to female resistance to such sensory traps. There is evidence for this in some fish (Garcia and Lemus 2012), and it would be interesting to explore this idea in other species. Finally, sensory traps involved in mating may be exploited by predators. Recent work by De Serrano *et al.* (2012) has shown that some predatory prawns have orange spots on their pincers, and that these seem to resemble the orange spots on males preferred by female guppies (*Poecilia reticulata*). This study raises important areas of future work to determine how often sensory traps and biases involved in mating are co-opted by predators to lure prey.

8.1.2 Where do Biases Come From?

Evidence that sensory exploitation is an important aspect of communication is steadily growing, but it is often still unclear why animal sensory systems have biases. However, there are several possibilities. Broadly, pre-existing biases may result from either sexual or natural selection, or processes such as genetic drift (Endler and Basolo 1998). Neither the pre-existing bias nor sensory exploitation models specify what causes preferences to arise in the first instance. The sensory trap model asserts that biases or preferences in the receiver's processing arose under selection in a one task and this is then exploited in other circumstances. Although biases may in some cases be non-adaptive, generally we would expect them to be maintained by selection given that sensory systems are used in many vital tasks, such as foraging, navigation, communication (Basolo and Endler 1995). Sensory biases could also evolve as neutral by-products of the way that sensory systems are constructed, and so they have also been termed *hidden preferences*. Artificial neural network simulations have been effectively used to investigate the evolution of such preferences (Arak and Enquist 1993; Enquist and Arak 1993; Hurd *et al.* 1995). For example, Enquist and Arak (1993) trained a network to distinguish between males of two species, one of which had longer tails. After the network could do this, they found that new stimuli could elicit stronger responses than any of the training stimuli, including those with exaggerated forms such as longer tails (supernormal stimuli; see below). Enquist and Arak (1993) concluded that their model showed that female preference for exaggerated traits in males can result simply from the need to recognize males of the correct species. Overall, Arak and Enquist (1993) argue that recognition systems are never perfectly tuned to any task, because sensory systems cannot perfectly code for all the vast range of signals and cues that they may be faced with over evolution. Consequently, biases will arise to new stimuli because the potential will always exist for new stimuli to elicit greater responses in the receiver than existing stimuli. Overall, biases may exist for a wide range of reasons and these need not be the same across species.

8.1.3 The Relationship Between Perceptual Exploitation and Mimicry

Mimicry and sensory exploitation are often treated separately, yet in many cases they could be related processes. In addition, it may sometimes be difficult to distinguish between their predicted effects on signalling systems (such as what type of signal evolves). This can have important implications for understanding the evolution of different strategies used by animals in prey capture and mating, and how such signals actually work.

Organisms may resemble each other for a variety of reasons, including shared ancestry, and convergent evolution owing to similar selection pressures. Many organisms also directly mimic another, or mimic the signals other organisms make. Traditional explanations describe how the mimic evolves to resemble another organism, the model, when the two species encounter the same receiver, such that the receiver will make recognition errors and mistake the mimic for the model. Generally, mimicry this thought to involve a misclassification of a stimulus by the receiver, so that it mistakes it for something else. As such, unlike most (but not all) forms of camouflage that prevent detection and discrimination from the background, mimicry also involves a cognitive element in terms of how the receiver categorizes an object. Naturally, this distinction (between hindering detection and hindering identification) is not entirely clear-cut. Nonetheless, the various types of mimicry should involve misclassification by the receiver, over and above simply stimulating the sensory system in an exaggerated way or in a manner that resembles something else.

Schaefer and Ruxton (2009) recently proposed that because perceptual biases are likely to be common in animals, a widespread underlying cause for apparent mimicry in nature may be perceptual exploitation. They argue that perceptual biases may often act as a precursor for true mimicry to evolve, and may bridge the gap between non-mimicry and full mimicry, allowing the presence of intermediate 'imperfect' forms (which are currently hard to explain by conventional mimicry theory). In addition, when similar sensory processing mechanisms exist between species, this may also select for convergence in signal form across animals if similar signals that are most effective at stimulating features of the sensory system independently evolve in different groups. Therefore, apparent mimicry may simply reflect similar sensory processes in different receivers. An important comparison between the perceptual exploitation and mimicry models is that perceptual exploitation relates to the underlying properties of the perceptual system (often sensory biases), whereas mimicry evolution traditionally is thought to require a key role of learning and generalization (i.e. cognition; Schaefer and Ruxton 2009).

Perhaps one of the largest areas of overlap between mimicry and perceptual exploitation involves sensory traps. Sensory trap models often assume a model-mimic relationship (Christy 1995). In fact, various examples are given by Christy (1995) as instances of sensory traps that are often traditionally regarded as instances of mimicry. However, a clear distinction is made by Christy *et al.* (2003b) with regards to female fiddler crabs (see above). If the crabs gain the same benefit (refuge from predators) from approaching both natural objects (the model) and 'hoods' built by males near the burrow (the mimic), then there need not be selection for females to discriminate between these. This is quite different from many other types of mimicry system whereby there should be selection for the receiver to discriminate between the model and mimic.

Finally, a related issue concerns aggressive mimicry (see below), where the predator mimics another object or organism to remain unnoticed by or to lure its prey. It is possible that many apparent cases of aggressive mimicry are really a type of sensory trap, or started off that way. For example, angler fish lure their prey (smaller fish) by waving a 'rod' over their head, the end of which is thought to resemble a small prey item (Gudger 1946; Pietsch and Grobecker 1978). The lure may simply work by attracting the interest of the prey by stimulating a feeding response, and in some cases may be especially effective when in an exaggerated form. Subsequently, in some cases the lure may evolve to more effectively mimic a small animal. Therefore, whether it is a sensory trap or a true mimic may also depend on the stage that evolution has reached in this system.

8.2 Supernormal Stimuli

Under sensory exploitation theories, it is often argued that stimuli will be favoured that elicit a stronger perceptual response in the receiver. This leads to the idea of supernormal stimuli, being exaggerated versions of an existing stimulus and that elicit a stronger response than the normal stimulus in the receiver (Tinbergen 1948). For example, some birds will preferentially incubate eggs much larger than those naturally occurring (see Tinbergen 1951). As discussed in Box 8.2, some male butterflies prefer to court females with flicking stimuli at much higher rates than the natural wingbeat frequencies of females. Rowland (1989) has shown that female sticklebacks (*Gasterosteus aculeatus*) presented with models of males prefer larger models even though they exceed the size of the largest natural males by 25%. In addition, some female birds have been shown to prefer songs significantly longer than normal male songs (Neubauer 1999), and some male moths are more attracted to supernormal synthetic pheromones over the naturally occurring female substance (Jaffe *et al.* 2007). Thus, preference for supernormal stimuli may be widespread across modalities. Supernormal stimuli need not convey a strategic message of quality but may simply provide greater sensory stimulation. This could lead to exaggeration of signal form without changes in receiver preference (Rowland 1989). However, interest in supernormal stimuli has not been common lately, and more work is needed in this area to gain a wider understanding.

8.3 Anti-Predator Defences: Mimicry and Exploitation

Almost all animals are targeted by predators, and so it is not surprising that the natural world is full of anti-predator strategies. Many of these exploit features of the predator's perceptual systems, including early stage processing mechanisms and through resembling other objects.

8.3.1 Masquerade

Some animals achieve camouflage by resembling an inedible or uninteresting object in the environment, such as a leaf, twig or bird dropping (Figure 8.1; Skelhorn *et al.* 2010a). This is called masquerade, and although predators may detect these animals they do not recognize them as a prey item and so ignore them (Stevens and Merilaita 2009b). Although masquerade has sometimes been described as similar to Batesian mimicry it does not require that the model is toxic, just that the model is of no interest to the receiver. Although there are many putative cases of masquerade in nature, only recently have experiments provided evidence for this concept. These involved presenting domestic chicks (*Gallus gallus domesticus*) with twig-resembling caterpillars and models (branches) that had been modified in some way, and showed that the chicks misclassify the caterpillars as twigs (Skelhorn *et al.* 2010b; Skelhorn and Ruxton 2010). Following this, the authors have shown that the benefit of masquerade decreases when the proportion of masqueraders becomes higher compared to the models. This is because when masqueraders become harder to find the predators' motivation to search declines (Skelhorn *et al.* 2011).

8.3.2 Batesian Mimicry

Some animals with dangerous, unpleasant, or otherwise unprofitable attributes advertise this to potential predators with conspicuous signals, so that predators avoid attacking them; they are aposematic (see Chapter 9). Not all bright colours are, however, honest signals of aposematism, and many animals falsely convey a risk to a predator. Batesian mimicry is usually taken as involving one harmless animal (the mimic) resembling another unprofitable animal (the model), such that the receiver is deceived into confusing the two (Edmunds 1974; Figure 8.2). Generally, a Batesian mimic should degrade the protective value of a model's aposematic markings, since predators that sample a Batesian mimic will not learn to avoid the pattern and may subsequently attack the model species. As such, a Batesian mimic is often expected to be particularly protected, and least costly to its model, when it is relatively rare compared to the model.

Figure 8.1 Stick insects resemble twigs, grass, and other objects in the environment so that potential predators fail to classify them as a food item.

Despite a long history of theoretical work, surprisingly few studies have investigated Batesian mimicry from the perspective of the receiver's sensory system and this is a major avenue for future studies. Early work suggested that some juvenile lizards, *Eremias lugubris*, from southern Africa mimic the colour and gait of toxic beetles known to spray an acidic fluid when attacked (Huey 1977). It was suggested by Wallace (1889) and recently confirmed that some jumping spiders resemble ants, and that this may help them to avoid avian and spider predators that often avoid ants that are unpalatable (Huang *et al*. 2011; Nelson and Jackson 2006). Another longstanding idea is that many birds

appear to mimic the vocalizations of other individuals or species. An interesting study concerns burrowing owls (*Athene cunicularia*) that nest in ground squirrel burrows, and make a hissing sound when disturbed, similar to that of a rattlesnake shaking its rattle. Work by Rowe *et al*. (1986) indicates that ground squirrels familiar with rattlesnakes will mistake the owl hissing sound for the sound of a real snake rattle. Thus, the owl vocalization may be an example of acoustic Batesian mimicry. Kelley *et al*. (2008; and Kelley and Healy 2011) define true vocal mimicry as involving the acquisition of sounds via learning during an individual's lifetime. They contrast this to similarities in calls or songs that

Figure 8.2 A hoverfly, *Episyrphus balteatus*, often assumed to be a Batesian mimic of harmful wasps and hornets.

arise due to evolutionary convergence resulting from similar selection pressures on species. Kelley *et al.* therefore argue that true Batesian acoustic mimicry must be learnt. This is conflicting with ideas of Batesian mimicry in other modalities, such as vision, where the mimic over evolutionary time adopts colour patterns that may increasingly resemble an unprofitable model. Limiting acoustic mimicry in birds only to cases where learning has occurred seems overly restrictive without good reason, and it is in conflict with our understanding of mimicry and its evolution in other modalities. One can readily imagine how acoustic mimicry could evolve over generations, such that the mimic better resembles the model and gains an increased benefit from this (e.g. in the case of the owls putatively mimicking snakes above). Indeed, work has recently shown that some moths are acoustic Batesian mimics of toxic species attacked by bats (Barber and Conner 2007). Here, the defence has arisen over many generations, including the use of specialized sound-producing organs in the moths. This example is directly comparable to Batesian mimicry in the visual sense.

8.4 Foraging and Prey Capture

While some prey animals have adopted deceptive or exploitative signals to reduce the risk of predation, many predators and foragers exploit competitors or prey. This subject presents a range of examples demonstrating how foraging and prey capture can involve genuine mimicry, camouflage, and luring and sensory exploitation.

8.4.1 Mimetic Alarm Calls and Kleptoparasitsm

It has been suggested for some time that different species of drongos mimic other birds. Recent work has shown that fork-tailed drongos (*Dicrurus adsimilis*) make mimetic alarm calls resembling those of other species directed at meerkats (*Suricata suricatta*), pied babblers (*Turdoides bicolour*), and glossy starlings (*Lamprotornis nitens*) that falsely indicate the presence of a predator (Figure 8.3). The drongos are kleptoparsites, and the false alarm calls cause individuals of the target species to drop the food they have just acquired, so that the drongo can

swoop down and steal it (Flower 2011). Of the instances of kleptoparasitsm, about 50% of these involved the drongo first making a false alarm, and playback experiments show that both meerkats and babblers are equally likely to drop food in response to real and false alarm calls, and more so than to non-alarm calls (Flower 2011). As with work on Batesian mimicry, the benefit to the mimic and the cost to the model is frequency dependent. Drongos safeguard against the target species from learning to ignore its calls by also making a number of honest alarm calls when a predator really is present. Thus, the drongos are not entirely costly to the targeted species (see also Radford *et al.* 2011).

8.4.2 Flower Attraction to Pollinators

Among the most tightly linked signaller and receiver systems occur between the vision of pollinating insects and the appearance of flowers. Many flower patterns appear to enhance detection by pollinators (Hempel de Ibarra and Vorobyev 2009). Work has shown that pollinators such as bees are attracted to more colourful and larger versions of flower signals (supernormal versions) than those they had originally been trained to associate with a reward, and this is even the case when such exaggerated displays do not equate to higher quality (Naug and Arathi 2007). Thus, exaggerated floral displays may benefit from biases in the pollinators' sensory system. The signals need not be costly to the pollinator, but likewise they may not be reliable indicators of flower quality. Other work has indicated that sensory biases in bumblebees (*Bombus terrestris*) to different flower colours may be driven by variation in flower properties, such as nectar production (Raine and Chittka 2007). For example, bee colonies that show stronger innate biases for some colours, such as violet, bring back more nectar than bees with biases towards other colours (e.g. blue) in areas where the violet flowers produce more nectar than blue flowers. Thus, the direction of sensory biases could be driven by ecological factors on a local scale.

However, some plants, especially orchids, are known to attract pollinators deceptively with visual and olfactory mimicry of food, mates, and oviposition sites, and may exploit sensory biases in pollinators (see Gaskett 2011; Jersakova *et al.* 2006). They

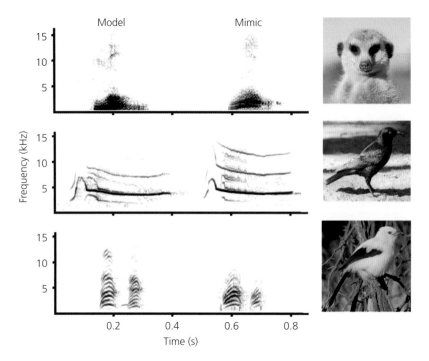

Figure 8.3 Sonograms showing the call structure of three model species that are targeted by kleptoparasitic drongos: meerkats (top), glossy starlings (middle), and pied babblers (bottom) and the mimetic drongo alarm call used towards each species. Sonograms and images reproduced with permission from Tom Flower.

may attract pollinators in a number of modalities, including using chemical (odour), colour, and pattern, shape, and tactile channels, of which chemical have been best studied (Gaskett 2011). Such cases are manipulative because the pollinators go unrewarded (they do not get nectar or pollen). For example, the orchid *Dendrobium sinense* from China mimics the alarm pheromones of honeybees, in order to attract hornets that often prey upon the bees, tricking the hornets into pollinating the flowers (Brodmann *et al.* 2009). Likewise, another orchid, *Epipactis veratrifolia*, seems to attract pollinating hoverflies by releasing compounds that mimic aphid alarm pheromones that the hoverflies use to locate aphids that their larvae eat (Stökl *et al.* 2011). Rather than mimicking one species of aphid, the orchid seems to be a generalist mimic of a group of aphids, presumably because most hoverfly larvae feed on a range of aphid species. Alternatively, the compounds could be a sensory trap. In contrast, work by Gaskett and Herberstein (2010) on four orchid species from the genus *Cryptostylis* shows that they accurately resem-

ble the colour of different parts of the female wasp they mimic (*Lissopimpla excelsa*) to the visual system of male wasps. Here, the male wasps are the only pollinator of the orchids, makings species-specific mimicry worthwhile. In other systems, there is high similarity between orchid chemical compounds and those of the model pollinator species (e.g. Schiestl *et al.* 2003). Sexual deception by flowers need not be limited to orchids, and recent work has shown that other plant groups utilize this tactic to attract pollinators, including petal ornaments aimed at male flies (Ellis and Johnson 2010). Recent work has also shown remarkably that some orchids seem to mimic fungus-infected parts of plants to attract flies, both visually with black mould-like spots and chemically (Ren *et al.* 2011)!

8.4.3 Luring Prey and Avoiding Predators in Spiders

Many spiders exhibit bright colour patterns on their bodies, including species that actively forage during the day. Such body coloration may work in camouflage, hiding the spider from its prey. Alternatively, the colour patterns may attract prey, increasing hunting success.

A range of work has been done in crab spiders to investigate spider camouflage and prey luring. For a long time it was assumed that crab spiders were an example of excellent camouflage against flowers in order to capture their prey by remaining hidden from them. This was, however, until recently based on human assessment. Théry and Casas (2002) modelled the coloration of crab spiders, *Thomisus onustus*, from France to both bee (prey) and bird (predator) vision and found that the spider coloration effectively resembled flower coloration to both visual systems. However, while some European crab spiders rely on camouflage to remain hidden from their prey, this does not appear to be the case for the Australian crab spider, *T. spectabilis*. Heiling *et al.* (2003) showed that while this species looks camouflaged to humans against the white daisy *Chrysanthemum frutescens*, it contrasts strongly against the flower petals in ultraviolet (UV) light, which European honeybees (*Apis mellifera*) can see (Figure 8.4). This high-UV contrast lures more bees than flowers without spiders. Heiling *et al.* (2005) found that spiders with reduced UV reflectance attracted fewer bees compared to flowers with UV-positive spiders. Interestingly, Australian honeybees (*Australoplebia australis*) are also attracted to flowers with spiders, but tend not to land on them (Heiling and Herberstein 2004). It is possible that coevolution between native bees and Australian crab spiders may have resulted in some defences arising in the bees. Comparative analyses indicate that UV reflection is found in Australian crab spiders, but not in European species (Herberstein *et al.* 2009). These studies also demonstrate the importance of considering the receiver's vision, rather than that of humans.

Many orb-web spiders wait in the centre of their web, and their coloration could act as camouflage against the general environment or to lure prey. Tso *et al.* (2006) compared these hypotheses in the orchid spider (*Leucauge magnifica*) with field experiments in Taiwan and by modelling spider and background colours to insect vision. The orchid spider has a brightly coloured opithosoma, being silver with

Figure 8.4 Top-left: A crab spider *Misumena vatia*, capturing a honeybee and camouflaged against the white flower. Image of *Misumena vatia* reproduced with permission from Roger Le Guen. Top-right: A spider, *Synema globosum*, camouflaged against a yellow flower and capturing a fly. Bottom images show an Australian crab spider, *Thomisus spectabilis*, camouflaged against a white daisy to human eyes but highly conspicuous when viewed in ultraviolet light. Images of *Thomisus spectabilis* reproduced with permission from Marie Herberstein.

Figure 8.5 Images of the orchid spider, *Leucauge magnifica*, (left) and the giant wood spider, *Nephila pilipes*, (right), showing their coloration that lures prey. Images reproduced with permission from I-Min Tso.

black stripes on the dorsum, and dark green with yellow stripes on the ventrum (Figure 8.5). Tso *et al.* compared the number of insects hitting the web per hour for webs with and without spiders. They found that webs with spiders caught approximately twice as many insects as webs without spiders. The authors also conducted an experiment whereby they altered various parts of the spider's coloration. They applied green paint to the yellow ventral stripes, the silver dorsal coloration, both these areas, or the green parts of the abdomen (control). They found that painting the yellow or silver colours with green (or especially both sides) decreased the amount of prey caught compared to the control treatment. Finally, vision modelling showed that while the green spider colour matched the background well, the yellow and silver patches had colours that were highly discriminable against the vegetation to insects. Thus, Tso *et al.*'s study provides good evidence that the spider coloration is effective by luring prey, probably by exploiting a pre-existing bias in the prey perceptual systems. Alternatively, the spider's coloration could be mimicking the colours of flowers (see discussion in Fan *et al.* 2009). Interestingly, Tso *et al.* (2006) suggest that the spider coloration may not be fully conspicuous but incorporate green camouflage colours because if the whole body was bright then the outline would be conspicuous and the prey may recognize the spider body shape and avoid it. Thus, the spider coloration may still utilize some aspect of camouflage.

Tso *et al.*'s (2006) study is also supported by other work. For example, Bush *et al.* (2008) investigated the yellow, white, and black bands on wasp spiders (*Argiope bruennichi*) in the UK. They found that naturally coloured spiders caught more than twice as many prey as spiders with their coloration painted black, or when the spider was obscured in the web by a leaf fragment. Colour analysis based on bee vision showed that the yellow bands are highly conspicuous against the background, although this was not the case for the leg banding coloration. Thus, the spider's coloration does seem to lure insect prey rather than camouflage the spider, but as with the orchid spider could still provide some disruptive camouflage benefits in concealing the spider's body outline. In other work, Tso *et al.* (2002, 2004) studied the giant wood spider (*Nephila pilipes*) found in East and South-East Asia (Figure 8.5). Tso *et al.* (2004) showed that the spider's colourful yellow stripes are highly visible to hymenopteran insects, and that the brightly coloured normal form captured significantly more prey than a melanic form (Tso *et al.* 2002). Later, Chiao *et al.* (2009b) took digital images of *N. pilipes* corresponding to the spectral sensitivity of the receptors of a bee. In addition, they Gaussian filtered the images to remove high spatial frequencies (fine-grained detail) to simulate a bee's spatial resolution at 10, 20, or 30 cm away from the spider. By doing so, they could investigate how the spider

coloration looks 'through a bee's eyes', showing that the spider markings are highly contrasting to a bee. The authors suggest that the pattern of the spider coloration is similar to those of some flowers the bees visit because the colour patches do not resemble the spider's body shape, but rather form clusters that look like a flower pattern. In addition, the patterns are also symmetrical, like many flowers, and bees have a visual preference for symmetry. These are interesting ideas, but remain somewhat speculative because the spider patterns were not quantitatively compared to real flowers. In addition, bilateral symmetry is found widely across nature and so is not strong evidence of mimicry. The crucial missing piece is to test if the spiders' colour patterns are mimicking flowers from the environment. It would be revealing to put flower parts (e.g. petals) on webs to see if these also increase attraction, and also to determine if bee species that do not encounter *N. pilipes* but that visit flowers are also attracted to the webs.

Overall, the majority of evidence indicates that spider coloration in these systems seems to be mainly for luring, but with a secondary camouflage benefit. Luring may also be an important function of spider coloration at night (e.g. Blamires *et al.* 2012). It is not yet clear whether the luring works by a general sensory exploitation, or by specifically mimicking flowers. Finally, spider coloration also seems to represent a balance between attracting prey but not predators that are also drawn to the bright coloration (Chuang *et al.* 2008; Fan *et al.* 2009).

8.4.3.1 Web Decorations

Some orb-web spiders build or add to their webs a range of conspicuous structures, called decorations (see reviews by Herberstein *et al.* 2000; Théry and Casas 2009). The function of these structures has not always been clear because they could work to camouflage the spider, attract prey, or act as mimetic decoys, drawing predator attacks away from the real spider. The various decorations are diverse, including prey remains, egg sacs, detritus, and silk patterns. Most evidence indicates that silk decorations work to lure prey, with limited support for the idea that they reduce predation risks (see discussion in Cheng and Tso 2007; Tan *et al.* 2010; Théry and Casas 2009).

Cheng and Tso (2007) conducted experiments on the silk decorations and body coloration of St Andrew's cross spiders (*Argiope aemula*). They manipulated the spider silk decorations (four cruciate 'zig-zag' like bands forming a cross shape from the centre of the web), and found that decorated webs intercepted 60% more insects than undecorated webs. As with the studies of spider coloration, the inference is that the spider's silk decorations may have a broad resemblance to flower signals (Cheng and Tso 2007). Other work has also shown that silk decorations in spiders attract insects to webs (Blamires *et al*. 2008), although findings are not always consistent or instead indicate an anti-predator function (Blamires *et al*. 2010; Nakata 2009). However, overall, the evidence for prey attraction by silk decorations is convincing. But why are prey attracted to them? A particularly impressive study by Cheng *et al*. (2010) sheds light on this in the *Argiope* group of spiders, which often build webs with either linear (usually single vertical lines) or cruciate (crosses with lines in diagonal directions) decorations (Figure 8.6). Cheng *et al*. (2010) hypothesized that cruciate decorations exploit preferences in insect vision to particular orientations, and are derived from an ancestral simpler linear form. Bees and other insects seem to have at least three specific orientation-sensitive visual channels, each encoding a particular orientation (e.g. Okamura and Strausfeld 2007; Srinivasan *et al*. 1994; Yang and

Maddess 1997). Cheng *et al*. (2010) predicted that the cruciate decorations should be more effective because they stimulate two of the channels, sensitive to diagonal orientations, whereas linear decorations would only exploit one channel sensitive to vertical arrangements. They first showed with a molecular phylogeny that the linear form is indeed ancestral in Asian *Argiope* spiders, evolving at least twice independently in the genus. Following this, they conducted field experiments where they manipulated the number and orientation of silk decorations. They presented pairs of webs side by side in the field (using either real decorations or artificial versions to control for total silk area). In one experiment, they compared prey interception rates of linear and cruciate decorations. In another they compared cruciate decorations in a natural orientation (two diagonal lines) with an arrangement rotated 45° (linear and vertical lines). They found that cruciate decorations attracted significantly more prey than linear forms, and that cruciate forms were twice as effective in their natural orientation than when rotated 45°. Thus, features of the prey sensory system could drive the evolution of specific decoration forms. One factor that should still be tested for, however, is whether the linear forms merge with the background environment more, due to the vertical nature of the vegetation, whereas the cruciate forms stand out more because their orientation contrasts with the general environment.

Recent work has also investigated the role of decorations comprising prey pellets and eggsacs wrapped in silk in the spider *Cyclosa mulmeinensis* (Tseng and Tso 2009). This study intriguingly found that while these decorations increase the likelihood of wasp attacks, the wasps were distracted to attack the decorations (which often resemble the colour and size of the real spiders; Gan *et al*. 2010). Thus, as is the case with studies of conspicuous lizard tails (e.g. Cooper and Vitt 1991; Cooper 1998), the cost of attracting higher predator encounters with conspicuous decorations may be offset by the benefits of escaping from attacks that occur. Decorations in this system are interesting because they seem to mimic the spider itself. This is essentially the inverse scenario to masquerade, the latter being where the prey species mimics an uninteresting or inanimate object in the environment (the models). With these decora-

Figure 8.6 Some spiders build silk 'decorations' in their webs, including these diagonal cruciate types. Research indicates that they may function in luring insect prey to the web.

tions, the model is the prey item, and the mimics are the decorations. We can call such decorations 'decoys' (Tseng and Tso 2009). There is also evidence that some birds may build false nests (decoys) that reduce the risk of nest predation (Galligan and Kleindorfer 2008). However, it is not certain that spider decorations have this function; other work on *C. mulmeinensis* has indicated that decorations function by making the spider cryptic against the decorations, reducing its detectability to insects, increasing prey interception rates (Tan and Li 2009).

Overall, evidence from both spider coloration and decorations indicates that luring prey is an important function of many of these traits, potentially in many species. However, there is also evidence for camouflage and decoys, and so spider coloration and decorations seem to have many functions. More work investigating what the primary function is in different species will be valuable.

8.5 Aggressive Mimicry

Unlike Batesian mimicry, where prey deceive the predator, aggressive mimicry involves the predator deceiving the prey. This is different from simply hiding, camouflage, or stalking prey, because the predator uses deceptive communication signals rather than concealing its presence. That is, the prey is expected to register the presence of the aggressive mimic, but to misidentify it as another non-threatening species or object. For example, the bluestriped fangblenny (*Plagiotremus rhinorhynchos*) resembles the cleaner wrass (*Labroides dimidiatus*), but instead of cleaning parasites from other fish, the fangblenny bites flesh and scales off the fish (e.g. Cheney 2012). One remarkable example of aggressive mimicry may exist in the cookie-cutter shark (*Isistius brasiliensis*). This shark bites characteristic 'plugs' of flesh from larger fast-swimming predatory species, such as tuna and porpoises. It also produces bioluminescence from its ventral surface, and Widder (1998) has suggested the pattern of this bioluminescence mimics the silhouette of a smaller fish, which normally produce downwelling bioluminescence to camouflage themselves when viewed from below against the light surface waters (counter-illumination). In this way the shark may lure its victims close

enough to attack them. Note, as above with angler fish, however, that whether this should be called aggressive mimicry or a sensory trap is unclear.

One of the most impressive examples of aggressive mimicry comes from the bolas spiders, *Mastophora spp.* Adult females of various species hang from a silk line strung between vegetation (a 'trapeze wire'). The spider produces a ball of sticky adhesive that it swings from a thread to capture moth prey that come close, with the liquid penetrating the scales and the underlying cuticle of the moths (Figure 8.7).

Eberhard (1977) showed that bolas spiders caught a comparable amount of prey to a similar sized orb web weaver (about 18% of the spider's total weight per day), and noted that moths approached the spiders from downwind with their antennae extended, beginning with wide arcs that become narrower as the moth gets closer to the spider. This is similar to the way that male moths approach females emitting mating pheromones and led Eberhard to suggest that bolas spider's success could stem from mimicking female pheromones. Later, Stowe *et al.* (1987) showed that the volatile emissions in *M. cornigera* contained compounds that are found in the female moth sex pheromones used to attract males. In another species, *M. hutchinsoni,* two prey species account for 93% of moths caught: the bristly cutworm (*Lacinipolia renigera*) and the smoky tetanolita (*Tetanolita mynesalis)* (Yeargan 1988). Of this, the bristly cutworm comprises approximately 40%. Gemeno et al. (2000) characterized the emissions of *M. hutchinsoni* and found that they contain the same two key components found in the real bristly cutworm moth sex pheromones of females: (Z)-9-tetradecenyl acetate and (Z,E)-9,12-tetradecenyl acetate. *M. hutchinsoni* not only mimics the composition, but also the blend ratio of these two components. Perhaps even more impressive is that bolas spiders can mimic the pheromone components of more than one prey type, and change the predominant components released to coincide with prey activity patterns. The bristly cutworm is active mainly before 22:30, whereas the smoky tetanolita is active mainly after 23:00, although there is some overlap. Work by Haynes *et al.* (2002) indicates that the spiders could attract both species during the night and seem to produce a compromise mixture of the two species' pheromones. However, the spiders also decrease

Figure 8.7 Image of a bolas spider (*Mastophora hutchinsoni*) and its bolas used to capture flying moths. Image reproduced with permission from Kenneth Yeargan.

emissions of the bristly cutworm pheromone components as the night goes on.

In some cases predatory spiders have developed adaptations to increase capture by mimicking the vibratory courtship displays of other spiders that they hunt. The spider *Portia fimbriata* eats a range of other species, including spiders of the genus *Euryattus*. Work by Jackson and Wilcox (1990) has shown that *P. fimbriata* from north-east Queensland in Australia use vibratory signals that mimic the courtship displays of male *Euryattus* to lure female *Euryattus* out from their nest of rolled up leaves. One of the interesting features of *P. fimbriata* is that it eats a range of prey and it will also invade the orb webs of other spiders. Here, it also uses aggressive mimicry by strumming the strands of the web with its legs and palps in a manner that mimics a prey item in the nest, luring the resident prey spider towards it (Tarsitano *et al.* 2000). Interestingly, the exact type of web strumming behaviour depends on the size of the prey. Against small spiders, *P. fimbriata* creates strong and steady vibrations that mimic the capture of an insect in the web, drawing the resident spider towards it. However, when the resident spider is large, and thus potentially a risk to *P. fimbriata*, it

produces faint intermittent vibrations that simulate an insect brushing the edge of the web. These are also ambiguous and keep the resident spider off the web so that it can be stalked and also reduce the level of predatory attack response from the resident spider. Finally, when *P. fimbriata* is walking on the web and towards its prey, it generates brief strong vibrations that simulate a large-scale disturbance of the web, such as wind or vegetation falling on the web, to mask its footsteps as it takes a lunge forward (a 'smokescreen') (Tarsitano *et al.* 2000). Spiders are not alone in having evolved the ability to invade other species' webs and lure the resident spider towards them for prey; as well as stalking, the assassin bug *Stenolemus bituberus* also has luring behaviour, using its forelegs to strum webs (Wignall and Taylor 2009, 2011).

Some species have evolved mechanisms to lure hosts to them and facilitate dispersal. The blister beetle, *Meloe franciscanus*, uses the solitary bee, *Habropoda pallida*, to transport it to the host nest. Larvae of the blister beetle hatch and form dark coloured masses of 120–2000 individuals that are attractive to male bees from a distance. Saul-Gershenz and Millar (2006) noted that male bees seem to approach the larvae aggregations from downwind, and hypothesized that the larval aggregations release compounds that mimic the sex pheromones of female bees to attract mates. They showed that male bees were attracted to larval aggregations but not towards unscented visual models of the aggregations. However, visual models with extracted chemicals of the larval aggregations attracted male bees at the same level as real larval aggregations (Figure 8.8). Analysis of chemical extracts of bees and larvae indicate a close correspondence between the chemicals of the female bees and the beetle larvae.

Male bees locate the source of the pheromone at close range by visual cues. The aggregation of larvae cooperate to mimic the size, colour, and perching location of female host bees (Hafernik and Saul-Gershenz 2000). Male bees then inspect the mass, and if contact is made the entire aggregation of larvae is transferred to the bee within seconds

(Figure 8.9). When the male bee next tries to copulate with a female, the larvae transfer to the female, who takes them to her nest where they feed and develop on the pollen, nectar, and bee egg in the nest (Saul-Gershenz and Millar 2006)!

8.6 Sensory Exploitation by Brood Parasites

8.6.1 Avian Brood Parasites

One of the most widely studied examples of exploitation involves brood parasites, such as many cuckoos (see also Chapter 9). These birds lay their eggs in the nests of other individuals (usually different species), so that the host parents rear the chicks to adulthood and the cuckoo avoids the cost of parental care. Cuckoo chicks are often bigger than those of the hosts, and so need more food. In addition,

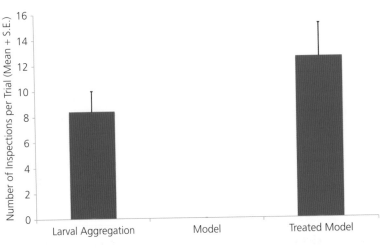

Figure 8.8 The number of inspections made by bees of real blister beetle aggregations, models of larval aggregations without chemicals, and models of aggregations treated with chemical extracts of the larvae. Data from Saul-Gershenz and Millar (2006).

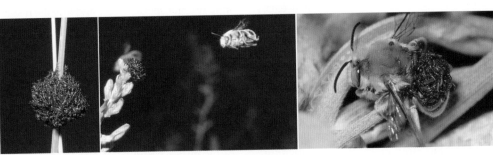

Figure 8.9 Images of a blister beetle, *Meloe franciscanus*, aggregation attracting a bee, and then moving onto the bee's abdomen. Images © Leslie Saul-Gershenz, published with permission from PNAS 2006, 103:38 14039–14044.

they are entirely unrelated to the hosts, and so should be completely selfish and extract as much care as possible. However, because many cuckoo chicks evict all host young from the nest, the host parents may lower their provisioning rate when faced with just one chick to rear (we expect hosts to reduce their provisioning rate to a lone chick to 'save' reproductive effort for future breeding opportunities). So how does the cuckoo induce the hosts to provide maximum care?

Davies *et al.* (1998) investigated food provisioning by reed warbler (*Acrocephalus scirpaceus*) hosts to common cuckoo (*Cuculus canorus*) chicks. They created three treatments: nests with 1) a single reed warbler chick, 2) a single blackbird or thrush chick, or 3) a single cuckoo chick. The reed warbler chick is much smaller (8 g) than the cuckoo or blackbird/thrush chicks (both about 26 g). Significantly more food was brought to the cuckoos than to the other two species, showing that large size alone is not enough for the cuckoo to maximize provisioning (Figure 8.10). Analysis of the call structure of the cuckoo shows that a single cuckoo chick vocalizes at a similar rate and amplitude to a whole brood of reed warblers. The cuckoo may therefore attract more care by mimicking a brood of host chicks. In a second experiment, Davies *et al.* (1998) placed a single blackbird/thrush chick in a warbler nest, and then played back begging calls of either a single cuckoo chick or an entire reed warbler brood through a speaker by the nest (or a control treatment with no playback). There was no difference in provisioning by hosts to nests with playbacks of either cuckoo chicks or a nest of warbler chicks. Thus, common cuckoo chicks seem to maximize provisioning by mimicking an entire brood of host chicks (Figure 8.10).

However, a follow-up study by Kilner *et al.* (1999) argues that common cuckoo chicks do not actually mimic the calls of the host brood *per se*. Instead, the cuckoo chick produces a highly exaggerated vocal display to make up for its deficient visual signal (gape). Reed warbler parents, when feeding their young, monitor the area of the brood gape displays in order to adjust feeding rate, as this correlates with brood size and the age of the chicks. In contrast, the vocal display of the brood allows parents to adjust their feeding to current hunger levels (Kilner *et al.*

1999). As such, both visual and vocal signals provide independent strategic information about required provisioning levels. As a cuckoo chick grows, the area of its gape becomes increasingly small compared to the combined gape areas of a whole brood of four host chicks. As such, a cuckoo chick needs an exaggerated call to compensate for its small gape area. Therefore, the cuckoo chick display need not be mimicry, but rather exploits features of the host begging signals that maximize provisioning. However, as Dawkins and Krebs (1979) discuss, the begging display of cuckoos in general may have a supernormal stimulus effect, exploiting the host's nervous system so that they bring more food. In such instances, both strategic and efficacy components may influence begging display form.

Some brood parasites use visual displays to induce parental care. The Japanese Horsfield's hawk cuckoo (*Cuculus fugax*) displays a bright yellow patch on the undersides of its wings when the host brings food to the nest (Figure 8.11). The wing patch looks like an extra gape and 'tricks' the hosts into bringing more food; sometimes the hosts will even try and place food into the wing patch instead of the mouth! Experiments by Tanaka and Ueda (2005) show that hosts (the red-flanked bluetail, *Tarsiger cyanurus*) bring less food to cuckoo chicks when the chick's wing patch is painted black, compared to un-manipulated controls or chicks with their wings painted with a transparent solution. The hawk cuckoo uses a visual signal because the hosts are ground nesting, and calling loudly would increase the risk of predation.

The common cuckoo discussed above has a bright red gape, which has often been suggested as a supernormal stimulus owing to its striking appearance. Manipulation of gape colours of host chicks to make them redder had no effect on parental provisioning, suggesting red coloration may not be a supernormal stimulus (Noble *et al.* 1999). However, these manipulations did not account for avian vision, and so may have been overly unnatural or unlike the cuckoo gape. In contrast, modelling of the hawk cuckoo wing patches with models of avian vision by Tanaka *et al.* (2011) indicates that the wing patches of the cuckoo are significantly brighter and reflect more ultraviolet light than do the host chick

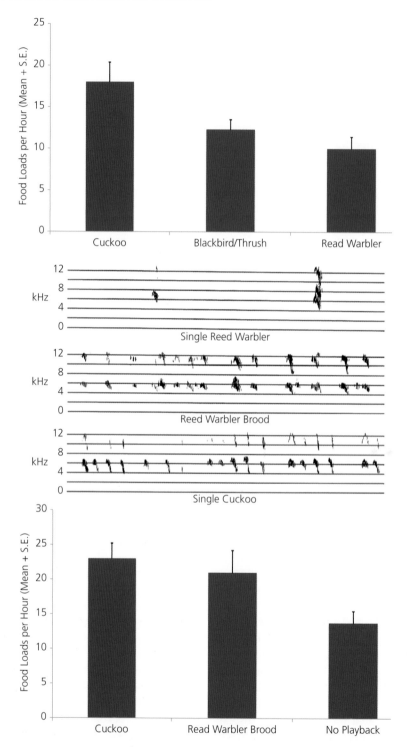

Figure 8.10 Top: The number of food loads brought per hour to single cuckoo, reed warbler, and blackbird/thrush chicks in a reed warbler nest. Middle: Spectrogram of the begging calls of a single reed warbler chick, a brood of four reed warblers, and a single cuckoo. Bottom: The number of food loads brought to single blackbird/thrush chicks when accompanied by playbacks of cuckoo or reed warbler brood begging calls, or silence. All data from Davies *et al*. (1998).

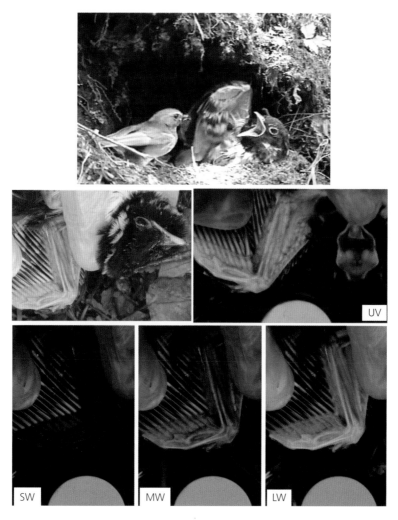

Figure 8.11 Top: A Horsfield's hawk cuckoo, *Cuculus fugax*, displaying its vivid orange wing patch to the host parent. Image of hawk cuckoo reproduced with permission from Keita Tanaka. The images below show the wing patch in human visible light, and then broken down into the information in ultraviolet (UV), shortwave (SW), mediumwave (MW), and longwave (LW) light, illustrating the intensity of the patch in UV reflectance.

mouth colours, and may thus be a supernormal stimulus to maximize parental care. The high ultraviolet component to the wing patch may be because the cuckoos utilize hosts in forested habitats at high altitude where there should be lots of scattered UV light in the environment.

8.6.2 Exploitation by Parasites of Social Insects

A range of butterfly and moth larvae infiltrate the nests of social insects and exploit their hosts for food and other resources (see also Chapter 9). Many species are known to mimic the cuticular hydrocarbons of the hosts (Dettner and Liepert 1994). The mountain Alcon blue butterfly (*Maculinea rebeli*) is 'adopted' and then fed by ant workers (Akino *et al.* 1999). While chemical mimicry explains the ability to infiltrate ant nests, Barbero *et al.* (2009b) note that ants also treat the butterfly larvae/pupae with an elevated status compared to their own young, including preferentially feeding the butterfly larvae. Queens in *Myrmica schencki* ant colonies produce sounds that induce workers to act in a benevolent way (attendance behaviour), reinforcing their queen status. Sounds produced by the butterfly larvae/pupae mimic those of the queen. The sounds cause more attendance to the butterfly larvae than towards workers, enabling the parasite to elevate its status and exploit the hosts to a high degree. Similar findings have been made in predatory *Maculinea* butterfly caterpillars and pupae (Barbero *et al.* 2009a), in parasitic wasps, whose larvae are visited and fed more frequently than host larvae (though how this manipulation in wasps is achieved is unknown; Cervo 2006), and in the death's head hawkmoth (*Acherontia atropos*), which invades the colonies of honeybees (*Apis mellifera*) and produces acoustic or vibratory signals (Moritz *et al.* 1991; Michelsen *et al.* 1986, cited in Moritz *et al.* 1991).

8.7 Future Directions

There is now good evidence showing that sensory exploitation and biases can be important in a range of systems in influencing behaviour. However, more work is needed to determine how widespread this is and where biases come from. Investigating where sensory exploitation fits in with other more 'traditional' explanations of mate choice also requires more work. The idea of supernormal stimuli has fallen from attention in recent times but there is enough evidence to suggest that it may be more widespread than currently presumed. It would be especially interesting to research how widespread supernormal stimuli are in exploitative systems, and if they are more likely to occur in some modalities than others. Finally, we need more empirical work to help distinguish between mimicry and sensory exploitation, and when the latter may lead to mimicry evolving. For example, it would be interesting to

know more about whether predators like spiders obtain prey with sensory exploitation, mimicry, or by being camouflaged, when each of these strategies arises, and the selection pressures involved.

8.8 Summary

Many signals in nature are dishonest, including various forms of mimicry. This involves organisms exploiting the communication systems of other species in order to deceive the receiver into mistaking an individual for something else. Likewise, many organisms exploit biases in the sensory systems of animals to manipulate the receiver's response. This can involve producing signals that are highly effective at stimulating aspects of the sensory system, or cause the receiver to respond inappropriately to an out of context signal. Sensory exploration and mimicry is widely found in mating signals and communication systems such as those of brood parasites, predators and prey, and pollination systems. It can lead to exaggerations in signal form and a wide range of complex interactions between organisms.

8.9 Further Reading

Readers should consult Schaefer and Ruxton (2009) and (2011) for discussion of perceptual biases, especially relating to plant-animal interactions, and Endler and Basolo (1998) for a general overview of many issues regarding sensory biases and exploitation. For various discussion of visual mimicry see Ruxton *et al.* (2004), and for a review of chemical mimicry see Dettner and Liepert (1994), and more recent discussions in later work.

PART 4

Diversification and Divergence

CHAPTER 9

Arms Races, Coevolution, and Diversification

Box 9.1 Key Terms and Definitions

Coevolution: A process involving pairs of species whereby changes in traits of individuals of one species causes reciprocal changes in the other species over evolutionary time. Diffuse coevolution can involve sets of species, whereby reciprocal changes still occur but specific pairs of species may be less tightly linked; species' traits may evolve in response to groups of other species.

Evolutionary Arms Race: The escalation of adaptations and counter-adaptations between two or more parties. Arms races may occur between individuals of the same species (e.g. rival males), or between species. Arms races are sometimes linked to coevolution, but need not involve reciprocal changes between two specific parties in response to specific adaptations by the other.

Polymorphism: Ford (1940) described genetic polymorphisms as the occurrence together in the same locality of two or more discontinuous forms of a species in such proportions that the rarest of them cannot be maintained merely by recurrent mutation. Ford's definition does not include continuous variation as he specifically states 'discontinuous'. However, while this strict definition is widely held, it can be useful to sometimes refer to 'polymorphisms' in species that show extreme variation in phenotype as long as these are regarded as phenotypic polymorphisms and not polymorphisms in the strict sense.

Red Queen Hypothesis: This is the idea that interactions between species can result in each party needing to consistently change specific trait values over evolution in order stay ahead or keep in touch with the other. Individuals in a species can be continuously changing over evolution but in relative terms stay in approximately the same place with regards to their fitness. The process can lead to oscillations or continuous changes in phenotype.

All organisms encounter other species during their lives, and the interactions between them are important factors in many species' evolution and ecology. Often, interactions occur as chance or intermittent encounters. However, relationships between species can also be much more intricate, involving regular and repeated interactions over evolution. For example, in Chapter 8 we discussed how organisms can develop sensory and signalling adaptations to exploit other specific individuals and species. Repeated interactions can help to explain the tremendous diversity in organisms that exists both within and across species. The aim of this chapter is to illustrate how two key concepts, evolutionary arms races and coevolution, can be a major driving force in generating both morphological and behavioural diversity associated with sensory information. For example, coevolution may occur between signals and sensory systems, as seems to be the case in bioluminescence producing fireflies and dragonfish (Chapter 4). Arms races and coevolution are substantial areas of research, and reviewing this field is beyond the scope of one chapter. Instead, we will focus on two subject areas (predators and prey, and brood parasitism) with a few specific areas in relative detail with respect to sensory systems and

Sensory Ecology, Behaviour, and Evolution. First Edition. Martin Stevens.
© Martin Stevens 2013. Published 2013 by Oxford University Press.

information, to illustrate important concepts and outcomes of these processes. In Chapter 2 we introduced the idea of signal detection theory, whereby a sensory system needs to detect and discriminate a stimulus of importance either from irrelevant background noise or from other stimuli. This is also highly relevant to both predators and prey and to brood parasites. For example, a predator may need to detect a camouflaged prey item against the background, or a host species may need to detect parasitic eggs against its own.

9.1 Arms Races and Coevolution

The processes of coevolution and arms races are often implicated as key mechanisms that can drive evolutionary change and diversification. The term coevolution was brought to common use in a paper by Ehrlich and Raven (1964) on plant–herbivore interactions. Although they did not define coevolution, we adopt the following definition:

Coevolution is a process of reciprocal adaptive change in two or more species over generations, caused by the selection pressures that each party imposes on the other.

Coevolution is not synonymous with the terms 'mutualism', 'symbiosis', and 'interactions' between species (although these can be coevolutionary), but must involve changes in both parties over evolution due to reciprocal selection pressure that each places on the other (Janzen 1980). Some of the best examples of coevolution associated with sensory ecology involve antagonistic coevolution between two or more species, where as each party improves its adaptations it reduces the fitness of individuals of the other species. A term sometimes also used is 'diffuse coevolution'. Here, there may not be a tight pairwise link between two species as in conventional coevolution, but rather interactions between a range of species occur that still lead to reciprocal changes, such as in groups of pollinators and flower colours.

Unlike coevolution, the idea of an evolutionary arms race is often less precisely defined. Generally, arms races involve the escalation of adaptation and counter-adaptation, either in two parties or across groups of organisms. The term 'escalation' is often used, yet it is not always defined as to what this means. In most cases, escalation will involve increased

investment in a specific line or type of attack or defence in each party. This may often also involve costs in terms of reduced performance in other functions. Arms races can be a feature of coevolution (e.g. 'coevolutionary arms race'; see below), but the two processes need not be the same or occur together. For example, arms races can occur within specific species, such as escalating adaptations in rival males or in sexual conflict between males and females. Furthermore, arms races can arise broadly between groups of species. For example, predators may develop sharper claws to attack a wide range of prey species, and prey species may develop thicker skins to fend off a suite of potential predators (note that distinguishing diffuse coevolution from this can be very difficult). In Dawkins and Krebs (1979) terminology, arms races can be either asymmetrical, involving defensive adaptations on one side and offensive adaptations on the other, or symmetrical, where each party becomes better at doing the same thing. This can be illustrated with an analogy to the Cold War. In symmetrical arms races, both parties may increase their stockpile of nuclear weapons. In asymmetrical arms races, one party may increase its nuclear weapons while the other improves its ability to detect an attack and take evasive action.

Coevolution may also be associated with another process: so-called Red Queen dynamics (van Valen 1973). Here, interactions between species can cause traits to change continuously as one party has to keep moving (over evolution) in order to keep one step ahead of its antagonist (Dieckmann *et al.* 1995; Marrow *et al.* 1992; van Valen 1973). Although two species may each undergo changes in their trait values (e.g. running speed), because both are changing, their relative fitness to one another may remain approximately the same.

Two different coevolutionary models are sometimes presented (Gandon *et al.* 2008; Gómez and Buckling 2011; Woolhouse *et al.* 2002). The first involves each species continuously gaining or improving adaptations without frequency-dependent selection occurring, termed 'arms race dynamics' (ARD). Here, accumulated adaptations and counter-adaptations in both populations occur, as in an arms race (Woolhouse *et al.* 2002). The alternative, 'fluctuating selection dynamics' (FSD) involves oscillations in the frequencies of traits in each party owing to frequency-dependent selection. This can involve Red Queen dynamics, with species changing over time but potentially staying in about the same place. These two models can lead to different outcomes over time (Gandon and Day 2009), although in practice many systems will contain a mixture of both dynamics (Gandon *et al.* 2008). In addition, the predictions of both models over short periods may be similar because both predict increasing patterns (escalations) of adaptation in the short term. However, over long time periods ARD predicts monotonic increases in adaptation, whereas FSD predicts fluctuations in frequencies and patterns of adaptation (Gandon *et al.* 2008), potentially with substantial diversification within species. While FSD could continue indefinitely (Gómez and Buckling 2011), under ARD coevolution we expect escalations in adaptations over time to eventually be countered by other selection pressures or physical limits.

How might such coevolutionary interactions end? A range of possibilities exist (see Dawkins and Krebs 1979; Dieckmann *et al.* 1995), including:

- *Extinction*—one side may drive the other extinct.
- A fixed *stable equilibrium* may be reached with unchanging phenotypes. Or
- A theoretically endless *limit cycle of temporally oscillating phenotypes* under Red Queen dynamics.

Such outcomes may result in stable polymorphisms, fluctuations (chaotic or cyclical) in trait frequencies over time (dynamic polymorphisms), or the fixation of traits (Woolhouse *et al.* 2002).

9.2 Predators and Prey

Interactions between predators and prey are often used as one of the main examples of arms races between species (Dawkins and Krebs 1979). Generally, most discussion centres on cycles in population structure over time, or general adaptations such as running faster to escape (see Abrams 2000). However, less work has focused on sensory aspects. Arms races between predators and prey have been implicated in driving the evolution of animal variation that exists today. The great expansion in diversity that occurred in the Cambrian radiation of life, where all major groups of animals existing today arose 543 million years ago, has been suggested to

be strongly linked and even driven by the evolution of sensory systems. This time period may have involved a substantial increase in the amount of information available and the ability of animals to acquire and process it (Plotnick *et al.* 2010). It has even been suggested that the evolution of complex eyes themselves was the driving force behind the Cambrian explosion (Parker 2003), leading to more intense predator–prey interactions with predators better able to locate prey, and prey needing better defences (e.g. armour). Although this hypothesis does not seem to be entirely consistent with the fossil record and other evidence (see Plotnick *et al.* 2010), it is a compelling argument that the development and refinement of sensory systems was a key driver of the diversity of animal life that has arisen. However, as we shall see, although prey animals have developed a staggering array of defensive strategies, evidence that there have been reciprocal evolutionary changes in predators' sensory systems in response to these is limited. At the end of this section we will discuss some potential examples, and why there seems to be a lack of response over evolution in predators; i.e. why predator–prey interactions involving sensory systems are good examples of arms races, but probably not true coevolution.

9.2.1 Defensive Coloration

Given that most animals face a significant risk of predation during their lives, it is no surprise that many species have evolved defences to reduce the likelihood of being eaten. In this section, we focus on anti-predator coloration to illustrate how diverse and widespread prey adaptations to reduce the risk of predation are, before discussing the issue of counter-adaptations in predators. The range of anti-predator defences is diverse, and they operate at various stages of interaction between predator and prey (see Box 9.2). Broadly, we can divide protective coloration into primary defences, which operate before a predator attacks, and secondary defences, which operate during or after a predator has attacked (Edmunds 1974). Primary defences include camouflage, warning signals, and mimicry, and reduce the likelihood that a direct interaction will occur between predator and prey. Secondary defences include tactics like startle displays and noxious secretions and reduce the likelihood that a prey animal will be captured, injured, or killed once an attack has occurred. Below, the main types of anti-predator coloration are outlined with some examples. This is not intended to be comprehensive by any means, owing to the large size of the subject area; indeed, several books have been entirely devoted to this subject alone (e.g. Ruxton *et al.* 2004)!

9.2.1.1 Camouflage

Camouflage is probably the most widespread form of protective coloration in nature, and works by reducing the likelihood that an object will be detected or recognized (Stevens and Merilaita 2009b). Several camouflage types are found in animals (Figure 9.1), and here we mainly cover recent studies.

Box 9.2 Defensive Coloration and Morphology

In this box we describe the main types of anti-predator defences involving coloration or morphological adaptations. Many of the definitions with respect to various types of camouflage are modified from Stevens and Merilaita (2011a).

Stage 1: Prevent Detection (Crypsis)

At this stage, the predator searches for prey in the environment, and the defences act to reduce the likelihood that the prey animal will be detected. All defences here are types of camouflage.

- *Background Matching:* This occurs when the object's appearance generally matches the colour, lightness, and pattern of the background.
- *Disruptive Coloration:* This involves the use of markings that create the appearance of false edges and boundaries and hinder the detection (or recognition) of an object's, or part of an object's, true outline and shape.
- *Obliterative Shading:* Many animals are countershaded and have a darker surface that receives higher light incidence. Countershading may lead to the obliteration of three-dimensional form.
- *Self-shadow Concealment:* Directional light, which would lead to the creation of shadows, is cancelled out by countershading. Some animals also have bioluminescent photophores that produce light that resembles downwelling light from above and cancels out their silhouette when viewed from below.
- *Silvering*: Here, an animal's body is highly reflective (like a mirror) making it difficult to detect when light incidence is non-directional (such as due to strong scattering by water-borne particles).
- *Transparency:* Some animals have part of their body transparent, reducing the likelihood that they will be detected.

Stage 2: Prevent Attack

Often a predator detects an object in the environment, but will not attack it. This may be because the prey warns the predator that an attack would be unprofitable, or because the predator fails to recognize it (i.e. it misclassifies it as something else).

- *Aposematism:* Many animals are dangerous, unpalatable, or toxic, and convey this by having bright conspicuous warning colours. Predators avoid such prey either due to a learnt avoidance during their life or an innate avoidance gained over evolution.
- *Batesian Mimicry:* In this instance, a harmless prey animal has an appearance that resembles an unprofitable model.
- *Decorations:* These are items from the environment that an animal adds to its body in order to avoid being recognized. They may also work to prevent detection.
- *Masquerade:* This is a type of camouflage that prevents recognition by resembling an uninteresting or unimportant object, such as a leaf or a stick.
- *Müllerian Mimicry:* Over evolution, many aposematic animals have adopted similar appearances. This is advantageous because they share the cost of educating naïve predators about their defences.

Stage 3: Halt or Inhibit an Attack from being Successful

Once a predator has initiated an attack, there are still tactics that prey use in order to halt or make the attack unsuccessful.

- *Confusion:* Some animal markings may inhibit the ability of the predator to pick out a prey item from the rest of the group. This may include silvering, where the bright reflectance of the fish body inhibits target segregation, or dazzle markings whereby the interacting effect of prey movement and markings makes capture difficult.
- *Deflection:* Some animals have conspicuous markings that are thought to deflect the attack of a predator to non-vital body regions.
- *Motion Dazzle:* These are markings that make estimates of speed and trajectory of moving prey difficult for the predator, preventing successful capture.

- **_Startle Displays:_** These are sudden conspicuous changes in appearance that cause the predator to pause its attack, allowing the prey to escape or re-conceal itself.

Stage 4: Prevent Subjugation and Consumption

Once an attack has been successful, some prey animals may have final adaptations that may prevent the predator from consuming them.

- **_Physical Defences:_** Some animals have strong spines, quills, and tough integuments that make it hard for the predator to ingest them, or increase the likelihood that they will survive an attack with non-lethal damage.
- **_Chemical Defences:_** Animals may secrete toxins or bad tasting chemicals when attacked, causing the predator to release them.

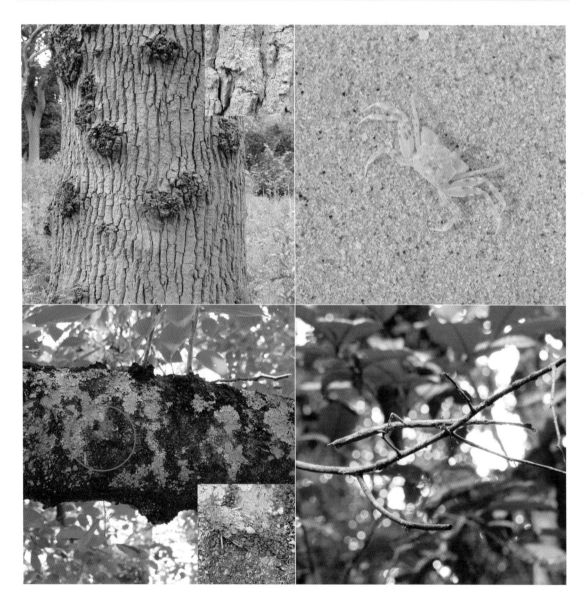

Figure 9.1 Examples of different types of camouflage. Top-left shows a moth matching the general appearance of tree bark, and top-right shows background matching to sand substrate in a juvenile horned ghost crab (*Ocypode ceratophthalma*). Bottom-left shows a cicada with potentially disruptive coloration breaking up its body shape against the tree and lichen colours, and bottom-right is an example of masquerade in a stick insect.

The most fundamental form of camouflage involves background matching. One of the earliest experiments to show a survival advantage of background matching was by Pietrewicz and Kamil (1977). Here, blue jays (*Cyanocitta cristata*) searched for camouflaged moths in photographic slides and found moths more difficult to detect when placed upon the appropriate background (i.e. a birch tree moth against a birch background), and in the natural resting orientation. In many vertebrates, background matching is also important. Work with rock dragon lizards (*Ctenophorus decresii* and *C. vadnappa*) in Australia by Stuart-Fox *et al.* (2003) has shown that different variants of the lizards vary in their level of conspicuousness to avian predators. Predation trials with plaster models resembling the lizards show that those individuals that match the background better are less likely to be attacked.

Some animals can also rapidly change their colour patterns, most notably chameleons and cuttlefish, and these are model systems to investigate camouflage expression (Stuart-Fox and Moussalli 2009). For example, cuttlefish (especially the European cuttlefish, *Sepia officinalis*) can rapidly adjust their body patterns using chromatorphoes: pigment sacs with radial muscles attached around the periphery that are directly controlled by motoneurones (Hanlon 2007). Work by Hanlon and colleagues has investi-

gated the visual features of the background that lead to different types of dynamic camouflage markings being expressed (see review by Hanlon *et al.* 2009; Figure 9.2). This shows that the substrate scale and contrast are key features in evoking different types of cuttlefish pattern responses (Chiao *et al.* 2007, 2009a), and that the colour patterns adopted show an effective match to the background in terms of the vision of fish predators (Chiao *et al.* 2011). The truly impressive abilities of cuttlefish to match the background patterning is matched by their capacity to change the three-dimensional properties of their skin using papillae, and the orientation of their body postures during camouflage, seemingly also driven by visual rather than tactile cues (Allen *et al.* 2009; Barbosa *et al.* 2012). Most cuttlefish work has used a range of artificial backgrounds to test when cuttlefish express different pattern types. However, Zylinski *et al.* (2011) used image analysis methods to show that the camouflage appearance of the Australian giant cuttlefish (*Sepia apama*) is a very good match to the natural backgrounds on which it is found.

Some colour-changing animals can also adjust their camouflage to different types of predators with different visual systems. Stuart-Fox *et al.* (2008) showed that dwarf chameleons (*Bradypodion taeniabronchum*) adjust their coloration when faced with either a bird (with tetrachromatic colour vision) or a

Figure 9.2 A great deal of work has been done to investigate the remarkable camouflage abilities of cuttlefish and the features of the environment that cause individuals to adopt different camouflage patterns. The top row shows field images of the giant Australian cuttlefish (*Sepia apama*). The bottom shows how the European cuttlefish (*Sepia officinalis*) can adopt camouflage with patterns ranging from relatively uniform through small mottled patterns to larger potentially disruptive markings. Images reproduced with permission from Roger Hanlon.

snake (a trichromatic) predator. The authors presented chameleons with either model snakes or model birds, and then analysed the chameleons' level of camouflage in terms of the visual systems of the two predator groups. Chameleons showed better camouflage when faced with birds, presumably because birds have more effective colour vision than snakes and so the camouflage needs to be better. Individuals can, therefore, identify the type of threat (snake or bird) and adjust their camouflage accordingly. Similarly, some ground squirrels also specifically direct infrared displays to deter snake predators, which unlike many other predators can see infrared (Rundus *et al.* 2007). In species where individuals cannot change their coloration, we would predict that in species that primarily face a threat from one type of predator that their camouflage would have evolved to be tuned to the visual system of this group.

A key drawback of background matching is that it still leaves the outline and edges of the body intact and potentially easy to detect. Therefore, some animals have evolved disruptive coloration, involving markings that break up the body shape and salient features like eyes and limbs (Stevens and Merilaita 2009a). Recently, Cuthill *et al.* (2005) created artificial 'moth' prey that resembled the colour and pattern of tree bark (to a predatory bird's vision) and pinned these to trees in woodland and monitored

'survival' over time. They found that targets with disruptive makings at the body edge survived significantly better than those simply matching the background pattern. Various subsequent experiments in the field, in aviaries, and with human 'predators' have added further support to these results (e.g. Cuthill and Székely 2009; Fraser *et al.* 2007; Merilaita and Lind 2005; Schaefer and Stobbe 2006; Stevens *et al.* 2006). Models of spatial vision indicate that disruptive coloration seems to work by creating high contrast markings at the body edge to destroy the coherence of the body outline, while creating 'false edges' away from the true object boundaries, so that the predator fails to detect the prey's shape (Osorio and Srinivasan 1991; Stevens and Cuthill 2006). Despite strong theoretical evidence in artificial systems, disruptive coloration has rarely been tested in real species, even though it is thought to occur in a wide range of animals.

As with edge detection and camouflage, animals may also seem to employ mechanisms to prevent successful estimates of motion. For example, in capturing fast moving prey the predator must correctly judge the speed and direction of the target in order to strike at the right time. Some characteristic patterns, such as stripes and zig-zags (similar to those found in snakes, zebras, fish, and insects; Figure 9.3) could interfere with motion detection and make it difficult for an observer to judge the speed and

Figure 9.3 Many animals have banding and striped patterns. One theory is that the patterns make it difficult for a predator to judge the speed and direction of moving prey, hindering capture. In zebra (*Equus burchelli*, pictured), the markings may also make it difficult for a predator to pick out an individual from the herd and to match the different body parts to specific individuals when the zebra are moving in the herd.

direction of a moving target (Scott-Samuel *et al.* 2011; Stevens *et al.* 2008b, 2011). However, how such patterns fool motion detectors is not yet well understood.

9.2.1.2 Aposematism

Aposematic animals advertise unpatability, toxicity, or general unprofitability with bright contrasting signals or structural adaptations to prevent both initial attacks by predators and promote learnt avoidance (Figure 9.4). Aposematism is found widely in invertebrates, but also in vertebrates like poison frogs, snakes, and even some mammals (Kingdon *et al.* 2012). Much work has shown that predators can have both innate and learnt avoidance of common warning colours, such as yellows, reds, and striped patterns (see review by Ruxton *et al.* 2004). These colours may be common in aposematism because they are very different from the green and brown colours found in many environments and so stand out strongly, and because simple pattern arrangements may be easier for predators to remember (Stevens and Ruxton 2012). They may also facilitate efficient detection by a potential predator before an attack occurs, preventing recognition errors (Guilford 1986). It is widely believed that warning colours should be conspicuous to be effective. However, there need not be a direct benefit to conspicuousness *per se*; an interesting idea is that conspicuousness may evolve in aposematism merely because it is very different from cryptic, undefended species, which cannot afford to be conspicuous because they lack defences. Sherratt and Beatty (2003) have demonstrated in experiments with humans that

defended prey may be avoided when they are cryptic if they are different in appearance from undefended prey (for example, with different patterns). Furthermore, experiments indicate that the characteristic zig-zag markings on snakes are distinctive enough to promote avoidance by predators without being especially conspicuous (Niskanen and Mappes 2005; Valkonen *et al.* 2011). In terms of visual processing, mechanisms like contrast enhancement, edge detection, and receptive fields (see Chapter 3) can have important implications for the evolution of communication signals, including warning coloration. For example, warning signals should be readily detectable and therefore often use high contrast and sharply defined boundaries to exploit features of visual processing (Stevens 2007; Stevens and Ruxton 2012). In general, more work is needed to understand why warning signals look the way they do and why they often share common features (markings and colours).

9.2.1.3 Startle Displays

A range of harmless animals have evolved startle displays, involving sudden conspicuous changes in appearance to scare away or confuse an attacking predator, allowing the prey to escape. Startle displays are relatively widespread in nature, from the bright hindwings of many moths to the displays of cuttlefish, frogs, and birds (Figure 9.5). They vary greatly in appearance, with different colours, patterns, and contrasts. Aviary experiments with blue jays and artificial stimuli show that jays hesitate for longer and show more startle responses when faced with novel, rarely encountered, and conspicuous

Figure 9.4 Examples of warning coloured lepidopteran larvae showing bright colours and conspicuous structures associated with warning signals.

Figure 9.5 Examples of startle displays. The moth (*Catocala nupta*) on the top-left exposes its hindwings when attacked (top-right). In other species (e.g. *Catocala fraxini*), the hindwing colour differs (bottom-right). Some startle displays involve eyespots, such as in the eyed hawkmoth (*Smerinthus ocellata*; bottom-left).

colours (Ingalls 1993; Vaughan 1983). Some startle displays are associated with conspicuous circular 'eyespots'. Vallin and colleagues (2005) have shown that eyespot displays effectively startle predators. They presented live peacock butterflies (*Inachis io*) to birds in aviaries with either their eyespots removed by drawing over them with a marker pen, or with the eyespots intact (including controls with other regions of the wings painted over). They found that of 13 out of 20 butterflies with their eyespots removed were killed, compared to only one out of 34 butterflies with their eyespots intact. Bioluminescence may even be used by some organisms in startle-like displays, or to distract the predator or deflect its attack elsewhere. Here, animals such as crustaceans, squid, jellyfish, and fish, release bioluminescent chemicals into the water producing clouds of light (see reviews by Haddock *et al.* 2010; Widder 2010). The bright flash of light may also act as a smokescreen for the animal to escape, or even attract secondary predators that eat the threatened animal's own predators (Haddock *et al.* 2010)! In general, startle displays should be used sparingly to prevent habituation by the predator since the prey is harmless.

9.2.2 Predator Counter-Adaptations

Prey clearly have a great variety in defensive strategies, but do predators respond with counter-adaptations? There are some examples, but usually not directly related to sensory systems. For example, some birds have undergone changes in bill shape and size to overcome pinecone structures that have arisen to reduce seed predation (see Benkman *et al.* 2003). In addition, snakes can evolve resistance to the toxins in their prey (Hanifin *et al.* 2008), or use

behaviours that manipulate prey escape responses to direct prey towards rather than away from the predator attack (Catania 2009). However, there are few clear examples of coevolution between predators and prey involving sensory systems, and in general, it is hard to find clear cases of sensory counter-adaptations in predators to capture prey with defences (Abrams 2000). There are several general reasons why this may be the case. The first relates to Dawkin's 'life-dinner principle' (Dawkins and Krebs 1979). Here, we expect prey to be ahead of predators in an arms race because if the prey loses an interaction it dies, whereas the predator only loses a meal. Second, prey will often have shorter generation times, potentially allowing them to evolve adaptations more quickly. Third, in many cases, predators may eat a range of prey types, for which they need various skills to capture them (see Vermeij 1994).

Another possible reason for why we rarely seem to find counter-adaptations in sensory systems to overcome prey defences is that sensory systems are generally used in many tasks, not just in finding prey, but also for communication with mates, detecting predators themselves, and navigation. Therefore, modifying a sensory/perceptual system to better find or discriminate between specific prey types may compromise other functions. In Chapter 4, we discussed how sensory systems face costs and trade-offs, and that there are only a few clear examples of sensory systems being specifically tuned for one task alone (let alone being modified to overcome a specific prey defence). However, sensory modalities other than vision may allow adaptations to better find prey that do not compromise other functions. For example, some animals are chemically camouflaged against the background (e.g. Silveira *et al.* 2012). In olfactory systems it may be possible for predators to have sensory receptors and neural processing that are dedicated to specific odours, such as to detect or discriminate chemically camouflaged prey (as happens in some species with regards to detecting potential mates; see Chapter 3). This may be a promising area for future work. In general, however, we first need to discover more about what different types of anti-predator defences exist outside of the visual sense, and how they work in order to understand if and how predators have overcome this.

Despite these issues, below we discuss three potential examples of predator perceptual counter-adaptations to prey defences involving vision.

9.2.2.1 Polarization Vision and Prey Transparency

Some animals, mainly from aquatic environments, are partially or largely transparent, and this may act as camouflage from predators. There has been a lack of experimental testing of how effective transparency is as a camouflage strategy (but see Zylinski and Johnsen 2011), with only a handful of mostly indirect studies of the phenomenon. However, transparency seems to have evolved multiple times and its occurrence is strongly influenced by the environment, which is consistent with its being a successful camouflage strategy (reviewed by Johnsen 2001; Ruxton *et al.* 2004). Under water, most light is polarized in a horizontal direction, and transparent prey bodies may either change the plane of polarization or depolarize the light. Some predators are sensitive to orientations of polarized light (see Chapter 2), and therefore may be able to break the camouflage of transparent prey, although little work has been done to test this idea and there is a lack of support for the hypothesis (Johnsen *et al.* 2011; Marshall and Cronin 2011). Ultraviolet sensitivity in some predators may also have a function in transparency breaking, but again there is a lack of experimental work (Johnsen 2001; Ruxton *et al.* 2004) and it is unlikely to have arisen in species specifically for this function.

9.2.2.2 Dichromatic Versus Trichromatic Colour Vision in Primates

A range of platyrrhine primate species have polymorphic colour vision (see Chapter 4). Here, the presence of mediumwave-sensitive (MWS) and longwave-sensitive (LWS) pigments are controlled by multiple alleles at a single locus opsin gene on the X chromosome. This means that heterozygous females are trichromatic (LWS and MWS plus the autosomally controlled SWS opsin gene), whereas homozygous females and all males are dichromatic (either LWS or MWS plus SWS). Polymorphic colour vision was believed to be maintained via heterozygote advantage to trichromats. However, work by Melin and colleagues (2007, 2010) has shown that dichromatic capuchin monkeys (*Cebus capucinus*) seem better able to locate colour-camouflaged insect prey, especially under low light, than trichromats. Similar such findings have been recently found in other species (e.g. Caine *et al.* 2010). This is probably because colour signals can mask achromatic and textural differences between stimuli whereas improved luminance vision may be better at camouflage breaking. Accordingly, Saito *et al.* (2005) found that dichromatic primates were better at detecting colour-camouflaged stimuli than trichromatic primates. Likewise, dichromatic tamarins (*Saguinus* spp.) find greater proportions of more effectively camouflaged prey than do trichromats (Smith *et al.* 2012). This indicates that the type of colour vision present can be driven partly by the need to locate certain types of prey. Although it is unclear whether dichromatic vision is a specific counter-adaptation of predators to break the camouflage patterns of their prey, or rather a general adaption to living at low light levels, this example shows that features of sensory systems may be at least partly maintained under selection to defeat the defensive strategies of their prey. The advantage of trichromats, in contrast, may stem from being able to detect ripe fruit against green leaves, or even be maintained by mating (see Chapter 4).

9.2.2.3 Predator Search Images, Prey Camouflage, and Polymorphism

Lukas Tinbergen (1960) proposed that on encountering relatively common prey, predators formed search images; a perceptual change in the ability of predators to detect prey. This would enable them to find prey of the same appearance on future encounters more efficiently, especially if they concentrate on one prey type alone (Lawrence and Allen 1983). A knock-on effect is that the relative abundance of prey can affect predator selection, so that predators concentrate on prey that is relatively abundant and overlook prey that is scarce; prey choice would therefore be frequency dependent (Allen 1988). When one morph of a species is more common than another, predators may prey more upon the common morph relative to its frequency in the population, compared to that of the rarer morph; this is apostatic or negative-frequency selection (Clarke 1962). This means that generally, if two different morphs are equally cryptic, the rarer morph will have an advantage. Thus, polymorphism could arise owing to predator visual search (see review by Bond 2007).

While various studies have proposed to demonstrate the presence of search image formation in predators in increasing their capture of certain prey types, various alternative explanations exist, including that predators learn to visit a particular location where food is found, or that predators may increase in the rate of searching for food (Dawkins 1971; Guilford and Dawkins 1987). Alternatively, a different counter-adaption to crypsis could involve searching the environment more slowly, and allocating more time and attention to each patch searched.

Despite criticisms of the search image concept, evidence has recently been found that cannot easily be explained by other ideas, and under controlled laboratory experiments predators show attentional biases and limitations consistent with search image formation (Bond 2007). The most convincing evidence for the importance search images, apostatic selection, and prey polymorphism comes from a series of impressive experiments by Bond and Kamil. In an initial experiment, Bond and Kamil (1998) used blue jays foraging for three distinct computer-generated moth morphs over a series of generations. At each generation, moths that had been detected were removed from the population, which was then reformed based on the surviving individuals. Over 50 generations, the population oscillated with respect to the frequencies of the different morph types, consistent with apostatic selection and search-image formation. However, in these experiments, the morphs were fixed and could not evolve. Bond and Kamil (2002) then conducted a study with camouflaged prey that could evolve via a genetic algorithm. Prey items that were not detected by the jays were more likely to reproduce, and it was atypical cryptic morphs that were less frequently detected, leading to frequency-dependent selection over generations and the evolution

and maintenance of polymorphism and high levels of phenotypic variability (Bond and Kamil 2002). However, evidence of apostatic selection on crypsis and predator search image in natural systems is still lacking, and many instances of polymorphism in crypsis may stem from species occupying different backgrounds with different appearances under disruptive selection (Nosil and Crespi 2006; Pellissier *et al.* 2011; Rosenblum 2006; Vignieri *et al.* 2010; see Chapter 11), rather than apostatic selection and predator search image. A key question for the future is to what extent predator search images exist in nature and drive prey polymorphisms.

9.2.3 Bat–Insect Arms Races

Another model area to study predator–prey interactions is between bats and insects in the acoustic sense. Many species of bat use echolocation to navigate and capture prey, including adaptations in the sensory system and brain (Chapters 2, 3). This ability is often so sophisticated that bats can not only precisely pinpoint the location of an object in space and its motion, but also a range of other features including size, surface texture, and wingbeat frequency. Consequently, bats represent a major selection pressure on the many organisms that they prey upon. This includes both vertebrates and invertebrates, but especially insects, on which we will focus here.

Unsurprisingly, given the threat that bats present to many organisms, species have often evolved defences to fight back (Miller and Surlykke 2001). Despite its accuracy, echolocation offers prey a lifeline. If they can detect the echolocation calls, then they can take evasive action before the bat arrives. As such, many insects have evolved hearing organs (tympanal organs; Chapter 2) either specifically under selection pressure to detect bat calls, or modified from existing organs to improve sensitivity to frequencies used in echolocation (Greenfield 2002; Hoy and Robert 1996; Miller and Surlykke 2001; Yack *et al.* 2007). Recordings from the nerve cells that innervate these hearing organs show that the ears are generally sensitive to ultrasonic frequencies that correspond to bat calls (e.g. Yack *et al.* 2007). The tympanal organs found are diverse in terms of their physical structure and location on the body, resulting from multiple independent evolutionary origins (Greenfield 2002; Hoy and Robert 1996; Miller and

Surlykke 2001). In line with hearing organs having evolved under selection pressure to detect bats, some prey populations that are or have become isolated from bat predators (for example, by living or moving into bat free habitats/locations, or by moving from nocturnal to diurnal lifestyle), have reduced or lost their hearing sensitivity to bat calls (e.g. Fullard 1994; Surlykke and Skals 1998). A range of insects have been shown to have behavioural responses (often referred to as 'acoustic startle' but these are really escape behaviours rather than a startle display aimed at the bat; see below), including turning away from the source of the sound, decreasing wingbeat frequencies, and dropping to the ground (e.g. Roeder 1967; Yack *et al.* 2007; Yager and May 1990; Yager *et al.* 2000; see review in Hoy *et al.* 1989). Some insects, which are found by bats that eavesdrop on the mating calls of the prey (Chapter 7), also stop singing when they detect bat calls (e.g. ter Hofstede *et al.* 2010). As an aside, some fish are sensitive to ultrasound and show negative phonotactic behaviour when presented with ultrasonic calls, and maybe even escape manoeuvres (Jones 2005). The implication is that these features may have evolved as a result of predation pressure from echolocating dolphins (see Chapter 2).

Many insects have also evolved an ability to produce sound, including a range of arctiid moths, which have tymbal organs on their metathorax. Here, special muscles buckle and distort the tymbal membrane, generating clicks of different amplitudes and pulses, often in the ultrasonic frequency range (Miller and Surlykke 2001). This opens up a number of potential anti-bat defences that insects could employ (Figure 9.6). First, moths may utilize acoustic startle displays, comprising sudden bursts of sound that startle or scare a bat, causing the bat to pause and allowing the moth to escape. Second, moths that are toxic or unpalatable may signal to the bats that they should be avoided: acoustic aposematism. Third, moths may produce clicks that interfere with the echolocation calls and processing by bats, termed 'sonar jamming'.

9.2.3.1 Startle Displays

There is some evidence that moths can startle bats, although few purpose-designed experiments exist. Bates and Fenton (1990) trained big brown bats (*Epte-*

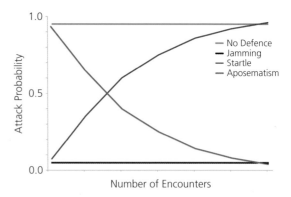

Figure 9.6 The relationship between attack probability by a bat over successive encounters of prey types with different defences, modified from Hristov and Conner (2005). With no defence (red line), bats attack at a constant high rate. When prey show a startle display (blue line), bats are initially startled and avoid the prey but habituate over time and begin to attack at high levels. With aposematism (green line), naïve bats attack the prey in initial encounters but then avoid moths as they learn to associate the display with toxicity (the reverse relationship to startle displays). In sonar jamming (black line), bats consistently have a low capture success of prey.

sicus fuscus) to fly to a platform where they were rewarded with a mealworm. They found that bats did appear to be startled by sudden clicking moth sounds, but that they rapidly habituated. Miller (1991) also found that naïve bats could be startled by moth clicks occurring in the terminal stage of an attack, but to varying degrees. More work into this potential defence and the response of bats would be valuable.

9.2.3.2 Acoustic Aposematism

Early work showed that bats could associate moth clicks with unpalatability (e.g. Bates and Fenton 1990; Surlykke and Miller 1985), but the clearest evidence for acoustic aposematism yet comes from a study by Hristov and Conner (2005). They used four species of arctiid moths that could either signal to bats (S+) with clicks or not (S−), and were either chemically defended and unpalatable (C+) or not (C−). They then presented naïve big brown bats with moths (and additional palatable controls) over a learning period of seven days and analysed, using high-speed infrared video recordings, the percentage of moths of each type that bats captured per day (Figure 9.7). They predicted that bats would only learn to avoid prey when they had both the chemical defence and the acoustic signal. As expected, the

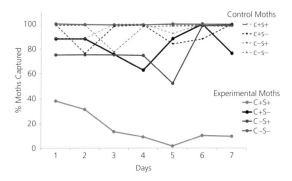

Figure 9.7 Results of Hristov and Conner (2005) showing that bats would only learn to avoid moths that had both the defensive chemicals and the acoustic signal (red line). In contrast, bats did not learn to avoid moths that lacked either the chemical or acoustic component or control moths. Data from Hristov and Conner (2005).

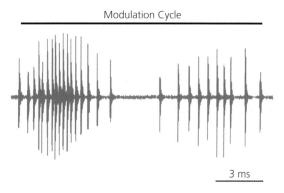

Figure 9.8 The clicks produced by the tiger moth *Bertholdia trigona* over time towards bats in sonar jamming. Data from Corcoran *et al.* (2009).

likelihood of bats attacking aposematic prey declined with successive encounters, but this did not occur when moths lacked either the signal or the toxin.

9.2.3.3 Sonar Jamming

The idea that some insects may produce sound that interferes with the echolocation calls of bats and prevents them from being located has been around for some time. Sonar jamming may work by 1) simulating multiple targets and creating *phantom echoes* of objects that do not exist (Fullard *et al.* 1994); 2) *interference* of range assessments, affecting how the bat's neural system encodes the timing of real echoes when defensive clicks overlap with or occur just prior to real echoes, diminishing the bats ability to assess range accurately; or 3) *masking* the echoes of a prey animal with abundant and intense clicks, thus hiding the prey animal entirely (Corcoran *et al.* 2011; Miller 1991; Tougaard *et al.* 1998). Each of these possibilities afford different predictions, with phantom echoes causing bats to detect multiple targets, interference causing imprecise target location estimates, and masking creating the impression of no target at all (Corcoran *et al.* 2011), although these and other possibilities are not mutually exclusive (Corcoran *et al.* 2009).

Initial work into sonar jamming, usually involving playing click calls to bats during hunting and testing the effect on range estimation, found little support for the theory (e.g. Surlykke and Miller 1985). However, Miller (1991) found that bursts of

clicks played in a 1.5 ms time window before echoes could severely interfere with range estimations, although this may be interpreted as the bats confusing clicks with echoes when the two are presented close together (Fullard *et al.* 1994). In line with the sonar jamming hypothesis, some moths also present clicks to bats during the terminal phase of the bat's attack, when jamming would be most effective (Fullard *et al.* 1994). In contrast, warning signals should be given in advance to prevent attacks from occurring at all. Later, Tougaard *et al.* (1998) showed that the clicks of moths could interfere with the neural responses of bats' auditory pathway for processing time delays between pulse and echo (in the lateral lemniscus), including a lack of neural response when clicks were presented close in time to a test stimulus. These cells are specialized for encoding echo timing, and therefore distance.

The clearest experimental evidence so far for sonar jamming has been found by Corcoran *et al.* (2009). They presented bats with the palatable tiger moth *Bertholdia trigona*, which produces rapid clicks by flexing and contracting its tymbal structure when a bat approaches (Figure 9.8).

In contrast to aposematism, where prey capture should decline with time (due to aversion learning), and startle displays, where capture should increase with encounters (habituation), sonar jamming would predict a constant low success rate even over repeated interactions. In accordance with this, Corcoran *et al.* (2009) found that big brown bats presented with moths in flight rooms successfully attacked far fewer of the clicking *B. trigona* moths than non-clicking control species over seven nights, and their success against *B. trigona* did not improve with time (Figure 9.9). Bats also seemed to try and adjust their echolocation calls in response to interference. Furthermore, when *B. trigona* moths were muted so that they could no longer click, this resulted in bats easily capturing them.

The clicks generated by *B. trigona* are different from those species that have warning signals in having a much higher duty cycle (sound production per unit time). A comparative analysis of tiger moth sounds from species in North and South America (Corcoran *et al.* 2010) indicates that sound producing species cluster into two groups with species either having few clicks and a low-duty cycle (probably used in aposematism), or a high-duty cycle and many clicks (possibly for sonar jamming). Comparisons of high-duty-cycle moth clicks with bat calls indicates that the more likely explanation for how they work is that they interfere with or mask the returning echoes to the bat, rather than creating phantom echoes. Further experimental work by Corcoran *et al.* (2011) found no evidence in support of phantom echoes (no capture behaviour observed towards erroneous targets). Furthermore, bats were able to track clicking prey items (*B. trigona*) in relatively close proximity and made echolocation calls consistent with attack phases, which is inconsistent with masking (where no target is detected). How-

ever, bats frequently missed the prey by distances that were consistent with the range interference hypothesis (*ca.* 15 cm), indicating that sonar jamming in this species of moth may work by preventing accurate assessments of target distance. Finally, note that sonar jamming is perhaps an acoustic analogue to dazzle coloration, both of which prevent effective localization and capture.

9.2.3.4 Bat Counter-Adaptations

It is clear that bats have remarkable sophistication in echolocation for prey capture, and that insects have evolved a range of adaptations to reduce the risk of bat attack. But have bats responded in turn? In order to counteract insect hearing and evasive behaviour, bats could shift the tuning of their echolocation calls away from the sensitivity of the insect hearing organs, or reduce the intensity or duration of their calls. However, although this may sometimes occur, evidence that shifts in call attributes are an adaptation to moth defences is mixed, and call modifications may represent adaptations for efficient prey capture instead (Miller and Surlykke 2001). However, recent work shows at least one bat species has developed 'stealth' echolocation. Goerlitz *et al.* (2010) found that the barbastelle bat (*Barbastella barbastellus*) emits calls that are 10–100 times lower in amplitude than other bats with similar hunting strategies, and that the calls should be hard to detect by moths until the bat is close to them. In addition, the barbastelle predominantly eats moths with ears, and is a specialist on prey that other bats may find hard to capture. However, this is not an ideal solution for the bats because it reduces prey-detection distances, adding support to the theory that stealth echolocation is a counter-adaptation specifically to evade prey defences rather than to facilitate capture efficiency. In addition, it does not seem a general adaptation to habitat or feeding mode because barbastelle's do not surface glean or forage in dense vegetation, where low amplitude calls may be favoured (Goerlitz *et al.* 2010).

9.3 Brood Parasites

Given that rearing young incurs considerable expenditure in terms of time and resources, it is not surprising that many species have become cheats:

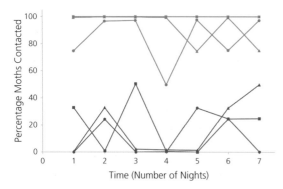

Figure 9.9 Results of Corcoran *et al.* (2009) showing capture success (contacts made) by individual bats over successive encounters over a period of nights towards either control moths (red lines) and *Bertholdia trigona* moths that produce sonar jamming clicks towards bats when attacked. Data from Corcoran *et al.* (2009).

they are brood parasites, exploiting others to rear their young. Brood parasitism is most studied, and seemingly most prevalent among birds and social insects, and here we focus on those as model systems to study coevolution in the context of sensory information.

9.3.1 Avian Brood Parasitism

Avian brood parasitism occurs in about 1% of birds, and has long been one of the most important systems to study antagonistic coevolution (Davies 2000; Rothstein 1990), and the subject has been well reviewed recently (Davies 2011; Kilner and Langmore 2011; Krüger 2007). Such interactions can be powerful generators of diversity, both within a species and also in terms of species richness. For example, parasitic cuckoos undergo higher rates of speciation (and extinction) than cuckoos showing parental care (Kruger *et al.* 2009). In Chapter 8 we learnt that brood parasites use a number of vocal and visual tricks to extract maximum provisioning from hosts. These adaptations are not the same as defences and counter-adaptations that evolve during coevolution that we discuss here but rather adaptations by the parasite once it has beaten the host.

9.3.1.1 Egg Laying and Nest Defence

A host's first line of defence is not to be parasitized in the first place. To this end, many hosts mob cuckoos near the nest. For example, reed warblers (*Acrocephalus scirpaceus*) mob common cuckoos (*Cuculus canorus*) with acoustic and visual displays and direct physical attack (Davies and Welbergen 2009; Welbergen and Davies 2009). Wallace (1889) noted that many parasitic cuckoos look similar to hawks, especially in their shape, direct flying style, and their barred underparts. Recent work shows that this mimicry of hawks by cuckoos reduces mobbing by hosts, with reed warblers less likely to approach and mob cuckoos with barred underparts than those without barring (Davies and Welbergen 2008; Welbergen and Davies 2011).

9.3.1.2 Rejection Behaviour and Egg Morphology

Once a parasite has succeeded in laying an egg in a host nest the most obvious host defence is egg rejection: throwing the foreign egg from the nest. In turn,

this has led in many systems to parasites evolving eggs that mimic the host's own eggs in appearance, in order to evade detection and rejection. Many brood parasites comprise several different host-races, or gentes, with each female cuckoo specializing on a particular host species. Brooke and Davies (1988) used museum collections to show that British common cuckoo host races differ in egg appearance (such as for the size and density of spotting, and egg darkness). They also performed egg rejection experiments, placing model eggs painted to look like different host eggs in the nests of different species, and showed that most hosts are more likely to reject less mimetic eggs. From their experiments, Davies and Brooke (1989a, 1989b) suggested that hosts learn the appearance of their own eggs, and then reject eggs that differ from these, rather than simply rejecting the odd one out.

These classic experiments showed clearly that egg rejection behaviour by hosts has led to the mimicry seen in cuckoo eggs across host species. However, much of this work was done from a human perspective, and humans and birds have substantial differences in their vision (see Chapters 2 and 4). Recent analyses of museum specimens using models of bird visual processing have shown that parasitic eggs can be a striking match in colour and luminance (lightness) to host eggs (Avilés 2008). In addition, host species that show greater egg rejection behaviour (more likely to reject non-mimetic eggs from their nest) have cuckoo host races with stronger mimicry for colour and pattern (Stoddard and Stevens 2010, 2011; Figure 9.10).

Avian vision models also accurately predict egg rejection behaviour in experiments (e.g. Avilés *et al.* 2010; Cassey *et al.* 2008; Spottiswoode and Stevens 2010, 2011; Yang *et al.* 2010). Analyses of the egg coloration and markings of both the common cuckoo (Stoddard and Stevens 2010), and the African cuckoo finch (*Anomalospiza imberbis*; Spottiswoode and Stevens 2010) and their hosts, show that egg appearance is composed of multiple independent attributes, such as colour, marking size, dispersion, and coverage. The fact that these attributes are uncorrelated may increase the amount of information that can be encoded about egg identity, as is predicted if egg markings act as signatures of identity (Davies 2011). In addition, recent egg rejection experiments with

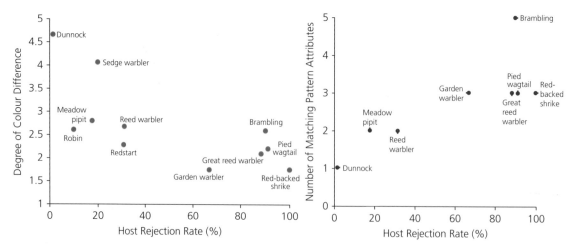

Figure 9.10 Analyses of museum eggs of common cuckoos and their various host species show that colour mimicry is better (difference in colour in terms of bird vision is lower) for cuckoos that have hosts with higher rejection behaviour (left). Likewise, the number of egg pattern components (dispersion, marking size, marking contrast, proportion of egg coverage, and marking diversity) mimicked by cuckoos is higher in host races with hosts showing stronger rejection. Data from Stoddard and Stevens (2010, 2011).

the cuckoo finch and its most common host, the tawny-flanked prinia (*Prinia subflava*) in Zambia have shown that hosts use multiple different aspects of egg appearance to discriminate foreign eggs from their own (Spottiswoode and Stevens 2010). Furthermore, the attributes of egg appearance used are also the cues that differ most between host and parasitic eggs, showing that hosts base their rejection decisions on those attributes that are most informative (convey most information) about egg identity.

Not all brood parasites have evolved mimetic eggs. In general, a lack of egg rejection behaviour (and consequently a lack of egg mimicry) is often attributed to evolutionary lag, where new hosts have not had enough evolutionary time to evolve defences (for example, in dunnocks, *Prunella modularis*; Brooke and Davies 1988). However, in some cases the absence of rejection behaviour may be the best strategy in hosts, and reflect a stable equilibrium (Dawkins and Krebs 1979) between the costs and benefits of rejecting (Davies 2011; Krüger 2007). For instance, if there is a relatively high chance that parasitic eggs do not hatch, the costs of re-nesting later in the season may outweigh the risk of accepting foreign eggs (Krüger 2011). In other situations, where host young are reared alongside the parasites (the parasites do not evict host eggs), parasites may 'punish' hosts for rejecting their eggs, so that acceptance is a better strategy than rejecting (e.g. mafia'

tactics; see Hoover and Robinson 2007; Soler *et al.* 1995). Furthermore, it may be better to accept parasitic eggs when multiple instances of parasitism per nest are common. This may be favoured if greater numbers of foreign eggs dilute the likelihood of host eggs being damaged by parasites in subsequent parasitism events (Gloag *et al.* 2012). Therefore, although counterintuitive at first, accepting parasitism can sometimes be the best strategy for hosts. Alternatively, some hosts may not have encountered relevant cues to reject eggs. For example, some hosts of Australian bronze-cuckoos (*Chalcites* spp.) accept dark olive-brown cuckoo eggs that are clearly not mimetic, and show no rejection behaviour. Work by Langmore *et al.* (2009) suggests that the cuckoo eggs may be dark in order to be effectively camouflaged against the lining of the dim host nest.

9.3.1.3 Escalation of Host Defensive Strategies at the Egg Stage

It is clear that many brood parasites have evolved egg mimicry as an adaptation to evade host defences. But once mimicry has arisen, how can hosts fight back? Hosts have at least two strategies for improved defence. First, they may improve their ability to detect a foreign egg and refine their rejection behaviour; this is likely to lead to improved mimicry by parasites, as seems to be the case in the common cuckoo. Second, hosts may evolve new egg appearances over time, in order to evolve away from or 'escape' the parasite, which we would predict would follow over evolution, as hypothesized by Swynnerton (1918). Theoretical modelling indicates that interactions between parasites and hosts can lead to continuous changes in egg appearance in both parties, and even lead to high levels of within species variation 'polymorphism' (Dieckmann *et al.* 1995; Liang *et al.* 2012; Mougi and Iwasa 2010; Takasu 2003, 2005), and cyclical changes over time (Dawkins and Krebs 1979). Here, in a very similar way to the example of polymorphisms in prey camouflage and predator search images, high levels of intraspecific egg polymorphism may arise under negative frequency dependent (apostatic) selection, where rare egg colours are at an advantage in hosts because there are no mimetic parasites that yet exist, making egg rejection easier. In general, intraspecific egg polymorphism makes it difficult for the parasite to match any given host egg appearance, and some hosts of the common cuckoo in China have, along with egg rejection behaviour, evolved different egg types (white and blue eggs) as an anti-parasite defence (Yang *et al.* 2010). Egg rejection experiments and modelling of bird vision have shown that two current hosts of the African cuckoo finch seem to have adopted different strategies to improve their defence (Spottiswoode and Stevens 2011; Figure 9.11). The tawny-flanked prinia has evolved extreme levels of egg polymorphism making effective mimicry difficult (one female always lays the same egg type, but individuals differ greatly), yet in absolute terms individuals need large differences in appearance between the foreign egg and their own to reject successfully. In contrast, another common host, the red-faced cisticola (*Cisticola erythrops*) has relatively low levels of egg polymorphism, but needs only very small differences in egg appearance to reject foreign eggs. It has developed refined rejection behaviour. Simulation modelling indicates that the two strategies have approximately the same payoffs in terms of rejection of mimetic parasite eggs. Thus, coevolution can take different routes even in closely related species with the same parasite. Modelling also suggests that hosts may be better able to fend off parasites when coevolution involves multiple traits ('multidimensional trait space'; Gilman *et al.* 2012), as is the case in the prinia. It would therefore be valuable to consider the relative benefit of the

two strategies above over longer periods of host–parasite interaction.

A key prediction in host–parasite interactions is the expectation for hosts to evolve new phenotypes to escape the parasite, which should track behind, analogous to the idea of the Red Queen hypothesis (van Valen 1973). In the case of the cuckoo finch and tawny-flanked prinia Spottiswoode and Stevens (2012) used a collection of eggs from the current time period and from predominantly 30 years ago from exactly the same location in Zambia to compare current eggs with those from the past. Despite the short timescale (in evolutionary terms), analyses of the eggs in terms of bird pattern and colour vision shows dramatic changes in appearance in both host and parasite. In terms of colour, eggs have greatly increased in diversity, and previously rare egg colours (olive host eggs and blue parasite eggs, to human eyes) are now common. Pattern has also changed in both parties. Furthermore, comparisons

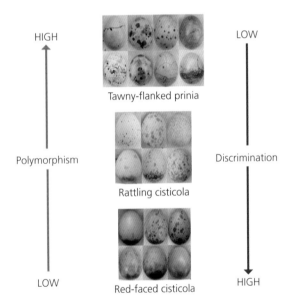

Figure 9.11 Hosts of the African cuckoo finch seem to have adopted different strategies to fight back against the parasite. Tawny-flanked prinia have extreme levels of egg variation across females but need large differences in appearance before they reject foreign eggs. In contrast, red-faced cisticola have low levels of intraspecific variation but have highly refined rejection behaviour, needing only small differences in appearance between their own and foreign eggs to reject. Another species, the rattling cisticola (*Cisticola chiniana*), which seems to be a former host, is intermediate for both rejection ability and egg variation. Images © Claire Spottiswoode and Martin Stevens.

of matching across time periods indicates that parasites are closely tracking host evolution, because they more closely match hosts from the same time period than across time periods, as has also been found in analyses of infectivity of *Daphnia* by its parasites in pond sediments (Decaestecker *et al.* 2007). In the cuckoo finch and prinia (and potentially other polymorphic species), the system probably began escalating in arms-race style coevolution (ARDs), with increasing parasite mimicry and host rejection. At some point, the system seems to have perhaps moved into a fluctuating selection dynamics (FSDs) model, with cycles in egg phenotype over time. It is quite possible that many brood parasitic systems may progress in this way, given sufficient evolutionary time. Overall, the system is also in accordance with theoretical models (Mougi and Iwasa 2010; Takasu 2003, 2005; Yoder and Nuismer 2010) suggesting that coevolution between host and parasite can generate extreme phenotypic diversity over short timescales. In accordance with the above findings that brood parasitism can drive increased phenotypic diversity, in populations of host species that are no longer targeted by brood parasites, there has been a reduction in egg diversity (Lahti 2005). Work into the genetics and inheritance of eggshell coloration in the above species would be valuable to better understand the basis of egg diversity and the rate of change in egg appearance.

9.3.1.4 Chick Rejection and Mimicry

In their early work, Davies and Brooke (1989a) found no evidence that host species of the common cuckoo discriminated against chicks of another species. However, recently, new studies show that chick rejection does occur, such as in some Australian host species. Langmore *et al.* (2003) found that superb fairy-wrens (*Malurus cyaneus*) parasitized by Horsfield's bronze cuckoos (*Chalcites basalis*) abandon nests with lone chicks. This simple rule of thumb makes sense because a lone chick has a high chance of being parasitic. Recently, Sato *et al.* (2010) found that the large-billed gerygone (*Gerygone magnirostris*) a host of the little bronze cuckoo (*Chalcites minutillus*) will actually throw (reject) cuckoo nestlings from the nest, also supported in later work in the mangrove gerygone (*G. laevigaster*) (Tokue and Ueda 2010). Chick rejection can also lead to mim-

icry. Langmore *et al.* (2011) showed that the skin colour, gape colour, and presence of downy feathers in several Australian cuckoo species matches that of their respective hosts (Figure 9.12).

So why has chick rejection and mimicry apparently only evolved in a few species? A recent theoretical perspective by Britton *et al.* (2007) indicates that the answer may partly lie in 'strategy blocking'. Here, a successful defence by the hosts at one stage reduces the need for subsequent later defences because the parasite is unlikely to get beyond this first line of defence. Therefore, proficiency in egg rejection by many hosts reduces the requirement for chick rejection. In many of the Australian hosts, egg rejection does not occur, possibly because parasitic eggs are cryptic and evade detection (Langmore *et al.* 2009) or because they are highly mimetic and have therefore 'won' at the egg stage (Langmore and Kil-

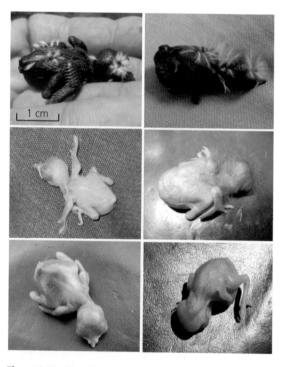

Figure 9.12 Skin coloration of bronze-cuckoo and host chicks. Top: little bronze-cuckoo (*Chalcites minutillus*) and large-billed gerygone (*Gerygone magnirostris*); Middle: shining bronze-cuckoo (*Chalcites lucidus*) and yellow-rumped thornbill (*Acanthiza chrysorrhoa*); Bottom: Horsfield's bronze-cuckoo (*Chalcites basalis*) and superb fairy-wren (*Malurus cyaneus*). Images reproduced with permission from Naomi Langmore.

ner 2009). Thus the arms race has escalated to the chick stage. We would also predict that selection on host defences should be strongest at the stages of parasite egg laying (e.g. mobbing) and mimicry, because successful defence here keeps the host's reproduction intact, whereas rejection at the chick stage does not (Dawkins and Krebs 1979; Langmore and Kilner 2010). In general, as Davies (2011) points out, three non-mutually exclusive explanations for why we see varied defences in different host species are 1) different defences are more appropriate for species with different life histories, 2) strategy blocking, and 3) the stage of coevolution.

As we have seen above, coevolution between hosts and parasites can have many outcomes. So, how might it end? One possibility is ongoing cyclical variations under frequency-dependent selection, or another is a stable equilibrium outcome. Alternatively, one side could become extinct. Here, the host may develop such good defences that the parasite is no longer able to beat them, or the parasite may become so prevalent that it drives the host population to extinction. The latter is less likely to occur because we would expect that when parasites become very common they would struggle to find enough hosts, and a classical predator–prey cycle of increasing and decreasing host and parasite numbers may arise. When the host wins, the parasite may switch to new host species. This may explain why potential host species exist that are not at present parasitized, yet show strong rejection behaviour (e.g. European blackbirds, *Turdus merila*), and why so many different parasite host races have arisen in different systems.

9.3.2 Brood/Social Parasitism in Insects

Many social insect colonies are invaded by brood parasites (often called social parasites) that prey upon the colony members, steal or solicit food, or even brood from the host nest, which are used subsequently to rear the parasites' own young. Parasites include other social insect species, as well as other insects such as beetles and butterfly larvae, of which many have evolved mimicry of the chemical cues their hosts use in recognition (Dettner and Liepert 1994). In these systems there are many analogues to avian brood parasitism, but in chemical and acoustic

sensory modalities instead of visual ones (see reviews by Brandt *et al.* 2005a; Cervo 2006; Kilner and Langmore 2011). Broadly speaking, two types of brood parasitism exist in social insects: slavemaker species that periodically conduct slave raids to capture brood from other colonies of ants, and inquilines that invade and live inside host colonies alongside the hosts (Brandt *et al.* 2005a; Foitzik *et al.* 2001).

9.3.2.1 Slavemaker Ants

Slavemaker ants steal the brood of other colonies (often of different species) and transport it back to their own nest (Hölldobler and Wilson 1994). The initial colony is set up when the slavemaker queen invades the nest of another species, kills or expels the resident queen and often the current workers, and then takes over the brood. The host brood then imprint on the new queen's odour and rear her young (Brandt and Foitzik 2004; Brandt *et al.* 2005a). The true workers of the slavemaking ants have just one task: to raid host colonies and steal their brood. During raids, they kill any opposing host workers with their formidable mandibles. In addition, many species also use chemical strategies, including exploitation of host nestmate recognition cues through chemical mimicry of cuticular hydrocarbons, or alarm pheromones that affect host behaviour (reviewed by Akino 2008). Some species have a greatly enlarged Dufour's gland, from which they spray chemicals onto the opposing host ants. These 'propaganda' substances can act as exaggerated alarm signals, causing workers from the host nest to flee and disperse (Regnier and Wilson 1971), or they can cause host workers to actually turn on and attack each other instead (Allies *et al.* 1986)! Alternatively, some species may use 'appeasement' substances to reduce aggression towards the invading ants (Akino 2008). The captured worker ants are reared by the slavemakers and then carry out many of the duties of the colony, including rearing the slavemakers' own brood. Some slavemaking species may utilize several slave species in different tasks, akin to the division of labour produced by different castes in non-slavemaking species (Hölldobler and Wilson 1994).

It is clear that slavemaking ants have adaptations to facilitate brood capture and slavery, but some hosts have evolved defences. A study by Foitzik *et al.* (2001; see also Brandt and Foitzik 2004) provides evidence for coevolution. They collected colonies of *Protomognathus americanus* and its host *Temnothorax longispinosus* from different regions of North America, and induced slave raids between host and parasite colonies from the same region and from different regions. By doing so, they showed that in regions where abundance and frequency of slavemaking raids is higher, *T. longispinosus* colonies were more successful in defending against the raids, including by being more likely to guard the nest entrance and showing greater aggression towards parasite scouts. In turn, *P. americanus* from the same region were more effective at raiding, including placing an individual near the host nest to help slave raiding ants to leave. Consequently, more host workers and host queens were killed during raids, and the raiders captured more brood.

However, adaptations and defence have not stopped there. Ants are well known for colonies sharing a common chemical signature, with individuals that differ from this being rapidly rejected or killed. Some slavemaker ants have chemical profiles that resemble their slaves (Kaib *et al.* 1993), and in general it should be beneficial for an invading queen when starting a new colony to mimic the profile of the host in order to reduce aggression towards her. *Temnothorax* hosts are capable of discriminating between invading slaves and free-living conspecifics. This defence, in terms of difference between chemical profiles to elicit aggression, is plastic, being stronger in regions and seasons when the risk of parasitism is higher (Brandt *et al.* 2005b). Likewise, recent encounters with slavemaker (e.g. scout) individuals can heighten defences in host colonies for several days (Pamminger *et al.* 2011). As with avian brood parasites, it was also widely believed that selection against slavemaker ants by hosts in social insects should occur only during parasite colony formation and during raids (analogous to egg laying). This was thought because as the host brood imprint on their new colony there is no selection pressure for them to detect that they have been enslaved; because the enslaved workers do not reproduce, there would be no fitness benefits to rebelling. Selection should be on preventing parasitism in the first place. However, Achenbach and Foitzik (2009) suggested that enslaved workers may

benefit from killing the slavemaker ant brood, thus reducing colony growth and the impact of slave raids, if the population structure was such that neighbouring host colonies were likely to be related to the enslaved workers. By killing the slavemaker brood, the enslaved ants could then increase their inclusive fitness by benefiting other host colonies. Achenbach and Foitzik (2009) found that enslaved *Temnothorax* workers would often remove *P. americanus* parasite queen and worker pupae from the nest chamber, and then either neglect and leave them to die, or directly kill them. Such behaviour and mortality was not found in normal host colonies. Experiments transferring brood between nests also show that host workers will kill pupae of different species, and the enslaved hosts were more likely to accept parasitic pupae than unenslaved host workers. In addition, parasites from the same locality have chemical profiles more like their hosts than from other locations (Achenbach *et al.* 2010).

9.3.2.2 *Formica* Ants and Cuckoo Bees

Slavemaker ants are by no means the only example of coevolution in social insect brood parasites. Martin *et al.* (2011) studied the chemical signature diversity and nestmate recognition behaviour in the ant *Formica fusca*, where colonies suffer high incidence of brood parasitism from other ants from various *Formica* species. Here, a queen takes over a colony and uses the host workers to rear her own workers, which gradually take over the nest. They found that Finnish ants, where parasitism is common, had excellent recognition abilities, compared to the UK population, where parasitism is largely absent. In addition, the UK ants had also lost key recognition compounds from their cuticular hydrocarbons. Thus, it seems that parasitism pressure has driven diversification in chemical recognition systems in social insects, similar to visual signatures in in avian brood parasite hosts. In addition, parasitized *Formica* species can also discriminate between the eggs of their own colony and that of parasitic eggs, and reject the foreign ones (Chernenko *et al.* 2011).

There are remarkable parallels to birds in other species too. For example, female cuckoo bees lay their eggs in the brood cells of other bee species, after which the cuckoo bee larvae destroy the host offspring and feed on the pollen supplies. Like many avian brood parasites, *Sphecodes* cuckoo bees are often secretive before parasitizing nests, waiting nearby for the host female to depart and only entering the nest when she is gone (Bogusch *et al.* 2006). In addition, they often possess strong stings, thick cuticle to resist attack, and sharp powerful mandibles (Martin *et al.* 2010). Remarkably, two species, *S. ephippius* and *S. monilicornis*, both exploit a range of host species, but a given female specializes predominantly on one host species, akin to the host races in avian systems (Bogusch *et al.* 2006). Of the various host species exploited, a range of defensive levels are shown, from no defence even when encountering the parasite in the nest, through to severe fights that can result in death (Bogusch *et al.* 2006). Other cuckoo bees *Psithyrus* (*Bombus*), parasitize bumblebees (*Bombus*) and have host specific mimicry of chemical signatures (Martin *et al.* 2010).

9.4 Future Directions

We need a greater understanding of if, when, and how predatory sensory systems change over evolution in response to prey defences. We would also benefit from a better understanding of anti-predator defences and predator counter-adaptations outside of the visual sense and bat-moth arms races. A particularly interesting recent study by Bura *et al.* (2010) shows the walnut sphinx caterpillars (*Amorpha juglandis*) produce ultrasonic sound by expelling air ('whistling') through enlarged spiracles on their eighth abdominal segment. This causes the birds to hesitate or abort the attack, indicating that they probably function as an acoustic startle display. What other types of anti-predator defence exist across other modalities needs further study. With regards to brood/social parasitism, a better understanding of when and why hosts tolerate parasites is important, and also how parasites exploit hosts in multiple modalities and the importance of these. In general, in antagonistic systems we could do with a better understanding of end points. For example, theoretical models indicate that stable equilibrium endpoints or small oscillations are more likely to arise when the prey or host defences are effective compared with the predator/parasite adaptations. In contrast, large-oscillations should occur when parasites or predators have effective adaptations compared to those of their host/prey (Mougi and Iwasa 2010). In many

systems, predators may be relatively generalist with a range of prey types and therefore lack specific adaptations to each prey type. In contrast, parasites are often host specific, and consequently have highly specialist adaptations to evade host defences. Thus, we may predict oscillations are more likely to occur in parasitic systems, and equilibrium endpoints in predators and prey (Mougi and Iwasa 2010). Empirical tests of these ideas would be valuable, albeit hard to come by owing to the difficulties of obtaining long-term data.

9.5 Summary

Interactions between species or groups of organisms, both cooperative and antagonistic can be powerful generators of biological diversity. Two key driving forces in this are arms races and coevolution. Predator–prey relationships provide clear examples of arms races, with predators having a range of general adaptations to effectively capture prey, which in turn have evolved varied defences. However, in many instances there is little evidence for genuine coevolutionary responses in the sensory systems of the predators. In contrast, coevolution seems widespread and diverse in brood and social parasites in birds and insects, and this has led to extraordinary defences and counter adaptations in both parasite and host in a range of modalities.

9.6 Further Reading

For an in-depth coverage of anti-predator defences see Ruxton *et al.* (2004). The sections on camouflage are now somewhat out of date and so readers could consult chapters in Stevens and Merilaita (2011b). Ruxton (2009) also reviews the evidence for camouflage working in non-visual senses. Schaefer and Ruxton (2011) cover various aspects of coevolution and the senses in their book on plant–animal communication, with many systems and examples not covered here. Readers are also encouraged to consult the gradually increasing number of studies that suggest a coevolutionary relationship between sensory systems and signals outside of predator–prey and parasitic systems, including occurring in mate selection and pollination (see Chapter 4).

CHAPTER 10

Adapting to the Environment

Box 10.1 Key Terms and Definitions

Atmospheric Absorption: During transmission, the loss of (e.g. acoustic) energy into other forms of energy.

Divergence (Spreading) Loss: The loss of information due to the divergence of energy with a wave-front from the source increasing with distance, resulting in a loss of amplitude.

Environmental Noise: The occurrence of abiotic-induced noise (e.g. wind) or biotic noise (e.g. other species calling) that can interfere with or mask information that would otherwise be available to an animal.

Masking: Occurs when the threshold for detection of a signal is increased due to the presence of interference, such as noise.

Refraction: The change in the direction or 'bending' of a wave due to a change in speed, often occurring when a wave passes from one medium to another with different properties.

Scattering: The process whereby particles or radiation such as wavelengths of light deviate from their original path due to interactions with the medium through which they are passing, such as by hitting other particles.

Sensory Plasticity: During development, individuals may invest more in senses that are of greater value based on prevailing environmental conditions, and less in senses that are likely to be less beneficial.

Organisms are found in a myriad of habitats, which can vary greatly in space and time even at a relatively fine scale. The environment plays a substantial role in the way that information is available, gathered, and used, including detecting and responding to different cues, the sensory modality that is most effective, and whether specific sensory modalities and communication signals are tuned to features of the environment. The aim of this chapter is not to go into detail about the physics of how different mediums affect signal/cue transmission. Instead, it is to focus from an ecological and evolutionary perspective on when and how organisms respond to different environmental features, and how different environmental characteristics lead to changes in the modality that is used and how it is tuned. This could occur in several ways. First, in

communication, animals could respond with changes in behaviour, such as choosing specific backgrounds to signal from/against, or selection could lead to changes in signal form. Second, sensory systems may vary between animals found in different environments. This may arise due to developmental plasticity, or over evolutionary time. In practice, there can be complex interactions between plasticity and longer-term changes in signals and sensory systems (see Fuller *et al.* 2010), and we should remember that sensory systems and signals may change in concert. Overall, we often expect selection to act on both signal generation and sensory systems to match to the environmental conditions (Endler 1992, 1993). The influence of the environment can also have profound effects in macroevolution. Many concepts such as sensory

Sensory Ecology, Behaviour, and Evolution. First Edition. Martin Stevens.
© Martin Stevens 2013. Published 2013 by Oxford University Press.

drive, divergence, and speciation follow the ideas presented here and are explicitly discussed in Chapter 11.

10.1 Signal Transmission and the Environment

Understanding the physics of how different mediums affect the transmission and extraction of information can be difficult enough, and so the task of understanding how complex biological habitats influence this is substantial! The aim in this section is to introduce some important ideas of information transmission, but otherwise key concepts will arise throughout the chapter with respect to specific examples. Both Dusenbery (1992) and Bradbury and Vehrencamp (2011) present in-depth discussions of the physics involved with transmitting information through different media. Readers are encouraged to consult those books in order to get a fuller understanding of the mechanics of information transmission in different modalities.

Here, we first use sound as an example to discuss important concepts relating to transmission. As sound travels through a medium, attenuation occurs as a reduction in the intensity of a sound as it travels from its source. This is due to geometrical spreading (also termed divergence loss, spreading loss, or spherical spreading), absorption (where sound energy is converted into another source of energy), and scattering. As sound passes through an environment it spreads out in space, meaning that the energy involved is spread over a wider area and to more molecules (divergence loss), and so the energy per unit area decreases. Although under some conditions animal sounds do attenuate according to divergence loss, in most natural environments a number of other factors are important in exacerbating attenuation (Brenowitz 1986). Sound energy is also lost through molecules colliding, resulting in loss of energy to frictional heat and absorption of the energy by the molecules themselves. This absorption depends upon the properties of the medium and the frequency of the sound. For example, the fraction of energy lost to the medium is greater for higher frequencies because the molecules are being forced to propagate the sound more quickly. Thus, higher frequencies tend to lose energy faster and travel

shorter distances. Sound also undergoes a range of scattering through a medium, via reflection, refraction, or diffraction from natural objects or environmental disturbance (e.g. wind), and this too can be frequency-dependent (in natural environments, high frequencies attenuate more than low frequencies due to scattering) (Brenowitz 1986). At the interface between two media of different properties, some sound will be directly reflected back to its source when the media have different acoustic impedances (the level of resistance presented to propagating a wave, such as a sound pressure wave). If the surface of an interface is textured, then sound will tend to be scattered in other directions, rather than reflected in the same path. Refraction arises when a change in speed occurs in the propagating wave, such as at the boundary between media, causing it to bend, and this can be influenced by temperature gradients and wind (Dusenbery 1992). The reflection of sound can also result in reverberation, whereby sounds generated cause echoes to come back from surrounding environmental features/objects, interfering with the temporal structure of sound signals.

Similar factors arise in light transmission. As light passes through the environment it is subject to absorption and scattering. As with sound, scattering can be frequency dependent. For example, small particles in the atmosphere tend to scatter shorter wavelengths of light more quickly than longer wavelengths, making the sky look blue on a cloudless day. Water properties also affect light transmission in a frequency-dependent manner. For example, in clear water the water molecules selectively attenuate longer frequencies and remove longer ('red') wavelengths more as you go deeper in the water column. Light can also undergo refraction, diffraction, and reflection.

In water, electric signals do not travel far because they are mostly limited by significant attenuation due to geometrical spreading, but unlike light and sound, electric signals do not undergo frequency dependent attenuation and there are few environmental constraints on temporal structure of electric signals (Brenowitz 1986). This means that electric fish can utilize a wide range of dominant frequencies and temporal structures. However, water conductivity can influence electric discharge transmission (see below).

While acoustic, seismic, and visual signals propagate through the environment in a wave-like manner, chemical dispersion occurs via bulk flow and molecular diffusion, with the latter mainly occurring over small spatial scales. As such, chemical information is lost through turbulent mixing (e.g. wind, water movements), diffusion, absorption, photolysis, and chemical transformation by other organisms (Atema 1995). Wind or water currents may lead to plumes of chemical information, and chemical stimuli often exist in patches. Generally speaking, animals that use chemical information to orientate and to find objects act relatively slowly, often travelling up chemical gradients that can have very low concentrations.

Given the above considerations, it is not surprising that different (micro)environments influence signal transmission and contrast (see Chapter 3) with the background. We now turn to some specific examples in real species. In vibratory communication, the substrate clearly has a major influence on how signals of different forms are transmitted. For instance, Bell (1980) produced artificial vibrations that resembled those of the tree cricket *Oecanthus nigricornis* from Canada through species of plant on which the species does and does not normally mate. Bell found that the plants that the crickets normally communicate on transmitted vibrations with less attenuation than plant species on which they are not normally found. Future work might explore further how different aspects of plant morphology drive changes in signal form. Another nice example of how the environment can affect seismic signal transmission is in the jumping spider *Habronattus dossenus*. This is found in the Arizona desert, USA, and males attract females with seismic signals: vibratory signals produced by tapping their legs and palps against the substrate. They tend to be found on three substrate types: sand/soil, leaves, and rocks (Figure 10.1).

Elias *et al.* (2004) used a technique called laser vibrometry to measure how these different substrate types attenuate different frequencies of vibration (between 10 and 2500 Hz). They found that rock substantially attenuated all frequencies, sand predominantly filtered out low frequencies, whereas leaves transmitted all frequencies found in the spider vibrations relatively well (Figure 10.2). In line with this, male courtship success when presented to females on the different substrates was three times higher on leaves than on rocks or sand/soil, showing that the substrate displayed on is an important component of mating success.

However, it is currently unclear if the properties of males' signals have evolved to effectively match a leafy background, how often males are found on each substrate, and whether males actively choose leaves to signal from in the wild. As with *H. dossenus*, the seismic signals of male wolf spiders, (*Schizocosa ocreata*) are more effectively transmitted on leaf litter than other substrates like sand and rock, and mating success is greater on leaves (*ca.* 85%) than on other substrates (*ca.* 30%) (Gordon and Uetz 2011). *S. ocreata* signals are multimodal, and males use significantly more visual signals when found on substrates that are less effective at transmitting seismic signals. Courting pairs are also more likely to select leafy substrates when given a choice with other substrate types. This shows that animals may adjust the substrate/background type against which they display, and also that multimodal signals may help as 'back-up' signals to cope with variable environments (Gordon and Uetz 2011; see Chapter 6).

Weakly electric fish living in tropical freshwater face variation in the level of conductivity of water over time, due to floods and rainfall, which could affect their electric organ discharge (EOD) characteristics. For example, in dry seasons, the concentration of ionic salts can be greater than in the wet season, increasing the conductivity of the water and meaning that EOD intensity may be less reliable as an indicator of distance (Brenowitz 1986). In some electric fish species, water conductivity may also affect the amplitude of EOD signals. However, this change is moderate and, in some groups at least, the effects do not seem large enough to destroy species-specific differences in EODs (Baier 2008). In general, relatively small changes in water conductivity may have only minor or short-term influences on recognition of electric signals (van der Sluijs *et al.* 2011). The discharge rate and amplitude of EODs can also be affected by ambient temperature, although the significance of this for communication in natural conditions is as yet unclear (Caputi *et al.* 1998; Dunlap *et al.* 2000). Weakly electric fish with pulse-type EODs (see Chapter 2) may tolerate low oxygen levels better and adapt to large changes in temperature

Figure 10.1 The jumping spider (*Habronattus dossenus*) communicates using seismic signals transmitted through the substrate. The transmission properties of the different substrates that the spiders are found on vary, influencing mating success. Images reproduced with permission from Damian O. Elias.

than wave-form fish, meaning that pulse fish are more likely to be found in slow-moving hypoxic water, and wave-fish in faster flowing habitats (Julian *et al.* 2003; van der Sluijs *et al.* 2011). This may arise due to the metabolic costs of 'scan swimming' (continuous forwards and backwards swimming and EOD scanning), which is found only in wave-type gymnotiforms and requires increased oxygen consumption, whereas different forms of EOD dis-

charge alone do not seem to influence oxygen consumption (Julian *et al.* 2003).

Finally, animals can also modify their behaviour in order to maximize signal transmission or contrast against the background. For example, Uy and Endler (2004) have shown that male golden-collared manakins (*Manacus vitellinus*) from Panama clear leaf litter from 'courts' against which they display to females, and that this increases the contrast of their

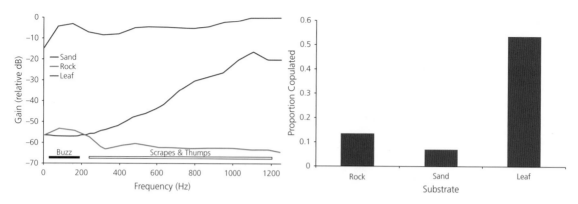

Figure 10.2 The left graph shows the relative transmission properties of the sand, leaf, and rock substrates on which *Habronattus dossenus* is found. Rocky substrates absorb strongly at all frequencies relevant to communication signals, and sandy substrates absorb low frequencies strongly, which are particularly relevant to the 'buzz' component of the signals. In contrast, signals on leaf litter show little attenuation at any frequencies. The right graph shows the proportion of male spiders that achieved copulations when presented to females on the different substrate types. Data from Elias *et al.* (2004).

plumage against the background (compared to uncleared regions).

10.2 Signal Form Under Different Environments

One of the key predictions in sensory ecology is that we expect signals to be modified by selection to improve their efficacy for transmission through the environment. This could be based purely on the type of habitat the species is found in, and to reduce the impacts of background/ambient noise that may interfere with the signal (see below).

Many birds sing to defend territories and to attract mates. Wiley (1991) investigated the association between habitat type (broad-leaved or deciduous forest, forest edge, parkland, grassland, shrub land, and marshes) and song structure in 120 species of North American oscine birds. In particular, there was a strong association between the temporal components of bird song (e.g. repeated elements) and habitat. Birds in open habitats used short song repetition periods, whereas birds in forest environments avoid these, presumably to decrease interference of reverberations from the foliage. In general, because vegetation attenuates and degrades song, foliated environments tend to lead to low-frequency tonal calls, whereas open environments favour repeated short-frequency modulated calls. Song can

also be tuned to the environment within a species. For example, green hylia birds (*Hylia prasina*) found in forests of sub-Saharan Africa, sing at lower frequencies when there is reduced canopy cover (Kirschel *et al.* 2009). In general, there now exists good evidence for song characteristics being linked to habitat features ('acoustic adaptation hypothesis'; see also Chapter 11).

Similar considerations apply in visual communication. In forest environments, the canopy comprises different light conditions and background types to the understory. Comparative analyses of bird plumage coloration show that birds found in different positions in the forest have different colours and pattern arrangements in order to either effectively signal to mates/rivals (i.e. to be conspicuous), or for camouflage (Gomez and Théry 2004, 2007). Two recent studies in fish illustrate that the ambient light and background are both important factors affecting signal form. In southern pygmy perch (*Nannoperca australis*) from Australia, males show more pronounced red coloration when in environments with more LW light (Morrongiello *et al.* 2010). This is likely because redder signals can take advantage of the abundance of LW light and become brighter, and perhaps also increase contrast against green vegetation. In comparison, Fuller (2002), investigating 30 populations of bluefin killifish (*Lucania goodie*) in Florida USA, found that when the light conditions were rich in UV and shorter wavelengths males with red anal fins were relatively common, whereas in environments with low transmission of UV and SW light, males with blue anal fins were more common. Here, the fish are utilizing different colours against the water column to have high colour contrast. Various other studies also show that water conditions and visibility can influence the coloration of fish in camouflage and communication (e.g. Eaton and Sloman 2011; Kekäläinen *et al.* 2010; Kelley *et al.* 2012; Seehausen *et al.* 2008; see Chapter 11).

How communication systems (both the signal and the response) can be modified in response to habitat properties is perhaps less well studied outside of visual, seismic, and acoustic senses, and we discuss a few examples below. In ticks, so-called assembly pheromones seem to play a role in creating aggregations of individuals, and this may aid in preventing water loss and reduce host localization behaviour when hosts are harder to encounter (Hassanali *et al.* 1989). At low humidity (25%), individuals assemble at the location of such pheromones, whereas at high humidity (85%) individuals may actually show a negative response (Hassanali *et al.* 1989). This change may be adaptive, as a pheromone that functions to prevent water loss and host localization should only function under appropriate conditions (e.g. low humidity). Humidity has also been shown to affect male response to female pheromones in the European corn borer moth (*Ostrinia nubilalis*). As humidity increases, the proportion of males taking flight and continuing to fly to the pheromone source declines (Royer and McNeil 1993). This may occur if increased humidity levels increase pheromone stimulation of receptors, causing adaptation or attenuation of sensory neurons, or due to interactions with the outputs of humidity detecting receptors (Royer and McNeil 1993). Although not related to signalling, moisture, such as from morning dew, may also decrease the ability of some spiders to detect and to show antipredator responses to cues of predatory spider species, such as silk and excretions (Wilder *et al.* 2005).

In electric fish, Kramer and Kuhn (1993) showed that in two closely related mormyrids, *Campylomormyrus spp.*, which have biphasic EODs, the resistance of water influences the duration and amplitude of the second phase of the EOD. While the changes to EOD form were not drastic, they could be important if they affect species recognition and mate selection. However, Kramer and Kuhn also reported that *Campylomormyrus* would adjust to changes in water conductivity over 30–48 hrs, partially restoring their EOD form, and illustrating that electric fish can have impedance-matching of their EOD signals to local environmental changes in the short term. Kramer and Kuhn suggest that when water becomes less conductive, fish may hormonally mediate increased channel proteins in the electrocyte cell membranes, strengthening the electric organ to produce greater voltage and current.

It is important to remember that signals may not always have been tuned over evolution to specific environments, even if this may appear to be the case. In contrast, animals may choose microhabitats or different temporal periods under which their signals will be most effective. It can sometimes be difficult to determine which is the case. For example, two species of tarantula spider from meadow environments in Uruguay (*Eupalaestrus weijenberghi* and *Acanthoscurria suina*) have similar behaviour and reproductive strategies, and males of both species court females with vibration signals generated from body movements that are very similar in form (Quirici and Costa 2007). The signals in both species seem to have converged on a similar structure because this is effective in courtship in the habitat where they are both found. Thus, selection could have modified signal form in both species to maximize efficacy. However, the alternative is that males use brief reproductive periods when receptive females are abundant and physical conditions are favourable, such as humid warm weather, as this may affect the transmission properties of the substrate, especially as humid soil is better at conducting seismic signals than dry soil (Quirici and Costa 2007). Of course, these explanations are not mutually exclusive, but elucidating which one plays a primary role is not always easy.

Finally, animals may also modify the type of signal that they use in communication. For example, when courting in dark conditions, male newts may use comparatively more olfactory than visual signals towards females than when courting in the daytime (Denoël and Doellen 2010). In Chapter 6 we also discussed how signals can be multimodal. Signalling in multiple modalities could allow animals to communicate effectively under a range of conditions when some modalities are less effective.

10.3 Sensory Plasticity and the Environment

We have seen above that animal signals are sometimes adjusted (plastically or over evolution) in order to cope with environmental features better. In addition, we also expect sensory systems to be tuned to improve the ability of animals to extract relevant information under the conditions under which they occur. Below we discuss when and how specific sensory systems can be tuned to the environment, but first we discuss how the value and use of different sensory modalities depends on the habitat in which the animal lives.

Animals may switch investment under different conditions to senses that are most useful. This could either be a response that occurs over evolution or resulting from developmental plasticity. For example, we find animals with electric senses in aquatic and moist environments but not in terrestrial systems because water conducts electricity much better than air. An electric sense would be useless for most terrestrial species. Likewise, vision is generally of most use when light levels are not very low, explaining why birds like cave swiftlets that live in caves have evolved a form of echolocation, and why many species of weakly electric fish are nocturnal or live in habitats where water visibility is poor.

Blind cave fish are a classic example of how sensory systems adapt to the environment, and can be lost when their functionality is reduced. The tetra *Astyanax mexicanus*, from Central America occurs in a surface-dwelling form with functioning vision, and a blind cave-dwelling form. Various populations were isolated a few million years ago and the cave populations have acquired a range of adaptations including more taste buds, loss of pigment and eyes, and changes in behaviour (Greenwood 2010). In darkness, fish swim towards water disturbances, such as clay particles that drop into pools or small invertebrates swimming to get food. Yoshizawa *et al.* (2010) showed that blind cave individuals showed this behaviour, whereas surface-dwelling individuals did not. This behaviour is advantageous for feeding success in that cavefish were better competitors than surface fish for food in the dark (but not in the light). The ability seems to be linked to the mechanosensory function of the lateral line, because fish responded to vibrations in the water in the region of 10–50 Hz, correspondent with the sensitivity of the lateral line of 20–80 Hz. Furthermore, cavefish lateral line neuromasts tend to be larger than in surface fish and have more sensory hair cells, and chemical inhibition of the neuromasts abolishes the attraction of the fish to vibrations (Yoshizawa *et al.* 2010).

Sensory systems are costly to produce and maintain, for example in terms of energetic requirements (Chapter 4), and animals should therefore not continue investment in sensory apparatus that are no longer useful. Instead, animals should invest most in senses that are likely to be more beneficial to them given current and future conditions. Such changes need not always occur over long periods of evolution. Sensory plasticity occurs when individuals compensate for sensory deprivation in one sense during development with an improvement in the performance of an alternative sense. For example, cats that are deprived of visual stimulation early in life show increased ability in auditory localization (Rauschecker 1995). A study by Chapman *et al.* (2010) with guppies (*Poecilia reticulata*) illustrates sensory plasticity. They reared guppies for 72 days under either low or high light conditions. Following this, they tested guppies from these two groups in their ability to locate food by either visual cues or chemical cues alone. Guppies reared in high light conditions were much better at finding food based on visual cues than guppies reared under low light (Figure 10.3). The situation was reversed when only chemical cues were present, with guppies reared under low light levels better at detecting the food than those reared under high light. Thus, guppies invest more in olfaction when vision is less valuable.

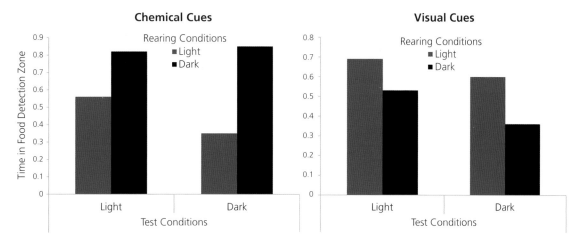

Figure 10.3 The left graph shows the success in finding food of guppies reared under high and low light conditions when presented with only chemical cues to the location of the food. When tested under either light or dark conditions, guppies reared in the dark were better able to find food than those reared in the light. The graph on the right shows the success of guppies in finding food based on visual cues only. Here, in both light and dark test conditions, guppies reared in the light were better at finding food than those reared in the dark. Data from Chapman *et al.* (2010).

10.4 Tuning of Sensory Systems to Habitats

It is clear that the value of different sensory modalities varies on a broad scale. However, individual sensory systems may also be tuned to finer aspects of the environment. As we shall see, whether this occurs or not can depend on several factors, relating to both the environment and the ecology of the specific species. In this section, we will focus on vision, because this has been particularly well studied with respect to tuning to environments.

10.4.1 Tuning of Receptor Sensitivity

In the deep sea, light is in short supply and also, due to wavelength-dependent attenuation, more dominated by shorter wavelengths. Even bioluminescence tends to be blue-green in nature. While many cavefish have reduced or lost vision entirely, deep-sea fish often have adaptations to facilitate vision under the dark shortwave- (SW) shifted light conditions, including large amounts of visual pigment in the retina and large photoreceptor outer segments (see Partridge *et al.* 1989). Most deep-sea

fish (almost 90% of species tested) also have a single visual pigment type that absorbs light most strongly around 470–490 nm (i.e. the SW part of the light spectrum). This is congruent with the spectrum of downwelling light and/or matches the peak emission of deep-sea bioluminescence (Douglas *et al.* 1995; Douglas and Partridge 1997; Partridge *et al.* 1989; Turner *et al.* 2009). Thus, the visual pigments reflect an adaptation to maximize sensitivity in the restricted light conditions (both to residual sunlight and bioluminescence). In relatively shallow-water fish, peak photoreceptor absorbance can also vary between different species occurring in different microhabitats with different backgrounds and light conditions (Cummings and Partridge 2001).

One of the best examples of how visual systems can adapt to different environments is in stomatopod crustaceans ('mantis shrimp'). These have remarkable diversity in both the number and spectral sensitivity of their visual pigments, and use their vision to actively hunt prey and in mate choice. In addition, they also have a large diversity in coloured pigments that filter out different wavelengths of light before it reaches the visual pigment, further modifying sensitivity. Different species of mantis shrimp are found across a wide range of habitat types and depths, where light conditions can change greatly. Cronin *et al.* (2000b) investigated vision in 12 species of mantis shrimp, found over a range of depths and light conditions, occurring from the intertidal down to below 50 metres. As depth increases, peak sensitivity of the peripheral photoreceptors from outside the midband (involved in spatial and motion vision), in the different species moves to shorter wavelengths (from 528 nm to 470 nm in the most extreme species). This enables individuals to retain effective colour vision within the restricted range of wavelengths available. Sensitivity can also vary within species. Cronin *et al.* (2001) investigated changes in sensitivity in the longwave- (LW) sensitive receptor of the mantis shrimp *Haptosquilla trispinosa* at Lizard Island, Australia. This species covers a range from the intertidal zone to about 30 metres in depth. They sampled individuals from 1 m and 15 m and found no difference in visual pigment across individuals, but there was a difference in the transmission properties of the coloured filters. As with the cross-species comparisons, peak sensitivity of the receptors is shifted to shorter wavelengths at greater depth (Figure 10.4). Cronin *et al.* calculate that at 15 m there is a 96% reduction in the amount of light present above 575 nm, and so this shift in sensitivity is essential for colour vision to still function effectively.

Cronin *et al.* also found that newly metamorphosed post-larvae have visual filters that resemble those of shallow-water adults. A lab experiment, rearing such post-larvae under either broad spectrum white light conditions or 'blue' light lacking wavelengths above 550 nm for three months, shows that individuals reared under blue light shift their sensitivity to shorter wavelengths (Figure 10.5). This is another example of sensory plasticity. It is unclear, however, if this phenotypic plasticity is brought about by changes in overall light levels or due to changes in light composition (spectra).

Cronin *et al.*'s (2001) study shows that sensory plasticity can allow tuning to the environment, but tuning will also often occur due to genetic changes over multiple generations. An experiment by Endler *et al.* (2001) with guppies shows that populations can respond to different environments with changes in their sensory system. Guppies have four types of cone: ultraviolet (UV), shortwave (SW), mediumwave (MW), and longwave (LW) sensitive, and have variation in the sensitivity of the LW cones across individuals. The researchers took six lines of guppies and selected them for their response to either 'red' (LW; 660 nm) light, or 'blue' (MW; 420 nm) light, and two unselected control lines. They tested the response of each line to different light compositions (predominantly red or blue) using an optomotor performance test. These generally involve chambers with moving stimuli (often

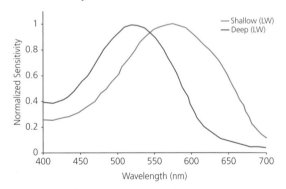

Figure 10.4 The spectral sensitivity of the longwave- (LW) sensitive receptor in the mantis shrimp *Haptosquilla trispinosa* is tuned to the light conditions at different depths owing to differences in the transmission properties of the coloured filters in the receptors. Individuals found at greater depth have sensitivity shifted to shorter wavelengths, consistent with the shift in wavelength composition in the light environment. Data from Cronin *et al.* (2001).

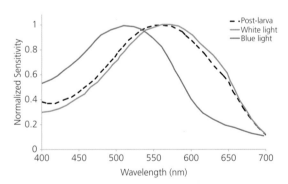

Figure 10.5 When *Haptosquilla trispinosa* individuals are reared under full spectrum white light they develop longwave-receptor visual sensitivity similar to that of shallow water individuals. However, when reared under blue light, sensitivity is shifted to shorter wavelengths, like individuals found in deeper water. Data from Cronin *et al.* (2001).

white and black bars), whereby individuals will follow (with their eyes, heads, or bodies) and orientate with respect to the stimuli, but will stop doing so when they can no longer see them. By gradually lowering the quantity of light within a given set of wavelengths, it is possible to determine how sensitive the visual system is to that light, based on when the fish stops responding. The lower the light intensity needed for this to happen, the more sensitive the fish is. At each generation, in each line, Endler *et al.* took the 40 most sensitive individuals (20 males, 20 females) after measuring at least 100 individuals, and allowed these to breed. They repeated this for seven generations and found a significant response to selection in both the red and blue lines, and that these changes were heritable. This study shows how sensory systems can respond to selection with altered sensitivity relatively quickly. An avenue for future study is to separate lines of individuals into different light environments and record changes in sensitivity over time and whether this affects behaviour (e.g. in foraging or mate choice).

Overall, in aquatic systems there is good evidence for tuning of vision to light conditions. However, the situation in terrestrial systems is less clear. Birds, for example, are a diverse and widespread group occurring in many habitat types, yet their visual pigments seem to be relatively constrained across species, habitat, and life history (see Chapter 4). This could be because light levels in terrestrial systems do not vary as much as in aquatic environments, and because the spacing of the cone types across the bird visual spectrum may be optimal to capture the range of different colours that birds may encounter in nature, associated with foraging, mate choice, and predator avoidance. The main exceptions in birds are penguins (that have shortwave-shifted vision to aid foraging in water), and nocturnal species (where colour discrimination is reduced to enhance night vision) (Hart and Hunt 2007). However, sensitivity in birds (and many other organisms) is not just driven by the pigments but also factors such as coloured oil droplets (that filter light before it reaches the visual pigment). It is possible that changes across species with different life histories may be found in the oil droplets and different proportions of receptor types, and this could tune colour vision to different

environments. More work is needed in this area before we can conclude that avian vision is as constrained as currently thought.

The situation in other terrestrial animals also needs more investigation. Recently, it has been shown that, unlike most mammals (except some rodents, bats, and marsupials), arctic reindeer (*Rangifer tarandus*) have a visual system sensitive to ultraviolet light. This probably does not come from an additional UV receptor, but from extending the sensitivity of the shortwave-sensitive receptor into the UV (Hogg *et al.* 2011). The authors suggest that the advantage of this may be that at high latitudes there is more UV light in the environment, as it is scattered more due to the sun being lower in the sky and because snow reflects UV strongly compared to most other natural substrates that absorb UV. The advantage of being able to see this is that it may enable reindeer to detect food sources like lichen by virtue of the contrast of this against the background in UV. Again, high latitudes represent relatively extreme ends of variation in terrestrial light environments (akin to aquatic systems), but raises the possibility for photoreceptor tuning within terrestrial systems. Overall, colour vision and spectral sensitivity in aquatic organisms often shows a match to the environment, whereas in terrestrial systems there is as yet little evidence for this and more work is needed. Currently, most evidence for photoreceptor tuning in terrestrial species seems to either occur with respect to specific tasks (e.g. foraging in primates or mating behaviour in fireflies) or owing to extreme conditions (high ultraviolet, dark conditions). See Chapter 4 for further discussion.

10.4.2 Eye Morphology

In contrast to tuning of sensitivity to different wavelengths, adaptations in eye morphology in terrestrial species with respect to the environment are more common. Comparative analyses in mammals and birds shows differences between diurnal, crepuscular, and nocturnal species in terms of eye morphology (Hall and Ross 2007; Schmitz and Motani 2010), although more work is needed to understand the specific selection pressures associated with different traits. Some species of bee, including the Indian carpenter bee (*Xylocopa tranquebarica*) have shifted from a diurnal to a nocturnal lifestyle. Here, adaptations have arisen to facilitate effective vision under the dark conditions. For example, *X. tranquebarica* has unusually large ocelli (1 mm diameter) that lead to reduced spatial acuity, and their apposition compound eyes (usually found in diurnal insects) are big and with large facets and wide rhabdoms, which makes them 27 times more sensitive to light than closely related diurnal bees (Somanathan *et al.* 2009). In addition, *X. tranquebarica* could also have photoreceptors with longer temporal summation to increase sensitivity further under the dark night conditions.

Differences in temporal niche across and within species of ants have recently been shown to strongly influence the morphology of compound eyes in ants. Narendra *et al.* (2011) investigated activity patterns in four species of *Myrmecia* ants and their workers in Australia, and sexually reproductive stages (alate) of females and males. In the different species, groups leave the nest at different times of the day or night. Generally, Narendra *et al.* found that eye area, facet size, and rhabdom size increased in castes that were active at night compared to those active during the day.

There is clear evidence of sensory system adaptation and tuning to different environmental conditions in terms of vision. More work is needed in other modalities to determine how widespread this is in other senses. In addition, as mentioned above, it is important to remember that changes in signals and sensory systems in accordance with environmental features may not be independent. For example, different environments can lead to changes in the form of bird songs in order to maximize transmission to the receiver (see Chapter 11). In some avian species, there have been corresponding changes in terms of auditory temporal and frequency resolution to best detect and discriminate these songs (Henry and Lucas 2008, 2010). It would be useful to know more when sensory systems have been tuned to the environment itself, as opposed to specific signals that work best in certain habitats.

10.5 Coping with Environmental Noise

One of the main factors in a habitat that may affect detection and information transfer is noise. Here,

we mean a source of interference in any modality that may prevent an animal from detecting relevant information or signals. Noise can arise through interference from other organisms, including their own communication signals, or from abiotic factors, such as wind. More recently, anthropogenic sources have arisen (see below). There are various recent studies of changes in signal form to cope with environmental noise (mainly acoustic) across a range of taxa, and we cover only a few examples here. Animals could respond to noise in a number of ways, including changes in the way that their sensory system filters out noise, or in communication by timing the use of signals to when noise interference is less, changing signal structure to make them more conspicuous, or by increasing redundancy in signal forms to increase the likelihood of detection/recognition.

Animals can change their signal form to increase conspicuousness and reduce overlap with noise profiles. It has been suggested for some time that wind-generated and insect noise could influence the evolution of song across species in different habitats (Ryan and Brenowitz 1985). Insect noise arises from cicadas, katydids, and crickets, and the abundance of these organisms in different locations can affect noise profiles. A playback study by Gillam and McCracken (2007) on the Brazilian free-tailed bat (*Tadarida brasiliensis*) in Texas shows that bats will adjust the frequency of their echolocation calls to reduce overlap with insect calls. The authors broadcast sounds of insect calls at different frequencies and measured the change in the bats' echolocation calls. Bats shifted up their peak and minimum call frequencies when presented with increasing frequencies of insect noise, in order to avoid overlap with the insect calls (Figure 10.6).

Similar to the bats, green hylia birds sing at lower frequencies when there is more noise from insect calls, in order to reduce overlap (Kirschel *et al.* 2009), and common marmosets (*Callithrix jacchus*) increase the amplitude of their calls when presented with white noise (Brumm *et al.* 2004). The concave-eared torrent frog (*Amolops tormotus*) from China uses calls in communication and has evolved an ability both to use and detect ultrasonic frequencies, which are distinct from the low-frequency background noise of moving water (Feng *et al.* 2006). Similar principles

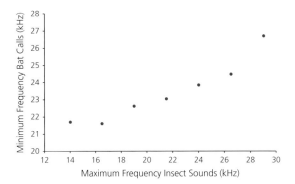

Figure 10.6 Brazilian free-tailed bats (*Tadarida brasiliensis*) have been shown to increase the frequency of their echolocation calls when presented with playbacks of the sounds of insect calls. Data from Gillam and McCracken (2007).

apply in aquatic environments. For example, humpback whales (*Megaptera novaeangliae*) will switch from underwater vocalizations when noise is low, to surface-based signals (breaching or pectoral fin slapping) when noise is high (Dunlop *et al.* 2010). In visual communication, many lizards signal to potential mates and rivals with displays involving movement. A potential problem is the background against which the signaller is seen is moving, for example due to wind-induced patterns in the vegetation. Work by Ord *et al.* (2007) with *Anolis* lizards in Puerto Rica shows that in environments with rapidly moving vegetation due to wind, individuals speed up their body movements to increase conspicuousness.

Animals may also have behavioural and timing changes in order to signal when noise levels are lower. For example, the acoustic signals used by Mediterranean gobies are matched in timing to 'quiet windows' in the ambient noise spectra of different habitats (Lugli 2010). Vibration-based communication is widespread in arthropods, including through plants, and wind can be a serious source of noise (Cocroft and Rodríguez 2005; Greenfield 2002). The treehopper, *Enchenopa binotata*, from the USA, is a plant-feeding insect in which males and females communicate (duet) with each other before mating. McNett *et al.* (2010), using analysis of signalling behaviour, measurements of noise in the field, and playback experiments in the laboratory with fans, showed that noise in the form of vibrations in the plant caused by wind could mask the treehopper communication, with females respond-

ing less to male signals when they are given in the presence of noise. However, in the field, most signalling occurred in the mornings and evening when wind velocity was low. In general, to reduce the cost of females not responding, males tended to signal to females in the periods in between noise occurrence, and signalled least when wind was of higher velocity. Variation in wind during the morning and evening lulls may favour calling at those times of day, whereas periodic lulls also give an opportunity to signal in gaps of noise (McNett *et al.* 2010).

Most work has been done on how signal structure or timing is modified to cope with noise in different habitats. However, a study by Witte *et al.* (2005) of cricket frogs (*Acris crepitans*) from southern and eastern USA shows that individuals from different habitats can have differences in their ability to filter out environmental acoustic noise. They studied frogs from either pine forest or grassland habitat sites and modelled the tuning of the basilar papilla in females. The level of noise is about the same in both habitats but the two environments affect the way that male calls are transmitted. Calls suffer greater attenuation and degradation in the pine habitat, making this environment more of a challenge for communication. Consequently, male frogs from the pine forest have advertisement calls with better transmission properties (higher dominant frequency, shorter duration, and faster pulse rate), than frogs from the grassland habitat, and these modifications are most advantageous in the forest environment. In addition, Witte *et al.* (2005) found that the auditory filter (tuning curve) of females from pine forest was better at eliminating noise than that of females from the grassland habitat. This increases signal reception in an environment where the signal does not transmit as far by reducing the impact of noise. It also illustrates that there have been changes in both the signal structure of male frogs and the receiver sensory system of females.

10.5.1 Altering Signal Components

Richards (1981) suggested that song structure in birds includes 'alerting' components. Instead of changing the structure of a signal to avoid detrimental effects of noise, alerting components convey no actual information (no strategic element) but attract attention with conspicuous alerts before giving the main signal. Richards followed others in noting that many animals have narrow-frequency, widely spaced notes that generally lack species-specific components that occur before the main song. Richards suggested that alerting components could overcome problems of environmental noise by being highly detectable and of high contrast with the other sounds in the environment. This then raises the detectability of the subsequent more complex message component. Alerting components may be expected in communication systems where information is rapidly transferred (we might not expect them to arise in chemical communication, for example). Because alerting components are conspicuous, they could attract unwanted attention from predators if used too often (see Chapter 7). Therefore, we may expect alerting components to be given only when really needed, such as under poor signalling conditions.

At present the idea of alerting signal components has received little experimental attention, but one study in the Yellow-chinned anole (*Anolis gundlachi*) in Puerto Rico by Ord and Stamps (2008) provides good evidence. In various species of *Anolis* the males advertise their ownership of a territory to neighbours with a head-bobbing sequence. In Yellow-chinned anoles, before the head-bobbing they present a four-legged push-up display. In their habitat, communication could be disrupted by low light levels, wind disturbed vegetation, and dappled light. As predicted, Ord and Stamps found that more alerts were given when the habitat was darker, at greater distances between signaller and receiver, and when motion-based noise was greater (Figure 10.7).

They also designed 'robotic' *Anolis* lizards resembling real males, but with modified displays. These would either give no alert before the head-bobbing, a conventional push-up alert first, or a novel alert in the form of a rapid dewlap extension not seen in the species. They found that the time for the receiver to orientate to the robotic display was shorter for both alerts than the no-alert control in dim light (Figure 10.8). However, there was no effect of an alert in high light levels.

10.5.2 Anthropogenic Noise

Finally, in the modern world, noise regularly arises due to humans. This idea has again been most stud-

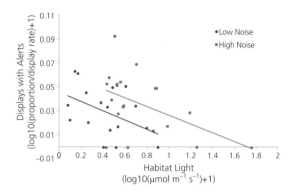

Figure 10.7 Yellow-chinned *Anolis* lizards sometimes produce an alert display consisting of a 'push-up' before they present their main signal of territory ownership to other individuals. Work by Ord and Stamps (2008) shows that individuals are more likely to produce alerting signal components when the light conditions are poor or when there are greater levels of wind-produced environmental noise. Data from Ord and Stamps (2008).

ied in acoustic senses (although this is changing), and a range of recent work shows that anthropogenic acoustic noise can affect breeding success,

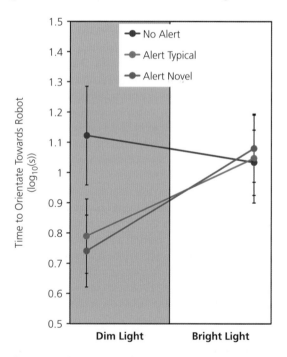

Figure 10.8 The time it takes for yellow-chinned *Anolis* lizards to orientate towards a robotic lizard performing a territorial display is shorter when the display contains an alert component (real or a novel form), but only when light levels are low. Data from Ord and Stamps (2008).

vigilance behaviour, foraging efficiency, predation levels, and spatial orientation across a variety of taxa (see Barber *et al.* 2009), and anthropogenic noise could be a major consideration for species conservation (Siemers and Schaub 2011).

It is well known that many birds sing to attract mates and defend territories. A major type of noise is from traffic. A study by Slabbekoorn and Peet (2003) in Leiden in the Netherlands shows that male great tits (*Parus major*) have songs with higher minimum frequencies in territories in urban locations with high levels of noise. In addition, songs in urban populations are shorter in duration than songs in forest individuals, and comparisons of a range of European forest and city populations show that urban populations have higher minimum song frequencies (Slabbekoorn and den Boer-Visser 2006). Some bird species have changed the time of day that they communicate to avoid noise. For example, European robins (*Erithacus rubecula*) sometimes sing during the night when human activity is lower when living in locations prone to high noise during the day (Fuller *et al.* 2007).

Siemers and Schaub (2011) have shown that traffic noise can reduce the foraging efficiency of greater mouse-eared bats (*Myotis myotis*), which capture ground-dwelling arthropod prey such as spiders, beetles, and centipedes, based by eavesdropping on the noise cues the prey make when moving. The authors used a large flight room and played back the sounds of prey rustling behaviour through a hidden speaker, and then played back noise corresponding to the profiles of a passing vehicle at different distances. They then recorded the ability of the bats to detect and localize the prey sounds. Noise from close proximity resulted in bats' search times being increased greatly, and could lead to a corresponding decline in foraging success. Interference was probably caused by acoustic masking of the prey rustling sounds

Species recognition in fish occurring via chemical communication can also be affected by anthropogenic disturbance. For example, females of the freshwater swordtail fish, *Xiphophorus birchmanni*, show strong preferences for conspecific males over another species when in clean water, but this preference is abolished leading to hybridization when the water is contaminated with sewage effluent and agricultural run-off (Fisher *et al.* 2006). This seems to occur due to the presence of humic acid in the water, which is a natural substance made more common in water due to human impacts stemming from degrading organic matter. It is not entirely clear how humic acid disrupts chemical communication, but one possibility is that it may chemically bind to the sexual pheromones of the fish, preventing the pheromones from triggering the chemoreceptors, or it may directly block or damage chemoreceptors (see Fisher *et al.* 2006). In the presence of chemical pollutants, crayfish can also behave inappropriately, orientating towards conspecifics that have been attacked by a predator and are releasing alarm signals, rather than fleeing from them (Wolf and Moore 2002). It is not yet clear how exactly this change is brought about.

10.6 Future Directions

The subject of how the environment influences sensory systems and signals is substantial and a great deal of work remains to be done. On a broad level, we know a lot more about how the environment influences transmission of acoustic and visual signals, but perhaps less in some other senses. In addition, while good examples exist as to how visual system properties may (or may not) be modified in response to environments, less is known in other sensory modalities. Following this, we need a more complete understanding of how common sensory plasticity is in animals and when it occurs (under what conditions one sensory modality should be favoured over another). We would also benefit from a better understanding of how environmental noise affects non-visual and non-acoustic senses. We do not yet know much about how different types of anthropogenic noise can affect the functioning of different sensory modalities, and whether other modalities can compensate for reduced function of another. Such information is important, as anthropogenic disturbance may be a major future issue for conservation. More work into how widespread alerting signal components are would be valuable, and why these evolve rather than modifications of the primary signal form. Studies investigating sensory integration (see Chapter 4) could also be important in understanding how using information from multiple modalities may overcome problems with environmental noise and uncertainty. Finally, research might consider how environmental noise is exploited by individuals to gain an advantage. For instance, some assassin bugs that capture spiders as prey use wind-induced vibrations of the web to hide cues of their own movement towards the spider (Wignall *et al.* 2011).

10.7 Summary

The environment significantly affects the way that organisms interact and communicate. This includes influencing the way that information is acquired and signals are transmitted, and the amount of noise present that could interfere with gaining relevant information. Organisms can deal with this by using different signal forms in different environments, switching between sensory modalities, or tuning their sensory systems to cope best with the environment (either over evolution or with phenotypic plasticity). Environmental noise, both natural and anthropogenic, can affect communication, and this can influence the evolution of signal form and create selection for conspicuous alerting signals and changes in behaviour.

10.8 Further Reading

For further discussions on the impacts of noise, including anthropogenic noise, and how this affects communication in fish see van der Sluijs *et al.* (2011), and in acoustic systems see Barber *et al.* (2009). Readers should also consult Chapter 6 of this book, Munoz and Blumstein (2012), and other references therein on if and how multimodal signals may enable animals to overcome environmental noise or to cope with changeable or unpredictable habitats.

CHAPTER 11

Divergence, Sensory Drive, and Speciation

> ## Box 11.1 Key Terms and Definitions
>
> **Adaptive Radiation:** The evolution of ecological and phenotypic diversity within a rapidly multiplying lineage.
>
> **Assortative Mating:** Individuals preferentially mate with partners that have a similar phenotype or genotype for a given trait (e.g. coloration).
>
> **Directional Selection:** Individuals at one end of a distribution have higher fitness than other individuals, such that over time, selection results in a directional shift in the mean trait value (e.g. to a larger body size).
>
> **Divergent and Disruptive Selection:** Selection operates in different directions in two populations, or favours extreme phenotypes at either end of a distribution (disruptive selection). Intermediate phenotypes have lower fitness (e.g. individuals do best either when small or large but not at intermediate sizes).
>
> **Elaboration:** Over evolution, traits change in value along the same axis of variation (e.g. brighter coloration, longer or louder calls of the same type).
>
> **Innovation:** This occurs when a new trait arises that is not down an already existing axis of variation, for example a new colour or call type used in mate choice.
>
> **Magic Trait:** A trait such as a colour pattern that is subject to divergent or disruptive selection and mediates assortative mating.
>
> **Sensory Drive:** The process where characteristics of the environment lead to (drive) selection pressure resulting in changes in the form of sensory systems and/or signals to be more effective under prevailing environmental conditions, and may potentially lead to divergence and speciation.
>
> **Stabilizing Selection:** Individuals with intermediate phenotypes have highest fitness whereas those with extreme phenotypes are selected against (e.g. intermediate body sizes do best).

In the last two chapters we have established some of the mechanisms that can lead to variation within and across species. However, perhaps the most fundamental questions in evolutionary biology relate to how and why populations diverge, and how new species arise. Until recently, the prevailing view of animal speciation was that it required geographical isolation (allopatric speciation; e.g. Mayr 1942). How- ever, increasing theoretical and empirical evidence now suggests that geographical barriers need not be an essential component of speciation, and that factors such as sexual selection and adaptation to local environmental conditions can lead to species divergence and formation with or without geographical isolation. In this chapter we largely focus on evidence for divergence and reproductive isolation caused by adaptation to local environments, sexual selection, and species recognition, because these examples have often placed more emphasis on signalling and sensory system evolution.

11.1 Divergence and the Environment

In Chapter 10, we discussed how sensory systems and signals may adapt to different environments. Here, we discuss examples that illustrate how differences in the local environment may result in divergence between populations.

11.1.1 The Molecular Basis of Camouflage Against Different Backgrounds

Following the development of powerful new approaches in molecular biology, we can now much more effectively understand the molecular basis for divergence in some animals. In Chapters 8 and 9 we discussed the different types of camouflage that exist and how they might prevent detection or recognition by predators. Work on mice has uncovered the underlying genetics of divergence in camouflage patterns associated with different habitats. For example, pocket mice (*Chaetodipus intermedius*) found in southern USA are light-coloured when found against light-coloured rocks, but dark (melanic) when found on larva rock substrates. Nachman *et al.* (2003) showed that mutations in the melanocortin-1-receptor gene, *Mc1r*, seem to underlie this change in coloration in some populations. Oldfield mice (*Peromyscus polionotus*) also show variation in coloration between subspecies associated with different habitats (backgrounds), with the mainland form dark in coloration on dark soil substrates and a subspecies, beach mice, being light on white sand environments. Again, some divergence is associated with protein coding changes from an amino acid mutation in *Mc1r*, but also with regulatory changes in gene expression in the Agouti signalling protein gene, *Agouti* (Steiner *et al.* 2007). The light colour pattern of mice largely results from epistatic interactions between these genes (and another gene, *Corin*), with *Mc1r* only having an effect when the derived *Agouti* allele is present (Steiner *et al.* 2007). *Agouti* also seems to be important in development (embryogenesis), with light-coloured beach mice

showing higher gene expression, and expression occurring higher up on the body flanks. Increased expression delays the maturation of melanocytes, meaning that pigment cannot be produced so quickly, resulting in a lighter coloration matching the white sands background (Manceau *et al.* 2011). Overall, the genetics in this system indicates that changes to relatively few interacting genes underlie changes in coloration (although we are a long way from understanding the genetic basis of more complex camouflage patterns). Recently, experiments using clay models of mice along field transects by Vignieri *et al.* (2010) have shown that the different colour patterns do provide a survival advantage on the appropriate background from visually hunting predators. However, as yet, no analyses have been undertaken as to the level of camouflage of different coloration with regards to predator vision.

Parallel work has shown that different co-occurring species of lizard in New Mexico USA have converged in camouflage, with morphs matching different background substrates in different environments, presumably under selection from predators (Rosenblum *et al.* 2004; Rosenblum 2006; Figure 11.1). Here, lizards occur in either a melanic form matching dark soil or a light form matching white sands, although different molecular mechanisms cause the changes in coloration across the species (Rosenblum *et al.* 2010). Interestingly, species showing more effective camouflage and greater divergence between the colour forms have higher levels of population structure and lower levels of gene flow (Rosenblum 2006). The changes in coloration in the lizard species are also caused by changes in melanin pigmentation, driven at least in part by mutations in *Mc1r* (Rosenblum *et al.* 2010; see also Hughes 2010). Overall, work on mice and lizards demonstrate that similar phenotypic changes can be based on both similar or alternative molecular mechanisms across species, and that the rates of adaptation to match the background (and therefore reduce the risk of predation) are influenced by degrees of gene flow between populations.

11.1.2 Avian Plumage and Signalling

Evidence that differences in environment can produce changes in plumage signals across species

Figure 11.1 Images of three species of lizard and the match between different morphs and the backgrounds on which they are found: eastern fence lizard (*Sceloporus undulates*), lesser earless lizard (*Holbrookia maculata*), and little striped whiptail (*Aspidoscelis inornata*), top, middle, and bottom pairs of lizards respectively. Images reproduced with permission from Erica Rosenblum.

comes from *Phylloscopus* warblers in Kashmir. Marchetti (1993) investigated eight species of *Phylloscopus*, each of which have different numbers of colour patches, found on the wings, crown, and rump. In these species, individuals temporarily flash bright colour patterns during intraspecific displays. Marchetti measured the quantity of light in the different habitats where each species is found, as well as the number and reflectance of the various plumage patches. She found that warbler species from darker habitats had brighter patches and higher numbers of plumage patches (Figure 11.2). The patches are important in defending territories. When Marchetti conducted an experiment with *P. inornatus*, whereby she either increased the conspicuousness of males by enlarging the wing bar (with green paint) or reduced the conspicuousness, she found that males with larger more conspicuous displays gain and hold on to larger territories than controls or those with reduced wing bars. Adding a novel colour patch (a crown stripe) also increased territory size.

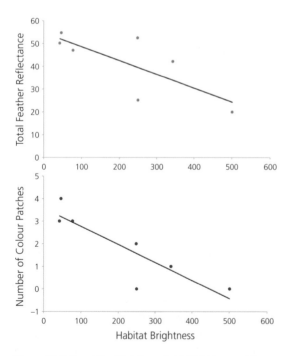

Figure 11.2 The relationship between habitat brightness and the total feather reflectance ('brightness') and number of different colour patches in *Phylloscopus* warblers in Kashmir. Species living in darker habitats tend to have brighter plumage and more plumage patches for communication. Data from Marchetti (1993).

Marchetti's experiments provide good evidence that variation in the physical environment can cause divergence in signals in particular directions. However, it is important to conduct similar studies analysing how conspicuous the signals are in terms of avian vision.

A comparative study by McNaught and Owens (2002) analysed the plumage patterns of Australian birds in allopatry and sympatry to investigate whether differences in plumage coloration is the result of selection for reducing hybridization in sympatric species (reproductive character displacement), or better explained by differences in light environment. They found little evidence for the first theory; species in sympatry were often more similar than those in allopatry. In contrast, there were significant associations between plumage coloration (colour and brightness) and habitat type (open or closed). Thus, selection against hybridization may have influenced other modalities like acoustic signals (see below), or other specific behaviours, but in Australian birds differences in plumage seem to be driven more by habitat structure.

11.2 Elaboration versus Innovation

It has been suggested for some time that differences in mating displays can maintain reproductive isolation. For example, Stratton and Uetz (1983) showed that vibratory signals used during mate attraction are key to maintaining reproductive isolation in two species of wolf spider (*Schizocosa rovneri* and *S. ocreata*). Isolation is maintained by females, which will only mate with males of the correct species based on their seismic-sound signals, with this display and behaviour differing between the two species. Males of both species court females indiscriminately and forced matings between the two species resulted in the production of egg sacs and offspring, suggesting that isolating mechanisms are pre-zygotic. Likewise, Uy and Borgia (2000) showed that two allopatric populations of the Indonesian Vogelkop bowerbird (*Amblyornis inornatus*) show strong differences in male bower structure but are otherwise morphologically similar, and that corresponding female preferences for those displays exist. However, how might differences in sexual displays arise in the first place?

Endler *et al.* (2005), in a study of bowerbird signal evolution, formalized the idea that new colour patterns could be considered elaborations of existing traits (such as changes from light blue to dark blue coloration), or innovations to produce novel traits (such as changing from red to blue). One way to think about this is if variation runs along a clearly defined axis then it is reasonable to conclude that a new value along that line is an elaboration, whereas a phenotype away from the line is an innovation (Figure 11.3). The concept may provide a helpful framework for understanding changes in signal form. However, defining something as an innovation or elaboration may sometimes be difficult or subjective, and may sometimes be better based on molecular mechanisms and changes rather than phenotypic measurements. The key issue is at what point we consider a change sufficient to switch from being elaboration to an innovation (i.e. how far away from the axis does a trait have to be to become an elaboration)? Furthermore, the concept of elaboration and innovation does not account for categorization: the grouping of stimuli into specific classes by the observer based on cognitive processes. It may be, for example, that some traits move along the axis of variation (i.e. elaborations) but change sufficiently to be put into another category. However, these problems aside, the idea of elaboration and innovation is potentially a productive way of thinking about how traits may diverge and arise. At the moment, only a few studies have explicitly tested the idea.

Bowerbirds use visual signals in sexual selection and perhaps species recognition, whereby males build bowers to attract females. These often comprise a cleared area and constructions of sticks, often decorated with brightly coloured objects from the environment, such as flowers and even man-made objects. Endler *et al.* (2005) quantified ornament evolution in Australian bowerbirds, using models of bird colour vision. They found that the ornaments used by bowerbirds to decorate their bowers were not elaborations of the species' own plumage patterns, but rather often innovations. In addition, species often showed clear evidence of efficiency innovation, because their ornaments were very different from the visual background and this acts to increase conspicuousness. Endler *et al.* suggest that the transfer of conspicuousness to bower ornaments may allow males

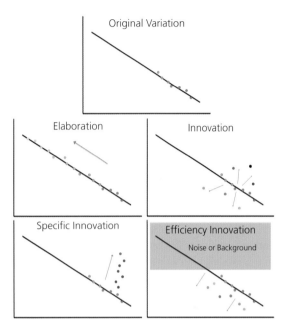

Figure 11.3 An illustration of the idea of elaboration versus innovation in signalling traits, modified from Endler *et al.* (2005). In elaboration, new phenotypes fall along the same axis of variation (e.g. brighter plumage, longer tails, louder and longer calls), whereas in innovation new components fall off the axis in another direction (e.g. new plumage colour or call type). This may be in any direction, or in one or a few directions only (specific innovation), or could occur in a direction that avoids background noise or enhances contrast with the background (efficiency innovation).

to circumvent some of the costs normally associated with signalling owing to increased predation risk by retaining some degree of camouflage in their plumage (see Chapter 7). Subsequent work (Endler and Day 2006), again considering avian visual processing, has shown that bowerbirds prefer coloured objects that differ from their own plumage, and that of the bower and surrounding background.

A study by Ryan *et al.* (2010) with túngara frog (*Engystomops pustulosus*) mating calls shows that females have preferences for both innovations and elaborations. Male túngara frogs call to females with either a simple call, comprising a whine only, or a complex call, being the whine plus 1–7 chuck components, with females preferring complex calls (see Chapter 7). Ryan and colleagues synthesized a range of calls consisting of whines plus chucks of various types, including heterospecific and conspecific calls, chucks with increased duration and amplitude modulation, human-made sounds, and other natural sounds. They found that females showed latent preferences for many of these complex calls over simple ones. Interestingly, Akre *et al.* (2011) have shown that the benefits of increased elaboration in male calls (greater numbers of chucks) has a diminishing benefit, with female preferences for elaborated calls declining with additional chucks. This seems to be because female responses follow something called Weber's law. Here, the ability to discriminate between two stimuli depends on their proportional rather than their absolute difference, meaning that as stimuli become more exaggerated, a greater difference is needed in order to be able to discriminate between them; i.e. it is easier to discriminate between male calls with either one or two chucks than it is to discriminate between calls with three or four chucks. Weber's law also applies in other contexts. For example, perceived visual contrast (e.g. brightness differences) between two stimuli also seems to depend more on the ratio of values between stimuli rather than absolute differences. Interestingly, predatory bats that eavesdrop on the male frog calls also show a response to increasing chuck components in line with Weber's law. This suggests that the benefits of adding multiple chucks for males is constrained not by greater risk of predation, but rather by perceptual limitations in females in discriminating between calls of increasing elaboration (Akre *et al.* 2011). The authors make the interesting point that such limitations on elaboration may favour the evolution of innovative signals occurring along a different perceptual axis. This idea would be valuable to investigate further in other systems.

11.3 Examples of Divergence and Speciation

In the following sections we focus on examples showing how a range of factors can lead to divergence in visual, electric, and acoustic signals, contributing to reproductive isolation and speciation.

11.3.1 Aposematism, Mate Choice, and Polymorphism in Poison Frogs

Poison frogs from Central and South America have long represented some of the most striking exam-

ples of toxicity and warning signals (aposematism) across a range of species (e.g. Darst *et al.* 2006; Summers and Clough 2001). The group is especially interesting because not only is there variation across species, but many species are also highly polymorphic for colour and pattern. That colour morph is highly labile in some species is interesting because it is not what conventional aposematism theory predicts. On the contrary, aposematic signals should be constrained because predators will have fewer colour patterns to learn to avoid. Recently, it has become clear that the colour signals of the frogs are also used in mate choice (Summers *et al.* 1999). Therefore, the group makes an interesting system to explore the evolution of intra- and interspecific signal diversity.

A phylogenetic study by Wang and Shaffer (2008) of the extremely polymorphic strawberry poison-dart frog, *Oophaga* (*Dendrobates*) *pumilio*, with morphs ranging from bright red, yellow, blue, to dull green (Figure 11.4), found in different geographical locations, used mtDNA sequence data to investigate the evolution of polymorphism. They found that shifts in coloration seem to occur regularly, including changes to relatively cryptic coloration, and that similar selective forces may drive convergence in some of the similar morph types (see Brown *et al.* 2010 for similar findings).

Summers *et al.* (1999) and Maan and Cummings (2008) have shown that female *O. pumilio* often prefer to mate with males of the same colour morph as themselves (assortative mating), based on the male dorsal coloration; although Maan and Cummings (2009) have also shown that female preferences may be directional in some populations, favouring, for example, brighter males. The conspicuousness of the frog coloration in terms of bird (predatory) vision is also correlated with the degree of toxicity across individuals from different populations in the species (Maan and Cummings 2012). Noonan and Comeault (2009) have demonstrated, using artificial clay models resembling morphs of *D. tinctorius* in the field, that immigrant (i.e. novel) phenotypes have a higher risk of predation than aposematic and cryptic phenotypes found in the local geographic area; i.e. evidence for stabilizing selection within populations but divergent selection between them. Thus, the coloration of poison frogs could act as a

Figure 11.4 Examples of some of the remarkable variation found in morphs of the poison frog *Oophaga pumilio* from different regions (Bocas del Toro archipelago and mainland and Escudo de Veraguas) of Panama. Middle left, bottom-left, and bottom-centre images © Laura Crothers, all other images © Martine Maan.

Box 11.2 Models of Speciation

Ultimately, divergence in populations can lead to the formation of new species. A crucial issue is how populations become reproductively isolated, that is, what mechanisms evolve that limit the exchange of genes between emerging species. These may act before or after fertilization, and so researchers talk about pre-zygotic and post-zygotic isolation. Pre-zygotic isolation can be further divided into pre-mating barriers, including for example prevention of mating owing to different behaviours, being active at different times of day, or being found in different geographical locations, and into post-mating barriers, which include the prevention of fertilization (for example, through incompatible gametes) and zygote formation once mating has occurred. Post-zygotic mechanisms include hybrid sterility or highly maladaptive behaviour in offspring.

Traditionally, biologists have defined three types of speciation relating to geography: *allopatric speciation*, where new species evolve in geographical isolation, *parapatric speciation*, where new species evolve in a geographically continuous population, and *sympatric speciation*, where new species evolve within the same geographical range. Note, however, that these represent more of a continuum and in practice it can be difficult to distinguish between them. For example, barriers to gene flow could operate on a very local scale within the general habitat/community. In addition, it is now realized that a diversity of geographic scenarios are likely to characterize speciation, and recent work focuses more on the mechanisms leading to reproductive isolation (Rundle and Nosil 2005). A common process leading to speciation is likely to be *ecological speciation*, whereby adaptation to different

continued

Box 11.2 *Continued*

ecological niches (e.g. competition or predation avoidance, or improved food acquisition) results in reproductive isolation between divergent populations (see reviews by Rundle and Nosil 2005; Schluter 2001).

Allopatric speciation has generally been regarded as the most likely route to species formation. It can be broken down into two types. *Peripatric speciation* occurs when two populations become isolated in space due to dispersal and colonization of new habitats. An example is the founder effect; for example, when a group of individuals colonize a new island separated from the mainland. Here, because the population size of the colonizing individuals may be low, random genetic drift may be important. In contrast, *vicariant speciation* occurs when the existing range of a species is separated due to the formation of a geographical barrier, such as a mountain range or a new river formation. If two populations come back into contact but are not yet completely reproductively isolated, hybrids may be formed. If these hybrids have lower fitness, selection may act to reduce the likelihood of the parent forms mating in a process known as *reinforcement*. Thus, reinforcement may complete the speciation process in sympatry.

In parapatric situations, populations do not significantly overlap in their range but are immediately adjacent to one another. Thus, in contrast to allopatry, emerging species can exchange genes but not to the extent of sympatric populations. Natural selection may favour different phenotypes in different areas, and so populations diverge. Again, where the two populations meet and interbreed, selection may act to strengthen pre-zygotic barriers if hybrids between the incipient species have reduced fitness (i.e. reinforcement).

Sympatric speciation has been a controversial topic in evolutionary biology as it presents a number of theoretical obstacles, and because it can be very difficult to show that two sympatric species did not evolve in allopatry and then come back together later. However, it could theoretically occur via two main routes. The first is by ecological speciation, where two populations adapt to different niches. Assortative mating may evolve as an incidental by-product. For example, selection might cause a population host-plant shift in phytophagous insects that mate on specific plants. Alternatively, assortative mating may limit individuals interbreeding with different morph types that would otherwise result in offspring with low fitness, in a process

analogous to reinforcement. For instance, two morphs may each be well camouflaged against two different background types, whereas hybrids match neither background well. The other type of sympatric speciation may occur through sexual selection, where female preferences for different male traits in a population could lead to assortative mating and the divergence of increasingly extreme male phenotypes, for instance under Fisherian runaway selection (van Doorn *et al.* 2004). Sexual selection models of speciation that include runaway selection or sexual conflict, whereby selection arises from interactions between the sexes alone, do not fall under ecological speciation (Rundle and Nosil 2005). Note that assortative mating is not synonymous with sexual selection because the latter involves 'competition and mate choice that generate variation in mating success among individuals of the same sex' (Maan and Seehausen 2011).

Regardless of the geographical arrangement, it is currently debated as to whether sexual selection, on its own, can lead to speciation without some ecological adaptation too (e.g. sensory drive, see Box 11.3 and main text). The idea, however, seems intuitive because sexually selected traits affect mate choice and could reduce interbreeding between diverging populations, and because species with high levels of sexual selection seem to diverge quickly and show high variation. For example, evidence that sexual selection might play a key role in divergence, diversification, and speciation has been found, for example, in spiders (e.g. Masta and Maddison 2002) and crickets (e.g. Mendelson and Shaw 2005). Such reproductive isolation may require only small differences in genomes between populations. Indeed, Darwin noted that many closely related species are similar in many characteristics but differ in terms of sexually selected traits. Evidence exists from modelling approaches, genetic studies, and comparative/phylogenetic work, but there is less direct empirical evidence and the relationship to ecological speciation may be complex and natural and sexual selection may often operate together in speciation (see reviews in Arnegard and Kondrashov 2004; Kraaijeveld *et al.* 2011; Maan and Seehausen 2011; Panhuis *et al.* 2001; Ritchie 2007; van Doorn and Weissing 2001; van Doorn *et al.* 2004; Weissing *et al.* 2011).

Overall, what makes understanding speciation so complex in real species is that a range of mechanisms may be operating at the same time and interact with each other, and many logically discrete theories may represent continuums in nature. Indeed, divergent selection (ecological selection) is likely to be important in many of the above systems, but that does not discount other factors, such as genetic drift (Rundle and Nosil 2005; Sobel *et al.* 2009).

Character Displacement

Character displacement is a process that occurs when two similar species that occur in sympatry evolve divergent traits owing to the advantages of reduced competition (see review by Pfennig and Pfennig 2009). Broadly, we can consider two types of character displacement. *Ecological character displacement* involves divergence in behavioural, morphological, or ecological traits to exploit different resources, whereas *reproductive character displacement* involves divergence in reproductive traits to reduce the likelihood of sexual interactions. Character displacement can prevent competitive exclusion from occurring, that may otherwise prevent similar species from occurring in sympatry. Character displacement should only happen when species co-occur, leading to the prediction that, all else being equal, populations in sympatry should differ more than populations in allopatry (Pfennig and Pfennig 2009). Character displacement could also lead to speciation if it finalizes reproductive isolation between incipient species.

Reproductive character displacement (RCD) on mating signals may be a particularly important evolutionary process. Closely related species in sympatry often diverge in signal form used in reproduction, under selection to reduce potentially costly reproductive interactions between heterospecifics, such as hybridization or simply attempting to mate with the wrong species (Crampton *et al.* 2011). Reinforcement, often considered a type of RCD, can occur when character displacement is driven by the costs of hybridization (Pfennig and Pfennig 2009). In addition, we also may expect to find *facilitated RCD*, where character displacement occurs in two

different species that do not produce hybrid offspring, in order to prevent wasted time, energy, or gametes, during sexual interactions between heterospecifics (Crampton *et al.* 2011). Furthermore, masking interference, where overlap between signals of different species reduces the ability of individuals to communicate effectively with conspecifics (Amézquita *et al.* 2006; Chapter 10) may be prevented by RCD in signal form. Masking interference need not involve direct contact or mismating, and often involves distantly related species. RCD is different from other mechanisms of signal divergence that may occur in sympatry, such as sexual selection, sensory drive, and predation pressure, as these do not arise under selection to prevent mismating and hybridization (Crampton *et al.* 2011).

Magic Traits

The idea of 'magic traits' (traits that are subject to divergent selection and contribute to non-random mating based on pleiotropic effects of the same gene(s)/allele(s); Servedio *et al.* 2011), has become a hot topic in evolutionary biology—albeit sometimes a controversial one. They are an attractive proposition as to how speciation could occur even with gene flow (e.g. in sympatry) because a trait (or traits with a common genetic basis) could influence both divergent selection and non-random mating (see Servedio *et al.* 2011; Smadja and Butlin 2011). Usually, the divergent selection is thought to be ecological (e.g. based on variation in predation pressure, or foraging niche competition), and this also affects mating cues or mate preferences. Magic traits could also be implicated in speciation by sensory drive, when selection pressure under different environments leads to changes in sensory systems to operate better in particular environments (in tasks such as foraging or navigation), and this also influences mate perception and preferences (Maan and Seehausen 2011; Servedio *et al.* 2011).

magic trait if it both influences assortative mating and is under divergent ecological selection (stemming from predation; Reynolds and Fitzpatrick 2007). These factors may select against immigrants of different appearance, resulting in greater genetic isolation between such populations with different colour morphs, rather than specifically based on

geographical distance alone (Wang and Summers 2010).

In *O. pumilio,* Wang and Summers (2010) suggest that the situation may be a case of incipient speciation. Mating between morphs could lead to offspring with reduced fitness if they have intermediate phenotypes or they exhibit inappropriate mate preferences. This could lead to reinforcement preventing mating between different morphs (Richards-Zawacki and Cummings 2011) because changes in coloration through hybridization may have two costs. First, intermediate forms may be unfamiliar to predators, and therefore not avoided if the predator does not associate the coloration with toxicity, and second they may not be attractive to potential mates. In populations where more than one morph is found, Richards-Zawacki and Cummings (2011) demonstrated that females had stronger preferences for their own morph type than females in populations with only one morph present, consistent with intraspecific reproductive character displacement. Conventionally, conspicuous mating displays are often constrained by predation pressure and selection for concealment (Chapter 7). However, the relationship between sexual signalling and aposematism could combine to contribute to the high diversity in poison frogs and other species because both types of communication may be enhanced by conspicuousness. More work is still needed in this and other systems discussed below into what drives the increased diversity. Dual-function signals seem to occur often in polymorphic or species-rich groups. This could be if these systems facilitate strong associations between ecological selection and mate choice, which is needed for sustained (in the face of geneflow) reproductive isolation. This could allow quick responses to environmental heterogeneity with regards to an important ecological variable.

11.3.2 Mimicry, Mate Choice and Speciation in Heliconius Butterflies

The *Heliconius* butterflies, comprising about 40 species and hundreds of geographic variants across the Neotropics, have long been a classical example of Müllerian mimicry, and a system to study the genetics of adaptation and speciation (Brown 1981; Jiggins 2008; Joron and Mallet 1998; Joron *et al.* 2006). In this group, closely related species often differ greatly in appearance, while distantly related species often converge in mimicry form (Jiggins 2008). Work has shown that the dual role of coloration in mate choice and predation can lead to reproductive isolation via ecological speciation. For example, Jiggins *et al.* (2001) showed that two sister species,

Heliconius melpomene and *H. cydno*, which are found in Central America but differ in coloration and habitat use, have recently diverged to form mimetic relationships with two other species of *Heliconius* (the red, black, and yellow *H. erato,* and the white and black *H. sapho*, respectively). Experiments show that males use these divergent colour patterns during mate recognition, leading to assortative mating (Jiggins *et al.* 2001). Hybrids are rare in nature, but when hybridization does occur, the intermediate phenotypes may lack effective mimicry of either model, and are thus likely to be vulnerable to predation. The effects of colour pattern on both mating and natural selection via predation risk can therefore influence both pre- and post-mating reproductive isolation (Jiggins *et al.* 2001).

Other experiments have also shown that males recognize females of the same species to mate with based on the wing coloration, and that there seems to be a tight genetic association between wing colour and colour preference (Kronforst *et al.* 2006; Merrill *et al.* 2011b). Because the coloration is used in both mate selection and as a warning signal to predators, this tight genetic association means that divergent natural selection on coloration may also simultaneously drive divergence in the mate recognition signal and preference, facilitating rapid evolution (Kronforst *et al.* 2006). Evidence also indicates that when populations occur in sympatry, individuals are less likely to court members of the other species than individuals from allopatric populations (Jiggins *et al.* 2001). Similarly, reproductive character displacement is also observed at interspecific contact zones in parapatric *Heliconius* species, where males are more resistant to court heterospecifcs compared to those from populations further from the contact zone (Kronforst *et al.* 2007). Below the species level, a study of *Heliconius melpomene* subspecies shows that males from four recently diverged parapatric populations of different colour patterns, which hybridize freely in contact zones, are more likely to court females from their own colour patterns than females from other races (Jiggins *et al.* 2004). Thus, pre-mating isolation based on coloration occurs before divergence in other traits, such as habitat preferences or genomic incompatibility. Further work, looking at pairs of sister *Heliconius* taxa, indicate that pre-mating isolation through mate preference arises early in diverging populations, and continues to increase as differences between incipient species become greater (Merrill *et al.* 2011a).

More work is still needed to investigate mate choice using models of *Heliconius* vision, in particular to determine the level of colour pattern difference that may be discriminated. This could be important because recent work shows not only that *Heliconius* species have evolved wing pigments not found in close relatives, increasing their diversity in coloration, but also that they have a range of visual pigments in their eyes, including two distinct UV-sensitive visual pigments (peak sensitivities of 355 and 398 nm) (Briscoe *et al.* 2010; see Chapter 4). In addition, divergence in colour patterns and the high level of mimicry (to human eyes) of other species indicates predators also discriminate between fine differences in colour pattern. Currently, little modelling of predator vision has been done to determine what level of detail predators could theoretically discriminate, although one study (Bybee *et al.* 2012) indicates that some colour patches on the butterflies (especially UV-yellow ones) may be less discriminable to birds than to butterflies owing to the extra UV visual pigment in *Heliconius*. Finally, although mark–release–recapture experiments with different morphs of *H. erato* in different locations indicate that foreign morphs have lower survival than resident morphs (Mallet and Barton 1989), predator selection against interspecies hybrids has rarely been directly tested and remains to be experimentally demonstrated (Jiggins 2008; but see a recent study by Merrill *et al.* in press). Overall, however, *Heliconius* wing patterns are perhaps the best-supported candidate for a 'magic trait' (Servedio *et al.* 2011), promoting divergence and speciation in the face of ongoing gene flow.

11.3.3 Divergence and Speciation in Electric Fish

Recent work with electric fish illustrates the role of electric signals in divergence and speciation, and how multiple isolating mechanisms may be involved. Species flocks comprise a number of species that have diversified from a common ancestor in a geographically restricted area, often over a relatively short period of evolutionary time (Sulli-

van *et al.* 2002). Sullivan *et al.* (2002) described a species flock of mormyrid electric fish from Africa ('elephant fish'), where electric organ discharges (EODs; Chapter 2) play an important role in reproductive isolation between the species, and where rapid speciation may have occurred leading to high numbers of species in the same location. At present, just over 20 species of the Gabon species flock *Paramormyrops* (previously called *Brienomyrus*) have been reported (though many await formal taxonomic descriptions) from river systems in Gabon and Equatorial Guinea (Arnegard *et al.* 2010a). The group exhibits great diversity in the structure of their EODs, and much of the diversification may have occurred within two million years. The different EODs are species specific and play a role in mate recognition, indicating that EODs may have been important in the origin and maintenance of the species boundaries (Arnegard and Hopkins 2003; Sullivan *et al.* 2002). Furthermore, allopatric taxa show greater overlap in EOD form than sympatric taxa (Arnegard and Hopkins 2003). Sympatric populations of *Paramormyrops* with different EOD characteristics and morphology appear reproductively isolated, although the relationship between some individuals (especially in the so-called the *magnostipes* complex) is complex and reproductive isolation has not always been demonstrated (Arnegard *et al.* 2005, 2006, 2010a).

Recently, a study by Arnegard *et al.* (2010a), involving phylogenetic comparative assessments, as well as measures of EOD characteristics and ecological traits, provided evidence that rapid EOD signal evolution has been driven by sexual selection and has led to species radiations in this group. They measured the variation in EOD characteristics across 21 species and morphs, plus variation in three other ecological traits that have been important in other adaptive radiations: body size, body shape, and trophic ecology (feeding ecology). They found that closely related species could be as divergent in EOD signals as distantly related species, whereas the level of ecological divergence was reflected in the level of phylogenetic difference. In all, the rates of signal divergence are estimated to be three to four times higher than those for ecological traits in *Paramormyrops*, and EOD characteristics diverged early in the radiation and ecological traits later. As Arnegard *et al.* (2010a) discuss, the different EOD characteristics are not driven by ecological selection by resource competition; there have been no observed associations between habitat and EOD composition, or evidence that EODs are matched to specific habitat characteristics (see also Feulner *et al.* 2009a). Furthermore, the study area also lacks electrosensitive predators that might have otherwise selected for differences in EOD characteristics, allowing for a diversity of signals to evolve unconstrained by predation pressure (Arnegard *et al.* 2010a). Thus, the species flock has rapidly diverged in EOD signal form in the absence of correlated divergence in ecological traits, indicating that sexual selection and/or species recognition on signals has been the major driving force behind species radiations in this group and other mormyrids.

In another radiation of electric fish, diversification may have been driven by a combination of ecological speciation and reproductive isolation based on EODs. The African mormyrid genus *Campylomormyrus* shows a high diversity in EOD characteristics, including between closely related species (identified by molecular phylogenetic analysis) and these may be important in reproductive isolation between species in the lower Congo River (Feulner *et al.* 2008). Here, an adaptive radiation with assortative mating based on EOD characteristics may have occurred. Indeed, closely related species in the genus show greater differentiation in EODs than do more distantly related species, suggesting that EODs act in reproductive isolation (Feulner *et al.* 2008). Species also differ in morphology (shape of the trunk-like snout) and such morphological traits may be under disruptive selection associated with different foraging niches (Feulner *et al.* 2009a). This could lead to trophic niche segregation; sister species sometimes differ in both EOD and snout morphology. So far, differences in EODs between males and females have not been found, indicating that sexual selection is unlikely to drive EOD divergence in this group; instead, EOD characteristics may promote species recognition and thus isolation (Feulner *et al.* 2009a).

Feulner *et al.* (2009b) conducted behavioural mate choice experiments with *Campylomormyrus compressirostris*, and predicted that females of this species would prefer conspecific males over those of the closely related sympatric species, *C. rhynchophorus*, which shows a longer EOD form. However, they also predicted that females would not discriminate between conspecific males and those from a more distantly related species, *C. tamandua*, whose EODs are similar to male *C. compressirostris*. Two-way choice experiments were conducted in the dark, with females' preference for either a hetero- or conspecific male (behind a grid) tested, or when presented with only playbacks of EODs of the different species. As predicted, female *C. compressirostris* preferred conspecific males over those of *C. rhynchophorus*, but did not discriminate between *C. compressirostris* and *C. tamandua* (either the males or playbacks; Figure 11.5). Thus, Feulner *et al.* (2009b) suggest that the longer EOD duration in *C. rhynchophorus* may have promoted reproductive isolation with sympatric *C. compressirostris*. Again, such differences in EOD are also in parallel with differences in snout morphology and may be linked with trophic niche separation. It seems, overall, that divergence in snout morphology under disruptive natural selection was the driving force behind speciation (ecological speciation), and that EOD characteristics are now responsible for maintaining reproductive isolation (Feulner *et al.* 2009a).

In South American gymnotiforms (electric 'knife fish'), there is good evidence that EOD diversity can be driven not by species recognition or sexual selection, but by predation pressure. Although predation is usually thought to constrain communication signals due to the risk of eavesdropping, a study by Stoddard (1999) indicates that greater signal complexity in gymnotiforms may have arisen because some types of electric signal are less detectable to electrosensitive predators, such as the electric eel (*Electrophorus electricus*, a gymnotiform species) and catfish species (Siluriformes), which are then subsequently modified under mate selection. Stoddard showed that in pulse-type gymnotiforms, a switch from a monophasic (the ancestral state, according to Stoddard 2002) to a biphasic EOD is accompanied by a shift to frequencies above those to which the low-frequency-sensitive ampullary receptors (used in passive electrolocation; see Chapter 2) in the predators (and knife fish) are most sensitive. The EODs can still, however, be effectively detected by the tuberous receptors of the knife fish, which are

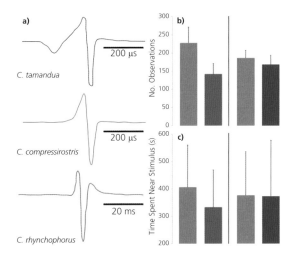

Figure 11.5 a) EODs of *Campylomormyrus compressirostris* (red line), *C. rhynchophorus* (green line), and *C. tamandua* (blue line). b) Experiments where *C. compressirostris* females could associate with either a conspecific male (red bars) or with a heterospecific male (green/blue bars) show that females associate more with conspecifics when paired with a closely related sympatric species (*C. rhynchophorus*) that have a longer EOD, but show no difference in association when presented with a heterospecific male of more distantly related species (*C. tamandua*) whose EODs are similar to male *C. compressirostris*. c) The results are similar when female *C. compressirostris* are presented with playbacks of EODs. Plots in B and C show means +SE for female responses to the EODs in A drawn with the corresponding colours. Data from Feulner *et al*. (2009b).

tuned to higher frequencies associated with EODs. The monophasic EOD has its peak power close to frequencies of zero Hz, similar to the peak sensitivity of the ampullary receptors. The addition of a second negative going phase in the biphasic EOD shifts the spectral energy above that of the ampullary receptors' peak sensitivity (Figure 11.6); so-called 'signal cloaking' (Stoddard and Markham 2008; see Chapter 7). This change from a mono- to a biphasic EOD appears to have evolved twice independently in pulse and once in wave-type families of gymnotiforms. In many gymnotiform electric fish, sexual dimorphism exists principally in the second phase of the EOD.

Stoddard (1999) showed that in playback trials with a predatory electric eel (*Electrophorus electricus*), which had been trained to approach electrodes producing electric playback signals accompanied with food, that biphasic signals of female *Brachyhypopomus pinnicaudatus* are less detectable to preda-

tors than monophasic signals (made by deletion of the second phase); the eel was half as likely to approach the biphasic signal than the monophasic signal. Finally, the three species of Gymnotiformes that have monophasic EODs are either a well-defended predator (the electric eel), occur in a region without electroreceptive predators, or may be a Batesian mimic of electric eels, producing EODs with very similar characteristics. In general, in Gymnotiformes there is little evidence that the evolution from a mono- to a biphasic EOD was driven by selection for heightened sensitivity for electrolocation, territorial defence, reproductive isolation, or through sexual selection, though the latter two factors may be involved in modification of the second phase once it arose (Stoddard 1999, 2002). However, recent phylogenetic work (Arnegard *et al*. 2010b) casts doubt on the hypothesis that a low frequency, monophasic EOD was ancestral in the Gymnotiformes, although electroreceptive predators in South America may nevertheless act as a significant selection pressure contributing to EOD evolution. Changes in EOD in some African mormyrids may have also occurred due to predation pressure from electroreceptive catfish (Hanika and Kramer 2000).

Finally, a recent study by Crampton *et al*. (2011), investigating seven species of *Gymnotus* electric fish from Brazil shows evidence for reproductive character displacement, with divergence of EOD form in signal space between sympatric species. Crampton and colleagues argue that the results cannot be attributed to avoidance of masking interference (ecological character displacement) from EODs of other species because the fish have a jamming avoidance response (Heiligenberg 1991), whereby pulse-repetition rate is altered to avoid coincident pulses with neighbouring fish and because EODs only function over very short ranges. Furthermore, the *Gymnotus* species seem not to have rapidly diverged and are genetically distinct species, with post-speciation divergence facilitated by reproductive character displacement. Crampton *et al*. also found that species with the most pronounced EOD sexual dimorphism also show the greatest signal displacement from heterospecifics, indicating that sexual selection and reproductive character displacement could interact to promote divergence in signals and female preference in this group. Simultaneously,

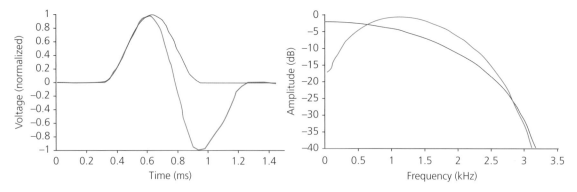

Figure 11.6 The left graph shows a modelled ancestral monophasic EOD (blue line) and a biphasic EOD (red line). The graph on the right shows the corresponding power spectra of these two EOD types, with the biphasic EOD having depressed amplitude below 0.6 kHz. The ampulary electroreceptors of potential predators are sensitive to about 8 Hz (Siluriformes) or 30 Hz (Gymnotiformes), frequencies that the biphasic EODs have lower amplitude compared to monophasic EODs. Data from Stoddard (1999).

that is, the signals could be under both selection to reduce the likelihood of mismating (leading to reproductive character displacement) and sexual selection (leading to sexual dimorphism). In some cases, EODs may facilitate the exploitation of different food resources and function in mate selection (either mating cues or sexually selected preferences), and could therefore provide examples of magic traits (Feulner *et al.* 2009a).

The diversity in EOD waveforms in many species of electric fish, both gymnotiform and mormyroid, seems to have been driven by a range of factors, including reproductive character displacement, predation pressure, and sexual selection, or a combination of those. Recent molecular work has also made exciting breakthroughs. Zakon *et al.* (2006) and Arnegard *et al.* (2010b) have shown that expression of the voltage-gated sodium channel gene, *Scn4a* (coding for the subunit of the ion channel, Nav1.4a), which is expressed in skeletal muscle of non-electrogenic teleost fish, has been lost from muscle but gained in myogenic (muscle derived) electric organs of electric fish. They found amino acid substitutions at sites involved with fast channel inactivation, and suggest that these changes may underlie some of the variation in EOD duration observed among species. Arnegard *et al.* (2010b) also found that similar genetic and molecular changes underlie parallel independent diversifications of electric organ discharges in gymnotiforms and mormyroids. Nav1.4a (*Scn4aa*) seems to have

been co-opted for restricted expression in electric organs during the independent origins of this novel organ of communication in the two groups, and that during the subsequent diversification of EODs, both groups exhibited amino acid substitutions in the same regions of the sodium channels related to ion channel kinetics (and, therefore, EOD timing). This shows that convergence of a novel form of communication in both groups is paralleled by similar independent genetic and molecular changes. Thus, there are interesting analogies here with changes in coloration for camouflage in mice and lizards (see 11.1.1), and with regards to changes in echolocation in bats and whales and infrared detection in snakes (see Chapter 2).

11.3.4 Harmonic Hopping in Bats

In Chapter 3 we discussed how different types of echolocation call in bats can be useful in different habitat types for prey detection. Work by Kingston and Rossiter (2004) in the large-eared horseshoe bat (*Rhinolophus philippinensis*) from south-east Asia and northern Australia shows that changes in echolocation associated with prey capture could lead to divergence between populations. The species comprises three sympatric size morphs (currently subspecies): large, intermediate, and small (nearly half the size of the large morph). These morphs are characterized by differences in the frequency of their echolocation calls, with large, intermediate, and small emitting frequencies of approximately 27, 40, and 54–kHz respectively. The frequencies of the small and intermediate morphs seem to correspond to different harmonics of the large morph's fundamental frequency (Chapter 3), leading Kingston and Rossiter to suggest that the morphs are switching between harmonics in their echolocation and that this could have a major influence on prey capture. This is because low frequencies travel further, increasing detection range, but also reflect poorly from small prey items reducing the number of small prey items that are detected. Thus, there may be a trade-off, with large morphs detecting larger prey types, and small morphs detecting smaller prey. The authors also suggest that the changes in call frequency would mean that each morph will become insensitive to the calls of the other morphs, and that if the calls are used in mate selection, then this could lead to reproductive isolation between the different morphs (i.e. the calls may be magic traits). In line with this they found significant genetic differences between the morphs. Finally, Kingston and Rossiter found that various species in the *R. philippinensis'* clade call at harmonics of the fundamental frequency of the large morph, and suggested that 'harmonic-hopping' may have contributed to the rapid radiation of these bats in the last five million years.

11.3.5 Acoustic Signals in Birds and Amphibians

As discussed in Chapter 10, there is good evidence that many features of animal acoustic signals are tuned to environmental characteristics, such as transmission properties and noise profiles. This idea is sometimes referred to as the acoustic adaptation hypothesis. In many groups, a range of factors may drive changes in song structure. For example, a comparative study of 163 species of neotropical antbird by Seddon (2005) shows evidence that changes in song structure are driven by changes in morphology and habitat transmission properties. Body mass and bill size were linked to frequency and temporal patterning respectively, and birds singing in the understory produced higher pitched songs than those in the midstory. The study also suggests that species recognition and character displacement have been important, with the songs of closely related sympatric pairs

more divergent than of those in allopatry. Overall, there is a steadily growing understanding of how song properties can contribute to reproductive isolation. Dingle *et al.* (2008) found that song divergence was related to ambient noise in two genetically distinct but hybridizing subspecies of the grey-breasted wood-wren (*Henicorhina leucophrys*), and that this possibly contributed to assortative mating or reproductive isolation upon secondary contact. Likewise, Qvarnström *et al.* (2006) have shown that the degree of similarity in song type predicts the likelihood of hybridization in two species of flycatchers.

The majority of work has looked at how environments can affect song characteristics, and the implications that it may have on reproductive isolation, but less so on receiver responses. However, Ripmeester *et al.* (2010) have shown that male European blackbirds (*Turdus merula*) show differences in their songs between urban and forest populations, and individuals from these habitats respond differently to playbacks of the same song forms. Seddon and Tobias (2010) have looked more specifically at the role of receiver perception. They tested mate recognition in two sympatric closely related species of Amazonian antbird, *Hypocnemis peruviana* and *H. subflava*. The two species' songs are very similar, but may contain potential information about individual identity. Playback experiments to females showed that they responded more to calls of their own mates than to conspecific strangers, but also responded more to the calls of conspecific strangers than to heterospecifc strangers. Therefore, females can discriminate between both species and individuals on the basis of very small differences between songs, even in densely vegetated environments with noise, allowing two species to co-occur without showing strong differences in song structure. However, without modelling the birds' acoustic systems, it is hard to determine whether there are any perceptual differences in the ability to detect different song structures rather than differences in motivation to respond. Work by Henry and Lucas (2008, 2010) shows that birds can undergo changes in their auditory systems associated with the use of different song types in different habitats, and this is an important area of future work.

In general, the extent of species differences in song varies greatly across different groups, and it is not yet clear why some species have evolved differences in song structure but others have not. Seddon and Tobias (2010) suggest that in recently diverged species, reproductive character displacement may be more pronounced because other mechanisms to prevent hybridization may not have arisen yet. In species like the antbirds, prolonged reinforcement could have also led to refined signal perception/response.

Overall, there is still only limited direct evidence that acoustic divergence in song has led to reduced gene flow and subsequent speciation (Dingle *et al.* 2008; but see Patten *et al.* 2004 for evidence in Californian song sparrows, *Melospiza melodia*). It is also important to consider that multiple interacting sensory and ecological factors may be driving differences in populations and isolation. For example, Darwin's finches from the Galápagos Islands are one of the most famous examples of natural selection and speciation, owing to differences in bill size and shape to cope with different food resources. However, differences in bill morphology also influence the ability of the bird to produce different song types, and this in turn can affect reproductive isolation. Podos (2001) has shown in the Galápagos finches that differences in bill morphology covary with song structure, and it seems that when the bills become modified to cope with different food types, that this also leads to the acquisition of distinct vocal signatures that distinguish them from other bill types; thus ecological adaption causes a change in signal form (Ryan 2001). Females should mate with males with similar bill types, because mating with dissimilar types could lead to intermediate bills that are less effective for either food resource, potentially leading to pre-zygotic isolation and speciation.

Comparable studies of song have also been conducted in amphibian calls. For instance, Ryan *et al.* (1990a) found evidence that two subspecies of the North American cricket frog, *Acris crepitans*, vary in their mate recognition calls owing to being found in different habitats with different sound transmission properties. As with the electric fish discussed above, a range of factors may influence call divergence and properties. The idea that sexual selection can lead to speciation, as opposed to mating signals diverging to prevent hybridization, is still somewhat controver-

sial. However, Boul *et al.* (2007) have recently found several lines of evidence suggesting that speciation is occurring in the Amazonian frog, *Physalaemus petersi*, perhaps owing to sexual selection on male calls and female preference. They looked at different populations of the frogs and analysed male call structure, performed phonotaxis experiments with females, and analysis of microsatellite loci. Neighbouring populations have strong behavioural isolation due to differences in male mating calls, coupled with female preferences for local call characteristics. In addition, a phylogenetic analysis based on mtDNA sequences showed that alternative call types (simple or complex types) have become fixed several times across the species range in different populations, and gene flow is 30 times lower between populations with different calls than between populations with similar calls separated by similar geographical distances. There was also no evidence for hybridization. Subsequent work has also shown that differences among populations in call structure (dominant frequency) are better explained by genetic differences than geographical distances or landscape features (Funk *et al.* 2009). Again, as with electric fish, different mechanisms seem to operate in different groups. For example, in contrast to sexual selection, species of *Pseudacris* frog in the USA show divergence in sympatric populations between species consistent with character displacement and reinforcement, including divergence in female preferences in sympatry resulting in reduced hybridization (Lemmon 2009). Subsequently, as a result of increased discrimination between species, heightened intraspecific discrimination on mating calls in sympatry seems to have led to increased sexual selection and the evolution of more energetically costly mating calls (Lemmon 2009).

11.4 Sensory Drive and Speciation

In Chapter 10 we discussed how sensory systems and signals can be influenced by properties of the environment. Sensory drive (see Box 11.3) describes how communication systems adapt to characteristics of the environment, and could even lead to reproductive isolation as a by-product of adaptive changes in behaviour and perception to certain environmental conditions (see review by Boughman 2002). Part of the appeal of sensory drive is that it provides an ecological explanation for differences between populations in specific features of signal form and sensory systems, and this is important for understanding how reproductive isolation arises (Boughman 2002). Divergence can occur owing to the different selection pressures that local environments place on signals and sensory systems, and how those systems respond over evolution. Speciation by sensory drive predicts that male mating signals will have differences in transmission under different habitats, and that sensory systems will be tuned to the characteristics of the local environment, altering the perception of the signal form of prospective mates.

Boughman (2001) provided some of the first evidence that sensory drive can lead to speciation in a study of six recently diverged populations of three-spine stickleback (*Gasterosteus* spp.) fish in Canadian lakes. The different lakes, and parts of the lake in which they occur, vary with regards to how red-shifted the ambient light is due to organic matter filtering out shorter wavelengths. In more redshifted water, red coloration would not stand out well from the background because it would blend in with the colour of the water column, whereas black would contrast effectively. Conversely, in clearer water red should be an effective signal. Boughman did not analyse the female visual systems directly, but instead performed an optomotor response test (Chapter 10). This exploits the behavioural response of many animals to follow a moving stimulus, usually stripes. By placing fish in an apparatus whereby they follow rotating stripes, and by illuminating the apparatus with red light, it is possible to gradually lower the total amount of light present and determine at what light intensity the fish are no longer able to see the moving stripes and respond to them. Fish with higher sensitivity for red light stop following the stripes at lower light intensities. Boughman found that fish showed differences in their red sensitivity in line with the amount of red light in the environment from where they are found. She also scored males on a scale of one to five in terms of the intensity and area of their red nuptial coloration. Ideally, this should have been done objectively or by modelling fish vision, but does still provide an assessment of the degree of redness. Boughman found that males from

these different environments varied in the amount of red coloration on their body. These differences were as predicted by the sensory drive model. Lakes with more redshifted conditions had males with less red coloration and more black coloration, and females from lakes with more redshifted light conditions showed a lower sensitivity to red light. This means that lakes with lower red light have males with redder coloration and females more sensitive to red light (and hence male red coloration). Finally, behavioural experiments showed that female preferences for red males were stronger when they were more sensitive to red light, and that females preferred to mate with males from the same population.

Boughman's (2001) study is consistent with sensory drive and speciation by sexual selection, but a key gap was the underlying mechanisms by which female preference had changed (i.e. a causal link between red sensitivity and red preference) and objective quantification of male coloration. To date, the most complete evidence for sensory drive leading to speciation is from recent studies with cichlid fish. Cichlids are highly diverse groups of fish that in the Great Lakes of Africa that have undergone rapid speciation in a relatively short space of time. In fact, they are a classical example of an adaptive radiation, with a huge diversity between species in terms of coloration and behaviour. For example, Lake Victoria is home to more than 500 cichlid species, many of which seem to have arisen in the last few hundred thousand years. Seehausen *et al.* (1997) suggested that directional selection in female mate choice for male coloration could lead to disruptive sexual selection and reproductive isolation, and this might partly explain speciation in cichlid fish. They found that lower cichlid species diversity is found when lakes are more turbid, destroying visual mate choice and reproductive isolation. Indeed, there has been various evidence showing that coloration is important in mate choice in cichlids and could promote speciation by sexual selection (e.g. Barlow *et al.* 1990; Couldridge and Alexander 2001; Elmer *et al.* 2009; Seehausen *et al.* 1997), and cichlids seem to have excellent colour vision involving a range of receptor types (e.g. Hofmann *et al.* 2010; O'Quin *et al.* 2010; Sabbah *et al.* 2010).

Work in Lake Victoria, Tanzania has been primarily done on two closely related sympatric cichlid

---

Box 11.3 Sensory Drive

Sensory drive is a concept that has been around since at least the early 1990s, but is now gaining increasing support as a process that could drive changes in sensory systems, signals, and communication in general. It is also a promising area where mate preferences and sexual signals may be implicated in speciation, even in sympatry. Endler (1992, 1993) and Endler and Basolo (1998) broadly described sensory drive as a process involving interactions between the environment, features and biases in sensory (and cognitive) systems, behaviour patterns, and signals. They discussed how this could influence communication systems and mate preferences to experience evolutionary change in particular directions. A range of interacting physiological, environmental, and behavioural factors are implicated, and changes to each one can influence the other. Some interpretations of sensory drive are somewhat hard to pin down, partly because they encompass much of communication and signalling theory (in terms of efficacy at least), making the idea of sensory drive somewhat vague and lacking in predictive power as a model. Here, we adopt an approach whereby specific features of the environment (e.g. transmission properties or noise) have a significant impact on how sensory systems evolve to be most effective in behavioural tasks (e.g. finding food), and/or on the specific forms of signals used in communication. Sensory drive predicts that unrelated species living in similar habitats would evolve similar sensory and communication systems (Endler 1993). Changes in sensory systems could, in turn, cause changes in perception and behaviour, including mate selection (either assortative mating for phenotypes that correspond to the sensory system found in individuals of that environment, and/or preferences

for increasingly exaggerated sexually selected signals that increase conspicuousness in certain conditions; see Maan and Seehausen 2010). It is therefore often argued that sensory drive could lead to reproductive isolation and divergence between populations, based on a combination of natural selection acting on visual performance in non-mating tasks and altered mate preferences resulting from this.

Note that our focus above is on how sensory drive could involve changes in sensory systems, and the knock-on effect of this on signal evolution and mate preferences. Characteristics of the environment also drive changes in signal form to best stimulate the sensory system of the receiver given prevailing environmental conditions (Chapter 10). Generally, we would not normally expect changes in signal form to influence changes in sensory systems, whereas we do expect changes in sensory systems to influence signal form (since sensory systems are often used in a great diversity of different tasks; e.g. eyes are not solely used to evaluate potential mates or detect food). However, there could be exceptions to this involving coevolution between signals and sensory systems involving reciprocal changes in both (as may have happened in some visual signals in *Heliconius* and fireflies; see Chapter 4). Endler (1992) and Endler and Basolo (1998) used sensory drive to encompass various other features of sensory systems, most notably sensory exploitation and bias (see Chapter 8). However, sensory exploitation models need not involve changes due to environmental conditions and biases can exist for a range of reasons, and so here we consider sensory drive and exploitation as separate models (although they may operate together). Finally, note that the effects of sensory drive are not restricted to mating, but may affect a wide range of communication systems (Figure Box 11.1).

---

species, *Pundamilia pundamilia* and *Pundamilia nyererei* (see review by Maan and Seehausen 2010). In these species, females look similar whereas males of *P. pundamilia* are blue and males of *P. nyererei* are red (Figures 11.7 and 11.8).

The species are broadly sympatric but live along rocky slopes, with *P. nyererei* occurring in deeper waters (4–7 m) and *P. pundamilia* in shallower water (0.5–2 m), although they overlap especially between 2 and 4 m. Maan *et al.* (2006) showed that as water depth increases, the light composition becomes relatively more biased towards longer wavelengths due to particulate matter in the water, with the environment being significantly redshifted where *P. nyererei* occurs (Figure 11.9).

Note that having redder fish under redshifted light conditions is the opposite direction to the stickleback example above, possibly if conspicuousness in cichlids is judged against the background substrate rather than the water column. Maan *et al.* also undertook an optomotor test, similar to Boughman's (2001) with sticklebacks, and tested the two species of cichlid for sensitivity to shortwave (436 nm) and longwave (656 nm) light. This showed that *P. pundamilia* females had higher sensitivity than *P. nyererei* to blue light, but that *P. nyererei* was more sensitive to red light. This is consistent with earlier work on several *Pundamilia* species by Carleton *et al.* (2005) showing that there is variation in the peak sensitivity of the longwave sensitive pigment

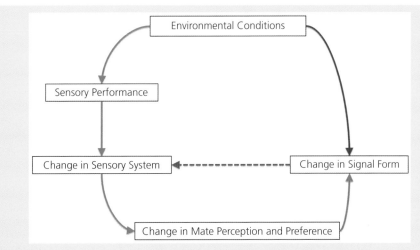

Figure Box 11.1 The process of sensory drive can lead to changes in both sensory systems and signal form under different environmental conditions, potentially with knock-on effects for mating and other communication systems. For example (red arrows), prevailing environmental conditions could influence the performance of a given sensory system in a particular habitat, leading to selection for changes in the sensory system in order to work more effectively. This could then influence the perception of potential mates, potentially altering mate selection or preferences, and in turn lead to changes in signal form, such that signal types change to more effectively stimulate the sensory system of the receiver. In some cases, coevolution may occur between the tuning of the sensory system and the specific form of the signal (dashed green arrow). Finally, the prevailing environmental conditions will often directly affect the specific signal form that evolves in order to facilitate signal transmission, detection, and recognition by a receiver (blue arrow).

across species, with species with red coloration having longwave sensitivity shifted to slightly longer wavelengths.

The results of Maan *et al.*'s (2006) study showed good indirect evidence for sensory drive, and a simulation model incorporating female choice, male

Figure 11.7 Images of *Pundamilia pundamilia* (top) and *Pundamilia nyererei* (bottom) showing the different colours of the males. Top image reproduced with permission from Oliver Selz/Eawag and the University of Bern, bottom image reproduced with permission from Ole Seehausen/Eawag and the University of Bern.

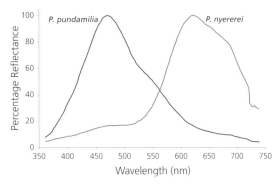

Figure 11.8 Normalized reflectance spectra of *Pundamilia pundamilia* and *Pundamilia nyererei* males, showing the differences in coloration resulting from greater longwave reflectance in *P. nyererei* and greater shortwave reflectance in *P. pundamilia*. Data from Maan *et al.* (2006).

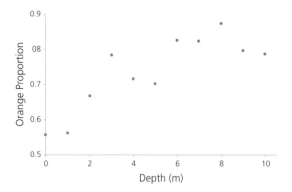

Figure 11.9 The proportion of longer wavelengths to shorter wavelengths increases with depth as the light conditions become more red-shifted. Data from Maan *et al.* (2006).

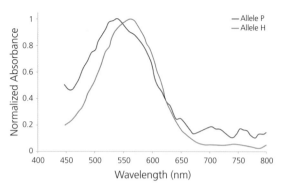

Figure 11.10 Normalized absorbance spectra from reconstituted longwave-sensitive pigments of the two alleles controlling sensitivity to different wavelengths of light. Allele P, predominantly found in *P. pundamilia*, is more sensitive to shorter wavelengths, whereas Allele H is more sensitive to longer wavelengths and predominantly found in *P. nyererei*. Data from Seehausen *et al.* (2008).

coloration, and visual sensitivity showed that speciation by sensory drive in cichlids was possible (Kawata *et al.* 2007). Subsequently, further evidence was presented by Seehausen *et al.* (2008), also on *P. nyererei* and *P. pundamilia* in Lake Victora, consistent with both the model and theory for sensory drive and speciation. They investigated the underlying alleles at the opsin genes, which control the tuning of the photoreceptors to different wavelengths of light (Chapter 2). They reconstituted the longwave sensitive pigments *in vitro* and measured the absorption spectra of these. Two alleles of the longwave sensitive cones seem to be of primary importance, and shifted in sensitivity by approximately 15 nm: Allele H, which is more sensitive to longer wavelengths and predominantly found in *P. nyererei* at greater depth, and Allele P, which is more sensitive to shorter wavelengths and found in *P. pundamilia* in shallower water (Figure 11.10).

In locations where there is a clear relationship between depth and coloration (e.g. where the water is not too turbid or light conditions do not change too rapidly with depth) mate choice experiments showed that female preferences were bimodally distributed, and the pattern of opsin divergence correlated with this. However, more work is needed to demonstrate that individual preferences are associated with specific opsin genes; i.e. whether females with Allele P prefer blue males and whether females with Allele H prefer red males. However, on a population level, fish in different visual conditions have differences in their visual sensitivity, and they in

turn prefer males of the 'correct' colour type. This is not to say that visual sensitivity alone would determine mate choice though, as other factors such as encounter rates of individuals of different types at different depths and sexual imprinting also seem to play a role (Kirkpatrick and Price 2008; Seehausen *et al.* 2008). The relationship between opsin allele type and male coloration also seems to be found in other cichlid species, and so sensory drive may have been an important factor in cichlid speciation in Lake Victoria in general. It may also be another example of a magic trait, with visual tuning affecting both ecological fitness and mate preference.

As discussed above, demonstrating an individual-level association between colour vision and colour preference in a mating context is still needed. In addition, as Maan and Seehausen (2010) discuss, one of the main pieces of evidence still lacking for a direct role of sensory drive producing speciation is confirmation of the assumption that females prefer male coloration that is more conspicuous, and that male colours have evolved to be more conspicuous to the respective female visual systems in their associated local habitat. Experiments carefully testing the role of male conspicuousness in female choice, in terms of the modelled visual system responses are needed, as is modelling of how conspicuous males are to each visual system in the different light conditions. In a different system, the Indonesian fish *Telmatherina sarasinorum*, different colour morphs (such as blue and yellow males) have different levels of conspicuousness in different habitats, and the locations where they are most conspicuous are also where their reproductive success is highest (Gray *et al.* 2008). Finally, it is not only genetics that may underlie visual system performance and preferences but also developmental plasticity when individuals are reared under different environments. This can affect the dynamics involved with sensory drive (see e.g. Fuller and Noa 2010).

As discussed above, differences in the environment can have a significant impact on the evolution of acoustic communication in birds and amphibians. As yet, there have been no explicit studies of sensory drive and speciation in acoustic communication, but one recent study provides some evidence for this in birds. Tobias *et al.* (2010) investigated the songs of Amazonian birds and compared closely related pairs of species in either bamboo or upland forest habitats. They found divergence in song between these pairs that was predicted by differences in sound transmission properties of the environment, rather than genetic divergence, ambient noise, or pleiotropic effects of bill morphology. The authors suggest that because songs can be important in reproductive isolation that these divergences could have been important in speciation. Unfortunately, this and many other studies of bird acoustic communication have not looked at changes in hearing sensitivity, and so we have little idea in most of those systems where there is good evidence that the songs have adapted to features of the environment whether acoustic systems have changed too. However, Henry and Lucas (2010) have shown that birds can undergo changes in their auditory systems (such as frequency and temporal resolution), that are consistent with the use of different song types to work more effectively in open or woodland habitats. Therefore, there is great potential to explore the concept of sensory drive both in terms of signal and sensory system in the acoustic sense. Finally, at present, little work has been done regarding sensory drive and speciation outside of visual communication in fish and song types in birds, but there is no reason why the concept should not be broadly relevant. For example, sensory drive may apply freely to areas such as vibratory communication in invertebrates (see discussion in Cocroft and Rodríguez 2005), given that different substrate types can have a substantial effect on signal transmission.

11.5 Future Work

Much of the subject matter in this chapter concerns ideas and theories that are still rapidly changing. There is much still to learn in many areas. For example, more work (especially outside of vision) is needed to determine how often there is divergence in sensory systems associated with habitats and behavioural tasks, and in turn do changes in sensory systems lead to divergence. Sensory drive is potentially a very important concept in divergence and speciation, yet we currently know little about this concept in terrestrial systems and in sensory modalities outside of vision. In general, a better understanding of how predation pressure and communication systems can lead to divergence and speciation in sympatry versus allopatry, and how, is a substantial area of future work.

11.6 Summary

Sensory and communication systems can be crucial in leading to divergence between populations. Within species, traits may become exaggerated versions or new innovations may arise and have important implications for how populations diverge. This used to be thought of as a relatively slow process only occurring in allopatry. However, increasing amounts of work illustrate this is happening both in sympatry and parapatry, and sometimes on a rapid evolutionary timescale. We are now starting to understand both the genetics underlying some of these changes, as well as the selective advantages incurred. Various factors can lead to divergence and subsequently speciation, including reproductive character displacement in both signals and behavioural responses, disruptive selection, reinforcement, and possibly so-called magic traits. In some systems, predation pressure alone seems to be a force leading to divergence. The environment can also cause divergence and speciation in some groups through sensory drive, although this has so far been studied little outside of vision.

11.7 Further Reading

Readers may wish to consult Schluter (2001) for a review of ecological speciation or Coyne and Orr's (2004) book. For a recent review of magic traits, see Servedio (2011), and for discussion on the role of sexual selection in speciation and interactions with ecological theories see Maan and Seehausen (2011). Boughman (2002) and Mann and Seehausen (2010) discuss important issues regarding sensory drive and speciation.

PART 5

Conclusions

CHAPTER 12

Concluding Remarks

Today, sensory ecology is one of the most inter-disciplinary subjects in biology. It is not simply concerned with how sensory organs work, but rather spans the way that sensory receptors acquire information, how nervous systems process it, how sensory information guides behaviour, through to the role of sensory systems and information in large-scale evolutionary change and speciation. One of the wonderful things about such a discipline is that there are many areas of work for the future. A key aim of this book was to discuss important behavioural tasks and evolutionary processes with examples and concepts across the range of sensory systems, rather than discussing each sensory modality in turn. Part of the reason for this was to understand and show when and how concepts and ideas operate across modalities. For example, many principles of communication, arms races and co-evolution, and divergence can be found across the different modalities. More discussion between scientists working on different modalities will surely be valuable, and researchers can learn a lot by considering work outside of their own modality and taxa of study.

12.1 Information is a Concept Linked to the Receiver's Sensory System

The idea of information is still central to much of sensory ecology, and we adopted an approach to define information in this book as leading to a reduction in uncertainty in the receiver. However, the idea of information still creates confusion, and information as an idea is all too often glossed over in many research areas and used in an imprecise manner. Information operates at many levels, and could be measured at various stages of stimulus form and sensory processing. Although a consensus on what information is and its importance in sensory ecology (such as in communication, where the idea is perhaps most controversial) may take time to arise, the field would be helped if researchers were more explicit about what they consider information to be and how they have measured it in their system.

Perhaps the most fundamental message of this book is that the sensory worlds of animals differ, and that we should not use our own perceptions to measure and interpret cues, stimuli, and signals aimed at other species. This may sound like an obvious point (and it is certainly not a new one), and there are numerous and increasing numbers of studies where researchers analyse or model how a stimulus may be encoded by sensory receptors or nerve circuits. Yet, it is also surprising how common it remains for experiments to either use human assessment or to not consider what the receiver may perceive. Studies of animal coloration are a good example of the range of approaches taken. Some researchers measure the stimulus features and incorporate what we (think we) know about the visual system of the receiver (e.g. the number and spectral sensitivity of photoreceptors) to analyse the coloration in terms of a model of visual processing. The approach has enabled great advances in how animals use visual information in guiding behaviour. Other studies analyse the shape of the reflectance spectrum from an object (how much light is reflected over a range of wavelengths). While objective, this approach does not consider that the receiver's receptors will not respond equally to all parts of the spectrum and that they may not even be sensitive to other spectral regions, or generally how spectral information is processed. Thus, it can be hard to interpret what aspect of a stimulus is most important and what the animal is responding to. Last, some work on coloration is even still undertaken based on humans assessing the colour patterns and interpreting the results of behavioural experiments on this basis. The great naturalists/biologists of the 1800s knew that the vision of animals is often different from our own. John Lubbock (1882), for example, showed that some animals could see ultraviolet light. It is time that assessments of animal coloration based on human assessment became a thing of the past. Acoustic stimuli provide a second example. With some notable exceptions (for example, a great deal of the work conducted on bats and barn owls), studies of sound are still largely restricted to analysing the structure of the sound itself (e.g. its frequency, amplitude, temporal structure), rather than how it stimulates the auditory system of the receiver. This remains especially common in communication work. This approach is objective, and certainly better than interpreting sound based on our own hearing. However, it is the acoustic equivalent of analysing a visual reflectance spectrum; analysing the structure of the sound without considering what part of the spectrum is relevant to the receiver's sensory system or how it is processed. Greater use of models of acoustic processing could revitalize studies of auditory information (for example, acoustic mimicry and communication).

The discussion above does matter. The information that an animal can extract from a stimulus is dependent on how their sensory system works. So, to understand how animals use information and the evolution of signals, we need to consider, as far as possible, what the animal can perceive. We are learning a lot more about how sensory systems work and developing new methods to study them and the stimuli they encode. Of course, we do not always know what our study species might hear or see, but where possible, we should not be afraid to move on and adopt new techniques.

A key caveat, however, is that we should also be aware of the assumptions of the models and techniques we use, and that there is still much to learn about sensory processing. We now have a good understanding of the fundamentals of many senses across broad groups of animal taxa. Yet, we must beware of overgeneralizations. For example, in Chapter 4 we discussed how the four cone types used in colour vision in birds seem to be relatively conserved in their sensitivity to light spectra across species. This seems broadly to be true, but less than 100 species have been analysed out of > 9000 birds. In addition, much flexibility in sensory systems stems not from receptor properties, but from subsequent neural processing. This area needs more work in many taxa and modalities, including how widespread different features of neural processing are across modalities and taxa (e.g. filtering, receptive fields), and how these mechanisms facilitate effective processing of particular types of stimuli.

12.2 Communication

Communication remains a vibrant and diverse field today. In recent years, research has more effectively considered not just concepts of signal reliability and ideas of signal strategy, but has begun to more fully consider signal efficacy too. Hopefully this will continue into the future because there is much about signal form and communication that cannot be understood from an economic and reliability perspective alone. We learnt that there are also a range of costs and trade-offs associated with both sensory systems and signals. For example, sensory systems are energetically costly, and signals can incur a range of physiological and energetic costs. As such, costs and trade-offs affect both the receiver, and the sender in communication systems and a better understanding of how the costs and trade-offs that exist for parties and how they constrain behaviour and communication would be valuable. A topic of increasing popularity is multimodal communication. This subject could do with a more consistently used framework to plan and interpret the results of research. In particular, understanding more fully when and why multimodal signals evolve in different species is required. Furthermore, closer links with work done on sensory integration and the advantages this brings would benefit research on multimodal communication. It would also be valuable to gain a better understanding of sensory integration outside of vision and olfaction and vision and hearing. Various other concepts and ideas in communication also require more work. For instance, the idea of private communication channels as a means to signal without the costs of eavesdroppers in as attractive one. Yet, work so far has rarely analysed the potential eavesdropper's response to the suggested private signal (modified and unmodified) and most of the work on this subject is based on modelling and theory. The idea of sensory exploitation has also received excellent work in some select systems, but how widespread this idea is, and how different types of sensory exploitation relate to phenomena like mimicry is unclear.

12.3 Life's Diversity

It is remarkable how diverse sensory systems and signals in nature are. Understanding what has driven this diversity (and convergence) across and within species is a major avenue for the future. Like many other areas of biology, sensory ecology has benefited greatly from modern advances in molecular work and phylogenetics. In particular, an increased understanding in recent years that molecular genetics, albeit important and interesting in its own right, has huge power when applied in combination with work on sensory physiology, behaviour, and evolution is providing many exciting advances in these fields. Some of the most interesting discoveries in sensory ecology have shown not just divergence in the molecular and genetic make-up of species, but, remarkably, instances of convergence in sensory systems that are underpinned by convergence at the molecular level. That this is being found in a wide range of modalities (e.g. infrared detection, electric fish, and echolocation) is exciting, and potentially has far-reaching consequences. No doubt many more exciting discoveries await as researchers delve more into the molecular basis of other sensory systems in many species, and more into the basis of signals too.

More work on how sensory systems and signals have resulted, and play a role in, evolutionary processes such as coevolution, arms races, divergence, and speciation is also needed. In particular, work outside of vision and hearing in these subjects would be especially valuable. Likewise, the role of the environment in influencing the evolution and function of signals and sensory systems should be explored further. A better understanding of how widespread sensory plasticity is across modalities is needed, as well as how sensory integration and

multimodal communication could overcome limitations of communication in particular environments. In the modern age, human impacts on the environment need special attention too. Finally, processes such as sensory drive and magic traits as routes towards divergence and speciation are attractive but our understanding of how important such mechanisms may be is currently limited.

12.4 How Best Can Sensory Ecology Develop from Here?

All in all, it is an exciting time to study sensory ecology, and there is much left to discover. One of the biggest broad challenges is to strengthen links with the study of animal cognition. This book has taken an evolutionary and behavioural approach to sensory ecology. An important issue has been to understand how the use of information, gathered by sensory systems, influences and directs particular behaviours. Many behaviours are 'innate' or 'hardwired', yet others require an important contribution from cognitive systems. Understanding how animals make decisions in such instances will often require knowing more than just how their sensory receptors and early processing work with regards to stimuli. In many instances, it is hard to fully understand certain concepts and behavioural outcomes without considering both sensory systems and cognition. For example, in Chapter 8 we discussed various examples of mimicry and sensory/perceptual exploitation. In both Batesian mimicry, where one harmless organism resembles a dangerous or toxic species, and masquerade, where camouflage is achieved by an organism resembling an uninteresting or unimportant object (like a dead leaf), the predator is thought to misclassify the potential prey item as something else. That is, these defensive strategies prevent recognition, rather than detection (the predator detects that an object is present, but fails to classify is correctly). Yet, for mimicry and masquerade to work, the prey animal must also resemble the model organism/object sufficiently closely in terms of the way that both objects stimulate the sensory receptors of the receiver. In short, distinguishing between detection and recognition is not easy (if possible at all), and both sensory and cognitive processes will dictate whether the mimicry is successful or not. Overall, therefore, a deeper understanding of animal decision-making could emerge if we had greater insight into how and why particular sensory stimulation patterns induce particular behaviours along with cognitive processing. Fortunately, work is now progressing in this area and will no doubt make many important advances.

I hope this book has demonstrated that there are many broad conceptual questions still to be tackled, and that scientists and students have much to gain from avoiding becoming too narrowly focused on a particular taxon or modality. Scientific communication seems particularly vital in sensory ecology, being a discipline that is so multidisciplinary and covers so many types of organism and environment. The discipline also certainly now seems sufficiently mature to merit its own dedicated journal. This could represent an ideal vehicle to allow those with interests in this field to keep abreast of developments across the full extent of sensory ecology. For similar reasons, there would be much to gain from international conferences dedicated to the subject. Likewise, I would be delighted if this book contributes in any small way to attracting bright young minds into a discipline that has substantial scope for development. It would be wonderful to see an increase in the number of universities offering their more advanced biology undergraduates and postgraduates a dedicated course in sensory biology, especially for those interested in more established related fields. Sensory ecology is fascinating and important in its own right, but its extraordinary broad scope, in both questions and techniques, makes it ideal to encourage students to increase their critical thinking and to reduce the risk of compartmentalizing their knowledge into isolated topics and subjects.

Bibliography

Able, K. P. & Able, M. A. 1995a. Manipulations of polarized skylight calibrate magnetic orientation in migratory bird. *Journal of Comparative Physiology A*, 177, 351–6.

Able, K. P. & Able, M. A. 1995b. Interactions in the flexible orientation system of a migratory bird. *Nature*, 375, 230–2.

Able, K. P. & Able, M. A. 1997. Development of sunset orientation in migratory bird: no calibration by the magnetic field. *Animal Behaviour*, 53, 363–8.

Abrams, P. A. 2000. The evolution of predator-prey interactions: theory and evidence. *Annual Review of Ecology, Evolution and Systematics*, 31, 79–105.

Achenbach, A. & Foitzik, S. 2009. First evidence for slave rebellion: enslaved ant workers systematically kill the brood of their social parasite *Protomognathus americanus*. *Evolution*, 63, 1068–75.

Achenbach, A., Witte, V. & Foitzik, S. 2010. Brood exchange experiments and chemical analyses shed light on slave rebellion in ants. *Behavioral Ecology*, 21, 948–56.

Aho, A. C., Donner, K., Helenius, S., Larsen, L. O. & Reuter, T. 1993. Visual performance of the toad (*Bufo bufo*) at low light levels: retinal ganglion cell responses and prey-catching accuracy. *Journal of Comparative Physiology A*, 172, 671–82.

Akino, T., Knapp, J. J., Thomas, J. A. & Elmes, G. W. 1999. Chemical mimicry and host specificity in the butterfly *Maculinea rebeli*, a social parasite of *Myrmica* ant colonies. *Proceedings of the Royal Society B: Biological Sciences*, 266, 1419–26.

Akino, T. 2008. Chemical strategies to deal with ants: a review of mimicry, camouflage, propaganda, and phytomimesis by ants (Hymenoptera: Formicidae) and other arthropods. *Myrmecological News*, 11, 173–81.

Akre, K. L., Farris, H. E., Lea, A. M., Page, R. A. & Ryan, M. J. 2011. Signal perception in frogs and bats and the evolution of mating signals. *Science*, 333, 751–2.

Alcock, J. 2009. *Animal Behavior, Ninth Edition*. Sunderland, MA: Sinauer Associates, Inc.

Allen, J. A. 1988. Frequency dependent selection by predators. *Philosophical Transactions of the Royal Society B: Biological Sciences*, 319, 485–503.

Allen, J. J., Mäthger, L. M., Barbosa, A. & Hanlon, R. T. 2009. Cuttlefish use visual cues to control three-dimensional skin papillae for camouflage. *Journal of Comparative Physiology A*, 195, 547–55.

Allies, A. B., Bourke, A. F. G. & Franks, N. R. 1986. Propaganda substances in the cuckoo ant *Leptothorax kutteri* and the slave-maker *Harpagoxenus sublaevis*. *Journal of Chemical Ecology*, 12, 1285–93.

Amézquita, A., Hödl, W., Lima, A. P., Castellanos, L., Erdtmann, L. & de Araújo, M. C. 2006. Masking interference and the evolution of the acoustic communication system in the Amazonian dendrobatid frog *Allobates femoralis*. *Evolution*, 60, 1874–87.

Ammermüller, J., Muller, J. F. & Kolb, H. 1995. The organization of the turtle inner retina. II. Analysis of color-coded and directionally selective cells. *Journal of Comparative Neurology*, 358, 35–62.

Ammermüller, J., Itzhaki, A., Weiler, R. & Perlman, I. 1998. UV-sensitive input to horizontal cells in the turtle retina. *European Journal of Neuroscience*, 10, 1544–52.

Andersson, J., Borg-Karlson, A.-K. & Wiklund, C. 2000. Sexual cooperation and conflict in butterflies: a male-transferred anti-aphrodisiac reduces harassment of recently mated females. *Proceedings of the Royal Society B: Biological Sciences*, 267, 1271–5.

Andersson, M. 1994. *Sexual Selection*. Princeton: Princeton University Press.

Andersson, M. & Simmons, L. W. 2006. Sexual selection and mate choice. *Trends in Ecology & Evolution*, 21, 296–302.

Andrews, K., Reed, S. M. & Masta, S. E. 2007. Spiders fluoresce variably across many taxa. *Biology Letters*, 3, 265–7.

Arak, A. & Enquist, M. 1993. Hidden preferences and the evolution of signals. *Philosophical Transactions of the Royal Society B: Biological Sciences*, 340, 207–13.

Archard, G. A., Cuthill, I. C. & Partridge, J. C. 2009. Light environment and mating behavior in Trinidadian guppies (*Poecilia reticulata*). *Behavioral Ecology and Sociobiology*, 64, 169–82.

Arikawa, K., Wakakuwa, M., Qiu, X., Kurasawa, M. & Stavenga, D. G. 2005. Sexual dimorphism of short-wavelength photoreceptors in the small white butterfly, *Pieris rapae crucivora*. *Journal of Neuroscience*, 25, 5935–42.

Arnegard, M. E. & Hopkins, C. D. 2003. Electric signal variation among seven blunt-snouted *Brienomyrus* species (Teleostei: Mormyridae) from a riverine species flock in Gabon, Central Africa. *Environmental Biology of Fishes*, 67, 321–39.

Arnegard, M. E. & Kondrashov, A. S. 2004. Sympatric speciation by sexual selection alone is unlikely. *Evolution*, 58, 222–37.

Arnegard, M. E., Bogdanowicz, S. M. & Hopkins, C. D. 2005. Multiple cases of striking genetic similarity between alternate electric fish signal morphs in sympatry. *Evolution*, 59, 324–43.

Arnegard, M. E. & Carlson, B. A. 2005. Electric organ discharge patterns during group hunting by a mormyrid fish. *Proceedings of the Royal Society B: Biological Sciences*, 272, 1305–14.

Arnegard, M. E., Jackson, B. S. & Hopkins, C. D. 2006. Time-domain signal divergence and discrimination without receptor modification in sympatric morphs of electric fishes. *Journal of Experimental Biology*, 209, 2182–98.

Arnegard, M. E., McIntyre, P. B., Harmon, L. J., Zelditch, M. L., Crampton, W. G. R., Davis, J. K., Sullivan, J. P., Lavoué, S. & Hopkins, C. D. 2010a. Sexual signal evolution outpaces ecological divergence during electric fish species radiation. *American Naturalist*, 176, 335–56.

Arnegard, M. E., Zwickl, D. J., Lu, Y. & Zakon, H. H. 2010b. Old gene duplication facilitates origin and diversification of an innovative communication system-twice. *Proceedings of the National Academy of Sciences of the USA*, 107, 22172–7.

Arnold, K. E., Owens, I. P. F. & Marshall, N. J. 2002. Fluorescent signaling in parrots. *Science*, 295, 92.

Arshavsky, V. Y., Lamb, T. D. & Pugh, E. N., Jr 2002. G proteins and phototransduction. *Annual Review of Physiology*, 64, 153–87.

Ashida, G., Abe, K., Funabiki, K. & Konishi, M. 2007. Passive soma facilitates submillisecond coincidence detection in the owl's auditory system. *Journal of Neuroscience*, 97, 2267–82.

Arthur, B. J., Wyttenbach, R. A., Harrington, L. C., Hoy, RR. 2010. Neural responses to one- and two-tone stimuli in the hearing organ of the dengue vector mosquito. *Journal of Experimental Biology*, 213, 1376–85.

Assad, C., Rasnow, B. & Stoddard, P. K. 1999. Electric organ discharges and electric images during electrolocation. *Journal of Experimental Biology*, 202, 1185–93.

Atema, J. 1995. Chemical signals in the marine environment: dispersal, detection, and temporal signal analysis. *Proceedings of the National Academy of Sciences of the USA*, 92, 62–6.

Au, W. W. & Benoit-Bird, K. J. 2003. Automatic gain control in the echolocation system of dolphins. *Nature*, 423, 861–3.

Avens, L., Wang, J. H., Johnsen, S., Dukes, P. & Lohmann, K. J. 2003. Responses of hatchling sea turtles to rotational displacements. *Journal of Experimental Marine Biology and Ecology*, 288, 111–24.

Avilés, J. M. 2008. Egg colour mimicry in the common cuckoo *Cuculus canorus* as revealed by modelling host retinal function. *Proceedings of the Royal Society B: Biological Sciences*, 275, 2345–52.

Avilés, J. M. & Soler, J. J. 2009. Nestling colouration is adjusted to parent visual performance in altricial birds. *Journal of Evolutionary Biology*, 22, 376–86.

Avilés, J. M., Vikan, J. R., Fossøy, F., Antonov, A., Moksnes, A., Røskaft, E. & Stokke, B. G. 2010. Avian colour perception predicts behavioural responses to experimental brood parasitism in chaffinches. *Journal of Evolutionary Biology*, 23, 293–301.

Baier, B. 2008. Effect of conductivity changes on the stability of electric signal waveforms in dwarf stonebashers (Mormyridae; *Pollimyrus castelnaui, P. marianne*). *Journal of Comparative Physiology A*, 194, 915–19.

Balasubramanian, V., Kimber, D. & Berry, M. J., II. 2001. Metabolically efficient information processing. *Neural Computation*, 13, 799–815.

Barber, J. R., Razak, K. A. & Fuzessery, Z. M. 2003. Can two streams of auditory information be processed simultaneously? Evidence from the gleaning bat *Antrozous pallidus*. *Journal of Comparative Physiology A*, 189, 843–55.

Barber, J. R. & Conner, W. E. 2007. Acoustic mimicry in a predator–prey interaction. *Proceedings of the National Academy of Sciences of the USA*, 104, 9331–4.

Barber, J. R., Crooks, K. R. & Fristrup, K. M. 2009. The costs of chronic noise exposure for terrestrial organisms. *Trends in Ecology & Evolution*, 25, 180–9.

Barbero, F., Bonelli, S., Thomas, J. A., Balletto, E. & Schonrogge, K. 2009a. Acoustical mimicry in a predatory social parasite of ants. *Journal of Experimental Biology*, 212, 4084–90.

Barbero, F., Thomas, J. A., Bonelli, S., Balletto, E. & Schonrogge, K. 2009b. Queen ants make distinctive sounds that are mimicked by a butterfly social parasite. *Science*, 323, 782–5.

Barbosa, A., Allen, J. J., Mäthger, L. M. & Hanlon, R. T. 2012. Cuttlefish use visual cues to determine arm postures for camouflage. *Proceedings of the Royal Society B: Biological Sciences*, 279, 84–90.

Barclay, R. M. R., Fenton, M. B., Tuttle, M. D. & Ryan, M. J. 1981. Echolocation calls produced by *Trachops cirrhosus* (Chiroptera: Phyllostomatidae) while hunting frogs. *Canadian Journal of Zoology*, 59, 750–3.

Bargmann, C. I. 2006. Comparative chemosensation from receptors to ecology. *Nature*, 444, 295–301.

Barlow, G. W., Francis, R. C. & Baumgartner, J. V. 1990. Do the colours of parents, companions and self influence assortative mating in the polychromatic Midas cichlid? *Animal Behaviour*, 40, 713–22.

Barlow, H. B. 1953. Summation and inhibition in the frog's retina. *Journal of Physiology*, 119, 69–88.

Barlow, R. B. 2009. Vision in horseshoe crabs. In: *Biology and conservation of horseshoe crabs* (Ed. by Tanacredi, J. T.), pp. 223–35. LLC: Springer Science & Business Media.

Barlow, R. B., Jr 1969. Inhibitory fields in the *Limulus* lateral eye. *Journal of General Physiology*, 54, 383–96.

Barlow, R. B., Jr 1983. Circadian rhythms in the *Limulus* visual system. *Journal of Neuroscience*, 3, 856–70.

Barth, F. G. & Geethabali. 1982. Spider vibration receptors: threshold curves of individual slits in the metatarsal lyriform organ. *Journal of Comparative Physiology A*, 148, 175–85.

Basolo, A. L. 1990. Female preference predates the evolution of the sword in swordtail fish. *Science*, 250, 808–10.

Basolo, A. L. 1995a. Phylogenetic evidence for the role of a pre-existing bias in sexual selection. *Proceedings of the Royal Society B: Biological Sciences*, 259, 307–11.

Basolo, A. L. 1995b. A further examination of a pre-existing bias favouring a sword in the genus *Xiphophorus*. *Animal Behaviour*, 50, 365–75.

Basolo, A. L. & Endler, J. A. 1995. Sensory biases and the evolution of sensory systems. *Trends in Ecology & Evolution*, 10, 489.

Bates, D. L. & Fenton, M. B. 1990. Aposematism or startle? Predators learn their responses to the defenses of prey. *Canadian Journal of Zoology*, 68, 49–52.

Bates, M. E., Simmons, J. A. & Zorikov, T. V. 2011. Bats use echo harmonic structure to distinguish their targets from background clutter. *Science*, 333, 627–30.

Beck, A. & Ewert, J. P. 1979. Prey selection by toads (*Bufo bufo* L.) in response to configurational stimuli moved in the visual field z,y-coordinates. *Journal of Comparative Physiology A*, 129, 207–9.

Bell, C. C. & Grant, K. 1989. Corollary discharge inhibition and preservation of temporal information in a sensory nucleus of mormyrid electric fish. *Journal of Neuroscience*, 9, 1029–44.

Bell, P. D. 1980. Transmission of vibrations along plant stems: Implications for insect communication. *Journal of the New York Entomological Society*, 88, 210–16.

Belwood, J. J. & Morris, G. 1987. Bat predation and its influence on calling behaviour in neotropical katydids. *Science*, 238, 64–7.

Benard, M. F. 2006. Survival trade-offs between two predator-induced phenotypes in pacific treefrogs (*Pseudacris regilla*). *Ecology*, 87, 340–6.

Benkman, C. W., Parchman, T. L., Favis, A. & Siepielski, A. M. 2003. Reciprocal selection causes a coevolutionary arms race between crossbills and lodgepole pine. *American Naturalist*, 162, 182–94.

Bernal, X. E., Rand, A. S. & Ryan, M. J. 2006. Acoustic preferences and localization performance of blood-sucking flies (*Corethrella coquillett*) to tungara frog calls. *Behavioral Ecology*, 17, 709–15.

Bernal, X. E., Page, R. A., Rand, A. S. & Ryan, M. J. 2007. Cues for eavesdroppers: do frog calls indicate prey density and quality? *American Naturalist*, 169, 409–15.

Bernard, G. D. & Wehner, R. 1977. Functional similarities between polarization vision and color vision. *Vision Research*, 17, 1019–28.

Bernard, G. D. & Remington, C. L. 1991. Color vision in *Lycaena* butterflies: spectral tuning of receptor arrays in relation to behavioral ecology. *Proceedings of the National Academy of Sciences of the USA*, 88, 2783–7.

Biro, D. 2010. Bird navigation: a clear view of magnetoreception. *Current Biology*, 20, R595–6.

Blamires, S. J., Hochuli, D. F. & Thompson, M. B. 2008. Why cross the web: decoration spectral properties and prey capture in an orb spider (*Argiope keyserlingi*) web. *Biological Journal of the Linnean Society*, 94, 221–9.

Blamires, S. J., Lai, C.-H., Cheng, R.-C., Liao, C.-P., Shen, P.-S. & Tso, I.-M. 2012. Body spot coloration of a nocturnal sit-and-wait predator visually lures prey. *Behavioral Ecology*, 23, 69–74.

Blamires, S. J., Lee, Y.-H., Chang, C.-M., Lin, I.-T., Chen, J.-A., Lin, T.-Y. & Tso, I.-M. 2010. Multiple structures interactively influence prey capture efficiency in spider orb webs. *Animal Behaviour*, 80, 947–53.

Bleckmann, H. 2008. Peripheral and central processing of lateral line information. *Journal of Comparative Physiology A*, 194, 145–58.

Blest, A. D. & Land, M. F. 1977. The physiological optics of *Dinopis subrufus* L. Koch: a fish-lens in a spider. *Proceedings of the Royal Society B: Biological Sciences*, 196, 197–222.

Blest, A. D. 1978. The rapid synthesis and destruction of photoreceptor membrane by and dinopid spider: a daily cycle. *Proceedings of the Royal Society B: Biological Sciences*, 200, 463–83.

Blest, A. D. & Day, W. A. 1978. The rhabdomere organisation of some nocturnal pisaurid spiders in light and darkness. *Philosophical Transactions of the Royal Society B: Biological Sciences*, 283, 1–23.

Blest, A. D., Kao, L. & Powell, K. 1978a. Photoreceptor membrane breakdown in the spider *Dinopis*: the fate of rhabdomere products. *Cell & Tissue Research*, 195, 425–44.

Blest, A. D., Powell, K. & Kao, L. 1978b. Photoreceptor membrane breakdown in the spider *Dinopis*: GERL differentiation in the receptors. *Cell & Tissue Research*, 195, 277–97.

Blest, A. D., Williams, D. S. & Kao, L. 1980. The posterior median eyes of the dinopid spider *Menneus*. *Cell & Tissue Research*, 211, 391–403.

Bogusch, P., Kratochvil, L. & Straka, J. 2006. Generalist cuckoo bees (Hymenoptera: Apoidea: *Sphecodes*) are species-specialist at the individual level. *Behavioral Ecology and Sociobiology*, 60, 422–9.

Boles, L. C. & Lohmann, K. J. 2003. True navigation and magnetic maps in spiny lobsters. *Nature*, 421, 60–3.

Bond, A. B. & Kamil, A. C. 1998. Apostatic selection by blue jays produces balanced polymorphism in virtual prey. *Nature*, 395, 594–6.

Bond, A. B. & Kamil, A. C. 2002. Visual predators select for crypticity and polymorphism in virtual prey. *Nature*, 415, 609–13.

Bond, A. B. 2007. The evolution of color polymorphism: crypticity, searching images and apostatic selection. *Annual Review of Ecology, Evolution, and Systematics*, 38, 1–25.

Booth, D., Stewart, A. J. & Osorio, D. 2004. Colour vision in the glow-worm *Lampyris noctiluca* L. (Coleoptera: Lampyridae): evidence for a green-blue chromatic mechanism. *Journal of Experimental Biology*, 207, 2373–8.

Borchers, H. W. & Ewert, J. P. 1979. Correlation between behavioral and neuronal activities of toads *Bufo bufo* (L.) in response to moving configurational prey stimuli. *Behavioural Processes*, 4, 99–106.

Boughman, J. W. 2001. Divergent sexual selection enhances reproductive isolation in sticklebacks. *Nature*, 411, 944–8.

Boughman, J. W. 2002. How sensory drive can promote speciation. *Trends in Ecology & Evolution*, 17, 571–7.

Boul, K. E., Funk, W. C., Darst, C. R., Cannatella, D. C. & Ryan, M. J. 2007. Sexual selection drives speciation in an Amazonian frog. *Proceedings of the Royal Society B: Biological Sciences*, 274, 399–406.

Bowdan, E. & Wyse, G. A. 1996. Sensory Ecology: Introduction. *Biology Bulletin*, 191, 122–3.

Bowmaker, J. K. 1980. Colour vision in birds and the role of oil droplets. *Trends in Neurosciences*, 3, 196–9.

Bowmaker, J. K., Dartnall, H. J. A. & Herring, P. J. 1988. Longwave-sensitive visual pigments in some deep-sea fishes: segregation of 'paired' rhodopsins and porphyropsins. *Journal of Comparative Physiology A*, 163, 685–98.

Bowmaker, J. K. 2008. Evolution of vertebrate visual pigments. *Vision Research*, 48, 2022–41.

Bradbury, J. W. & Vehrencamp, S. L. 2011. *Principles of Animal Communication, Second Edition*. Sunderland MA: Sinauer Associates, Inc.

Brady, P. & Cummings, M. E. 2010. Differential response to circularly polarized light by the jewel scarab beetle *Chrysina gloriosa*. *American Naturalist*, 175, 614–20.

Brandt, M. & Foitzik, S. 2004. Community context and specialization influence coevolution between a slavemaking ant and its hosts. *Ecology*, 85, 2997–3009.

Brandt, M., Foitzik, S., Fischer-Blass, B. & Heinze, J. 2005a. The coevolutionary dynamics of obligate ant social parasite systems—between prudence and antagonism. *Biological Reviews*, 80, 251–67.

Brandt, M., Heinze, J., Schmitt, T. & Foitzik, S. 2005b. A chemical level in the coevolutionary arms race between an ant social parasite and its hosts. *Journal of Evolutionary Biology*, 18, 576–86.

Brenowitz, E. A. 1986. Environmental influences on acoustic and electric animal communication. *Brain, Behavior and Evolution*, 28, 32–42.

Brill, R. L., Sevenich, M. L., Sullivan, T. J., Sustman, J. D. & Witt, R. E. 2006. Behavioural evidence for hearning through the lower jaw by an echolocating dolphin (*Tursiops truncatus*). *Marine Mammal Science*, 4, 223–30.

Brinkløv, S., Kalko, E. & Surlykke, A. 2010. Dynamic adjustment of biosonar intensity to habitat clutter in the bat *Macrophyllum macrophyllum* (Phyllostomidae). *Behavioral Ecology and Sociobiology*, 64, 1867–74.

Briscoe, A. D. 2001. Functional diversification of lepidopteran opsins following gene duplication. *Molecular Biology and Evolution*, 18, 2270–9.

Briscoe, A. D., Bybee, S. M., Bernard, G. D., Yuan, F., Sison-Mangus, M. P., Reed, R. D., Warren, A. D., Llorente-Bousquets, J. & Chiao, C. C. 2010. Positive selection of a duplicated UV-sensitive visual pigment coincides with wing pigment evolution in *Heliconius* butterflies. *Proceedings of the National Academy of Sciences of the USA*, 107, 3628–33.

Briscoe, A. D. & Chittka, L. 2001. The evolution of color vision in insects. *Annual Review of Entomology*, 46, 471–510.

Britton, N. F., Planque, R. & Franks, N. R. 2007. Evolution of defence portfolios in exploiter-victim systems. *Bulletin of Mathematical Biology*, 69, 957–88.

Brodmann, J., Twele, R., Francke, W., Yi-bo, L., Xi-qiang, S. & Ayasse, M. 2009. Orchid mimics honey bee alarm pheromone in order to attract hornets for pollination. *Current Biology*, 19, 1368–72.

Brooke, M. de D. & Davies, N. B. 1988. Egg mimicry by cuckoos *Cuculus canorus* in relation to discrimination by hosts. *Nature*, 335, 630–2.

Brown, J. L., Maan, M. E., Cummings, M. E. & Summers, K. 2010. Evidence for selection on coloration in Panamanian poison frog: a coalescent-based approach. *Journal of Biogeography*, 2010, 891–901.

Brown, K. S. 1981. The biology of *Heliconius* and related genera. *Annual Reviews*, 26, 427–56.

Brumm, H., Voss, K., Köllmer, I. & Todt, D. 2004. Acoustic communication in noise: regulation of call characteristics in a New World monkey. *Journal of Experimental Biology*, 207, 443–8.

Brunner, D. & Labhart, T. 1987. Behavioural evidence for polarization vision in crickets. *Physiological Entomology*, 12, 1–10.

Bruns, V. 1976a. Peripheral auditory tuning for fine frequency analysis by the CF-FM bat, *Rhinolophus ferrumequinum*; II: Frequency mapping in the cochlea. *Journal of Comparative Physiology A*, 106, 87–97.

Bruns, V. 1976b. Peripheral auditory tuning for fine frequency analysis by the CF-FM bat, *Rhinolophus ferrumequinum*; I: Mechanical specializations of the cochlea. *Journal of Comparative Physiology A*, 106, 77–86.

Bruns, V. & Schmieszek, E. 1980. Cochlear innervation in the greater horseshoe bat: demonstration of an acoustic fovea. *Hearing Research*, 3, 27–43.

Bruns, V., Burda, H. & Ryan, M. J. 1989. Ear morphology of the frog-eating bat (*Trachops cirrhosus*, Family: Phyllostomatidae): apparent specializations for low-frequency hearing. *Journal of Morphology*, 199, 103–18.

Bura, V. L., Rohwer, V. G., Martin, P. R. & Yack, J. E. 2010. Whistling in caterpillars (Amorpha juglandis, Bombycoidea): sound-producing mechanism and function. *Journal of Experimental Biology*, 214, 30–7.

Burr, D. & Thompson, P. 2011. Motion psychophysics: 1985–2010. *Vision Research*, 51, 1431–56.

Bush, A. A., Yu, D. W. & Herberstein, M. E. 2008. Function of bright coloration in the wasp spider *Argiope bruennichi* (Araneae: Araneidae). *Proceedings of the Royal Society of London B: Biological Sciences*, 275, 1337–42.

Bybee, S. M., Yuan, F., Ramstetter, M. D., Llorente-Bousquets, J., Reed, R. D., Osorio, D. & Briscoe, A. D. 2012. UV photoreceptors and UV-yellow wing pigments in *Heliconius* butterflies allow a color signal to serve both mimicry and intraspecific communication. *American Naturalist*, 179, 38–51.

Cade, W. 1975. Acoustically orientating parasitoids: fly phonotaxis to cricket song. *Science*, 190, 1312–13.

Caine, N. G. & Mundy, N. I. 2000. Demonstration of a foraging advantage for trichromatic marmosets (*Callithrix geoffroyi*) dependent on food colour. *Proceedings of the Royal Society B: Biological Sciences*, 267, 439–44.

Caine, N. G., Osorio, D. & Mundy, N. I. 2010. A foraging advantage for dichromatic marmosets (*Callithrix geoffroyi*) at low light intensity. *Biology Letters*, 6, 36–8.

Caldwell, M. S., McDaniel, J. G. & Warkentin, K. M. 2010. Is it safe? Red-eyed treefrog embryos assessing predation risk use two features of rain vibrations to avoid false alarms. *Animal Behaviour*, 79, 255–60.

Caputi, A. A., Silva, A. C. & Macadar, O. 1998. The electric organ discharge of *Brachyhypopomus pinnicaudatus*. *Brain, Behavior and Evolution*, 52, 148–58.

Carleton, K. L., Parry, J. W. L., Bowmaker, J. K., Hunt, D. M. & Seehausen, O. 2005. Colour vision and speciation in Lake Victoria cichlids of the genus *Pundamilia*. *Molecular Biology*, 14, 4341–53.

Carlson, B. A. & Arnegard, M. E. 2011. Neural innovations and the diversification of African weakly electric fishes. *Communicative and Integrative Biology*, 4, 720–5.

Carlson, B. A., Hasan, S. M., Hollmann, M., Miller, D. B., Harmon, L. J. & Arnegard, M. E. 2011. Brain evolution triggers increased diversification of electric fishes. *Science*, 332, 583–6.

Carr, C. E. & Konishi, M. 1990. A circuit for detection of interaural time differences in the brain stem of the barn owl. *Journal of Neuroscience*, 10, 3227–46.

Carr, C. E. & Boudreau, R. E. 1993. Organization of the nucleus magnocellularis and the nucleus laminaris in the barn owl: encoding and measuring interaural time differences. *Journal of Comparative Neurology*, 334, 337–55.

Cassey, P., Honza, M., Grim, T. & Hauber, M. E. 2008. The modelling of avian visual perception predicts behavioural rejection responses to foreign egg colours. *Biology Letters*, 4, 515–17.

Catania, K. C. 2009. Tentacled snakes turn C-starts to their advantage and predict future prey behavior. *Proceedings of the National Academy of Sciences of the USA*, 106, 11183–7.

Cervo, R. 2006. *Polistes* wasps and their social parasites: an overview. *Annals Zoologici Fennici*, 43, 531–49.

Chapman, B. B., Morrell, L. J., Tosh, C. R. & Krause, J. 2010. Behavioural consequences of sensory plasticity in guppies. *Proceedings of the Royal Society B: Biological Sciences*, 277, 1395–401.

Charlton, B. D., Reby, D. & McComb, K. 2007. Female red deer prefer the roars of larger males. *Biology Letters*, 3, 382–5.

Chatterjee, S. & Callaway, E. M. 2003. Parallel colour-opponent pathways to the primary visual cortex. *Nature*, 426, 668–71.

Chávez, A., Bozinovic, B., Peichl, L. & Palacios, A. G. 2003. Retinal spectral sensitivity, fur colouration, and urine reflectance in the genus *Octodon* (Rodentia): implications for visual ecology. *Investigative Ophthalmology & Visual Science*, 44, 2290–6.

Cheney, K. L. 2012. Cleaner wrasse mimics inflict higher costs on their models when they are more aggressive towards signal receivers. *Biology Letters*, 8, 10–12.

Cheng, R.-C. & Tso, I.-M. 2007. Signaling by decorating webs: luring prey or deterring predators? *Behavioral Ecology*, 18, 1085–91.

Cheng, R.-C., Yang, E.-C., Lin, C.-P., Herberstein, M. E. & Tso, I.-M. 2010. Insect form vision as one potential shaping force of spider web decoration design. *Journal of Experimental Biology*, 213, 759–68.

Chernenko, A., Helantera, H. & Sundström, L. 2011. Egg rejection and social parasitism in *Formica* ants. *Ethology*, 117, 1081–92.

Chiao, C.-C., Chubb, C. & Hanlon, R. T. 2007. Interactive effects of size, contrast, intensity and configuration of background objects in evoking disruptive camouflage in cuttlefish. *Vision Research*, 47, 2223–35.

Chiao, C.-C., Chubb, C., Buresch, K. C., Siemann, L. & Hanlon, R. T. 2009a. The scaling effects of substrate texture on camouflage patterning in cuttlefish. *Vision Research*, 49, 1647–56.

Chiao, C.-C., Wu, W.-Y., Chen, S.-H. & Yang, E.-C. 2009b. Visualization of the spatial and spectral signals of orb-weaving spiders, *Nephila pilipes*, through the eyes of the honeybee. *Journal of Experimental Biology*, 212, 2269–78.

Chiao, C. C., Wickiser, J. K., Allen, J. J., Genter, B. & Hanlon, R. T. 2011. Hyperspectral imaging of cuttlefish camouflage indicates good color match in the eyes of fish predators. *Proceedings of the National Academy of Sciences of the USA*, 108, 9148–53.

Chichilnisky, E. J. & Wandell, B. A. 1999. Trichromatic opponent colour classification. *Vision Research*, 39, 3444–58.

Chiou, T.-H., Kleinlogel, S., Cronin, T. W., Caldwell, R. L., Loeffler, B., Siddiqi, A., Goldizen, A. & Marshall, N. J. 2008. Circular polarization vision in a stomatopod crustacean. *Current Biology*, 9, 755–8.

Chiou, T. H., Place, A. R., Caldwell, R. L., Marshall, N. J. & Cronin, T. W. 2012. A novel function for a carotenoid: astaxanthin used as a polarizer for visual signalling in a mantis shrimp. *Journal of Experimental Biology*, 215, 584–9.

Chittka, L. & Menzel, R. 1992. The evolutionary adaptation of flower colours and the insect pollinators' colour vision. *Journal of Comparative Physiology A*, 171, 171–81.

Chiu, C., Xian, W. & Moss, C. F. 2008. Flying in silence: Echolocating bats cease vocalizing to avoid sonar jamming. *Proceedings of the National Academy of Sciences of the USA*, 105, 13116–21.

Chow, D. M. & Frye, M. A. 2008. Context-dependent olfactory enhancement of optomotor flight control in *Drosophila*. *Journal of Experimental Biology*, 211, 2478–85.

Christensen, T. A. 2005. Making scents out of spatial and temporal codes in specialist and generalist olfactory networks. *Chemical Senses*, 30 Suppl 1, i283–284.

Christy, J. H. 1995. Mimicry, mate choice, and the sensory trap hypothesis. *American Naturalist*, 146, 171–81.

Christy, J. H., Backwell, P. R. Y., Goshima, S. & Kreuter, T. 2002. Sexual selection for structure building by courting male fiddler crabs: an experimental study of behavioral mechanisms. *Behavioral Ecology*, 13, 366–74.

Christy, J. H., Backwell, P. R. Y. & Schober, U. 2003a. Interspecific attractiveness of structures built by courting male fiddler crabs: experimental evidence of a sensory trap. *Behavioral Ecology and Sociobiology*, 53, 84–91.

Christy, J. H., Baum, J. K. & Backwell, P. R. Y. 2003b. Attractiveness of sand hoods built by courting male fiddler crabs, *Uca musica*: a test of the sensory trap hypothesis. *Animal Behaviour*, 66, 89–94.

Chuang, C.-Y., Yang, E.-C. & Tso, I.-M. 2008. Deceptive color signaling in the night: a nocturnal predator attracts prey with visual lures. *Behavioral Ecology*, 19, 237–44.

Clark, D. L., Roberts, J. A., Rector, M. & Uetz, G. W. 2011. Spectral reflectance and communication in the wolf spider, *Schizocosa ocreata* (Hentz): simultaneous crypsis and background contrast in visual signals. *Behavioral Ecology and Sociobiology*, 65, 1237–47.

Clark, D. L., Roberts, J. A. & Uetz, G. W. 2012. Eavesdropping and signal matching in visual courtship displays of spiders. *Biology Letters*, 8, 375–8.

Clarke, B. 1962. Balanced polymorphism and the diversity of sympatric species. In: *Taxonomy and Geography* (Ed. by Nichols, D.). London: The Systematics Association.

Clutton-Brock, T. H. & Albon, S. D. 1979. The roaring of red deer and the evolution of honest advertisement. *Behaviour*, 69, 145–70.

Cocroft, R. B. & Rodríguez, R. L. 2005. The behavioural ecology of insect vibrational communication. *BioScience*, 55, 323–34.

Collin, S. P., Davies, W. L., Hart, N. S. & Hunt, D. M. 2009. The evolution of early vertebrate photoreceptors. *Philosophical Transactions of the Royal Society B: Biological Sciences*, 364, 2925–40.

Colonius, H. & Diederich, A. 2004. Multisensory interaction in saccadic reaction time: a time-window-of-integration model. *Journal of Cognitive Neuroscience*, 16, 1000–9.

Conner, W. E. 1999. 'Un chant d'appel amoureux': acoustic communication in moths. *Journal of Experimental Biology*, 202, 1711–23.

Cooper, W. E. J. & Vitt, L. J. 1991. Influence of detectability and escape on natural selection of conspicuous autotomous defenses. *Canadian Journal of Zoology*, 69, 757–64.

Cooper, W. E. J. 1998. Reactive and anticipatory display to deflect predatory attack to an autotomous lizard tail. *Canadian Journal of Zoology*, 76, 1507–10.

Corcoran, A. J., Barber, J. R. & Conner, W. E. 2009. Tiger moth jams bat sonar. *Science*, 325, 325–7.

Corcoran, A. J., Conner, W. E. & Barber, J. R. 2010. Anti-bat tiger moth sounds: Form and function. *Current Zoology*, 56, 358–69.

Corcoran, A. J., Barber, J. R., Hristov, N. I. & Conner, W. E. 2011. How do tiger moths jam bat sonar? *Journal of Experimental Biology*, 214, 2416–25.

Corneil, B. D., Van Wanrooij, M., Munoz, D. P. & Van Opstal, A. J. 2002. Auditory-visual interactions subserving goal-directed saccades in a complex scene. *Journal of Neurophysiology*, 88, 438–54.

Couldridge, V. C. K. & Alexander, G. J. 2001. Color patterns and species recognition in four closely related species of Lake Malawi cichlid. *Behavioral Ecology*, 13, 59–64.

Coyle, B. J., Hart, N. S., Carleton, K. L. & Borgia, G. 2012. Limited variation in visual sensitivity among bowerbird species suggests that there is no link between spectral tuning and variation in display colouration. *Journal of Experimental Biology*, 215, 1090–05.

Coyne, J. A. & Orr, H. A. 2004. *Speciation*. Sunderland: Sinauer Associates.

Crampton, W. G. R., Lovejoy, N. R. & Waddell, J. C. 2011. Reproductive character displacement and signal ontogeny in a sympatric assemblage of electric fishes. *Evolution*, 65, 1650–66.

Cronin, T. W. & Marshall, N. J. 1989. A retina with at least ten spectral types of photoreceptors in a mantis shrimp. *Nature*, 339, 137–40.

Cronin, T. W., Jarvilehto, M., Weckstrom, M. & Lall, A. B. 2000a. Tuning of photoreceptor spectral sensitivity in fireflies (Coleoptera: Lampyridae). *Journal of Comparative Physiology A*, 186, 1–12.

Cronin, T. W., Marshall, N. J. & Caldwell, R. L. 2000b. Spectral tuning and the visual ecology of mantis shrimps. *Philosophical Transactions of the Royal Society B: Biological Sciences*, 355, 1263–7.

Cronin, T. W., Caldwell, R. L. & Marshall, N. J. 2001. Tunable colour vision in a mantis shrimp. *Nature*, 411, 547.

Cronin, T. W., Shashar, N., Caldwell, R. L., Marshall, N. J., Cheroske, A. G. & Chiou, T.-H. 2003. Polarization vision and its role in biological signaling. *Integrative Computational Biology*, 43, 549–58.

Cronin, T. W. & Marshall, J. 2011. Patterns and properties of polarized light in air and water. *Philosophical Transactions of the Royal Society B: Biological Sciences*, 366, 619–26.

Cummings, M. E. & Partridge, J. C. 2001. Visual pigments and optical habitats of surfperch (Embiotocidae) in the California kelp forest. *Journal of Comparative Physiology A*, 187, 875–89.

Cummings, M. E., Rosenthal, G. G. & Ryan, M. J. 2003. A private ultraviolet channel in visual communication. *Proceedings of the Royal Society B: Biological Sciences* 270, 897–904.

Cummings, M. E., Jordão, J. M., Cronin, T. W. & Oliveira, R. F. 2008a. Visual ecology of the fiddler crab, *Uca tangeri*: effects of sex, viewer and background on conspicuousness. *Animal Behaviour*, 75, 175–88.

Cummings, M. E., Larkins-Ford, J., Reilly, C. R., Wong, R. Y., Ramsey, M. & Hofmann, H. A. 2008b. Sexual and social stimuli elicit rapid and contrasting genomic responses. *Proceedings of the Royal Society B: Biological Sciences*, 275, 393–402.

Curio, E. 1976. *The Ethology of Predation*. Berlin, Germany: Springer.

Curtis, C. C. & Stoddard, P. K. 2003. Mate preference in female electric fish, *Brachyhypopomus pinnicaudatus*. *Animal Behaviour*, 66, 329–36.

Cuthill, I. C., Stevens, M., Sheppard, J., Maddocks, T., Párraga, C. A. & Troscianko, T. S. 2005. Disruptive coloration and background pattern matching. *Nature*, 434, 72–4.

Cuthill, I. C. & Székely, A. 2009. Coincident disruptive coloration. *Philosophical Transactions of the Royal Society B: Biological Sciences*, 364, 489–96.

Czech-Damal, N. U., Liebschner, A., Miersch, L., Klauer, G., Hanke, F. D., Marshall, C., Dehnhardt, G. & Hanke, W. 2011. Electroreception in the Guiana dolphin (*Sotalia guianensis*). *Proceedings of the Royal Society B: Biological Sciences*, 279, 663–8.

Dall, S. R. X., Giraldeau, L.-A., Olsson, O., McNamara, J. M. & Stephens, D. W. 2005. Information and its use by animals in evolutionary ecology. *Trends in Ecology & Evolution*, 20, 187–93.

Darst, C. R., Cummings, M. E. & Cannatella, D. C. 2006. A mechanism for diversity in warning signals: Conspicuousness versus toxicity in poison frogs. *Proceedings of the National Academy of Sciences of the USA*, 103, 5852–7.

Darwin, C. 1859. *The Origin of Species by Means of Natural Selection*. London: John Murray.

Darwin, C. 1871. *The Descent of Man and Selection in Relation to Sex*. London: John Murray.

Davies, N. B. & Halliday, T. R. 1978. Deep croaks and fighting assessment in toads *Bufo bufo*. *Nature*, 274, 683–5.

Davies, N. B. & Brooke, M. de D. 1989a. An experimental study of co-evolution between the cuckoo, *Cuculus canorus*, and its hosts. II. Host egg markings, chick discrimination and general discussion. *Journal of Animal Ecology*, 58, 225–36.

Davies, N. B. & Brooke, M. de L. 1989b. An experimental study of co-evolution between the cuckoo, *Cuculus canorus*, and its hosts. I. Host egg discrimination. *Journal of Animal Ecology*, 58, 207–24.

Davies, N. B., Kilner, R. M. & Noble, D. G. 1998. Nestling cuckoos, *Cuculus canorus*, exploit hosts with begging calls that mimic a brood. *Proceedings of the Royal Society B: Biological Sciences*, 265, 673–8.

Davies, N. B. 2000. *Cuckoos, Cowbirds and Other Cheats*. London: T & A D Poyser.

Davies, N. B. & Welbergen, J. A. 2008. Cuckoo-hawk mimicry? An experimental test. *Proceedings of the Royal Society B: Biological Sciences*, 275, 1817–22.

Davies, N. B. & Welbergen, J. A. 2009. Social transmission of a host defense against cuckoo parasitism. *Science*, 324, 1318–20.

Davies, N. B. 2011. Cuckoo adaptations: trickery and tuning. *Journal of Zoology*, 284, 1–14.

Davies, N. B., Krebs, J. R. & West, S. A. 2012. *An Introduction to Behavioural Ecology, Fourth Edition*. Chichester: Wiley-Blackwell.

Dawkins, M. S. 1971. Perceptual changes in chicks: another look at the 'search image' concept. *Animal Behaviour*, 19, 566–74.

Dawkins, R. 1976. *The Selfish Gene*. Oxford: Oxford University Press.

Dawkins, R. & Krebs, J. 1979. Arms races between and within species. *Proceedings of the Royal Society B: Biological Sciences*, 205, 489–511.

de Bruyne, M. & Baker, T. C. 2008. Odor detection in insects: volatile codes. *Journal of Chemical Ecology*, 34, 882–97.

De Jong, G. 1982. Orientation of comb building by honeybees. *Journal of Comparative Physiology A*, 147, 495–501.

De Serrano, A., Weadick, C., Price, A. & Rodd, F. 2012. Seeing orange: prawns tap into a pre-existing sensory bias of the Trinidadian guppy. *Proceedings of the Royal Society B: Biological Sciences*, 279, 3321–28.

DeBose, J. L. & Nevitt, G. A. 2008. The use of odors at different spatial scales: comparing birds with fish. *Journal of Chemical Ecology*, 34, 867–81.

Decaestecker, E., Gaba, S., Raeymaekers, J. A. M., Stoks, R., Van Kerckhoven, L., Ebert, D. & De Meester, L. 2007. Host-parasite 'Red Queen' dynamics archived in pond sediment. *Nature*, 450, 870–3.

Denoël, M. & Doellen, J. 2010. Displaying in the dark: light-dependent alternative mating tactics in the Alpine newt. *Behavioral Ecology and Sociobiology*, 64, 1171–7.

Denton, E. J., Gilpin-Brown, J. B. & Wright, P. G. 1970. On the 'filters' in the photophores of mesopelagic fish and on a fish emitting red light and especially sensitive to red light. *Journal of Physiology*, 208, 72–3.

Denton, E. J., Herring, P. J., Widder, E. A., Latz, M. F. & Case, J. F. 1985. The role of filters in the photophores of oceanic animals and their relation to vision in the oceanic environment. *Proceedings of the Royal Society B: Biological Sciences*, 225, 63–97.

Dettner, K. & Liepert, C. 1994. Chemical mimicry and camouflage. *Annual Review of Entomology*, 39, 129–54.

Dieckmann, U., Marrow, P. & Law, R. 1995. Evolutionary cycling in predator-prey interactions: population dynamics and the Red Queen. *Journal of Theoretical Biology*, 176, 91–102.

Diederich, A. & Colonius, H. 2004. Bimodal and trimodal multisensory enhancement: effects of stimulus onset and intensity on reaction time. *Perception & Psychophysics*, 66, 1388–404.

Dingle, C., Halfwerk, W. & Slabbekoorn, H. 2008. Habitat-dependent song divergence at subspecies level in the grey-breasted wood-wren. *Journal of Evolutionary Biology*, 21, 1079–89.

Dominy, N. J. & Lucas, P. W. 2001. Ecological importance of trichromatic vision to primates. *Nature*, 410, 363–6.

Dominy, N. J., Garber, P. A., Bicca-Marques, J. C. & Azevedo-Lopes, M. A. d. 2003. Do female tamarins use visual cues to detect fruit rewards more successfully than do males? *Animal Behaviour*, 66, 829–37.

Dornhaus, A., Franks, N. R., Hawkins, R. M. & Shere, H. N. S. 2004. Ants move to improve: colonies of *Leptothorax albipennis* emigrate whenever they find a superior nest site. *Animal Behaviour*, 67, 959–63.

Doucet, S. M., Mennill, D. J. & Hill, G. E. 2007. The evolution of signal design in manakin plumage ornaments. *American Naturalist*, 169, S62–80.

Douglas, R. H., Partridge, J. C. & Hope, A. J. 1995. Visual and lenticular pigments in the eyes of demersal deep-sea fishes. *Journal of Comparative Physiology A*, 177, 111–22.

Douglas, R. H. & Partridge, J. C. 1997. On the visual pigments of deep-sea fish. *Journal of Fish Biology*, 50, 68–85.

Douglas, R. H., Partridge, J. C., Dulai, K. S., Hunt, D. M., Mullineaux, C. W., Tauber, A. Y. & Hynninen, P. H. 1998. Dragon fish see using chlorophyll. *Nature*, 393, 423–4.

Douglas, R. H., Partridge, J. C., Dulai, K. S., Hunt, D. M., Mullineaux, C. W. & Hynninen, P. H. 1999. Enhanced retinal longwave sensitivity using a chlorophyll-derived photosensitiser in *Malacosteus niger*, a deep-sea dragon fish with far red bioluminescence. *Vision Research*, 39, 2817–32.

Douglas, R. H., Mullineaux, C. W. & Partridge, J. C. 2000. Long-wave sensitivity in deep-sea stomiid dragonfish with far-red bioluminescence: evidence for a dietary origin of the chlorophyll-derived photosensitizer of *Malacosteus niger*. *Philosophical Transactions of the Royal Society B: Biological Sciences*, 355, 1269–72.

Duistermars, B. J. & Frye, M. A. 2008. Crossmodal visual input for odor tracking during fly flight. *Current Biology*, 18, 270–5.

Dunlap, K. D., Smith, G. T. & Yekta, A. 2000. Temperature dependence of electrocommunication signals and their underlying neural rhythms in the weakly electric fish, *Apteronotus leptorhynchus*. *Brain, Behavior and Evolution*, 55, 152–62.

Dunlap, K. D., DiBenedictis, B. T. & Banever, S. R. 2010. Chirping response of weakly electric knife fish (*Apteronotus leptorhynchus*) to low-frequency electric signals and to heterospecific electric fish. *Journal of Experimental Biology*, 213, 2234–42.

Dunlop, R. A., Cato, D. H. & Noad, M. J. 2010. Your attention please: increasing ambient noise levels elicits a change in communication behaviour in humpback whales (*Megaptera novaeangliae*). *Proceedings of the Royal Society B: Biological Sciences*, 277, 2521–9.

Dusenbery, D. B. 1992. *Sensory Ecology: How Organisms Acquire and Respond to Information*. New York: W. H. Freeman and Company.

Eaton, L. & Sloman, K. A. 2011. Subordinate brown trout exaggerate social signalling in turbid conditions. *Animal Behaviour*, 81, 603–8.

Eberhard, W. G. 1977. Aggressive chemical mimicry by a bolas spider. *Science*, 198, 1173–5.

Edmunds, M. 1974. *Defence in Animals: A Survey of Antipredator Defences*. Harlow, Essex: Longman Group Ltd.

Egughi, E., Nemoto, A., Meyer-Rochow, V. B. & Ohba, N. 1984. A comparative study of spectral sensitivity curves in three diurnal and eight nocturnal species of japanese fireflies. *Journal of Insect Physiology*, 30, 607–12.

Ehrlich, P. R. & Raven, P. H. 1964. Butterflies and plants: a study of coevolution. *Evolution*, 18, 586–608.

Eisthen, H. L. 2002. Why are olfactory systems of different animals so similar? *Brain, Behavior and Evolution*, 59, 273–93.

Elder, J. H. & Sachs, A. J. 2004. Psychophysical receptive fields of edge detection mechanisms. *Vision Research*, 44, 795–813.

Elias, D. O., Mason, A. C. & Hoy, R. R. 2004. The effect of substrate on the efficacy of seismic courtship signal transmission in the jumping spider *Habronattus dossenus* (Araneae: Salticidae). *Journal of Experimental Biology*, 207, 4105–10.

Elias, D. O., Lee, N., Hebets, E. A. & Mason, A. C. 2006. Seismic signal production in a wolf spider: parallel versus serial multi-component signals. *Journal of Experimental Biology*, 209, 1074–84.

Ellis, A. G. & Johnson, S. D. 2010. Floral mimicry enhances pollen export: the evolution of pollination by sexual deceit outside of the Orchidaceae. *American Naturalist*, 176, E143–51.

Elmer, K. R., Lehtonen, T. K. & Meyer, A. 2009. Color assortative mating contributes to sympatric divergence of neotropical cichlid fish. *Evolution*, 63, 2750–57.

Endler, J. A. 1978. A predator's view of animal color patterns. *Evolutionary Biology*, 11, 319–64.

Endler, J. A. 1980. Natural selection on colour patterns in *Poecilia reticulata*. *Evolution*, 34, 76–91.

Endler, J. A. 1991. Variation in the appearance of guppy color patterns to guppies and their predators under different visual conditions. *Vision Research*, 31, 587–608.

Endler, J. A. 1992. Signals, signal conditions and the direction of evolution. *American Naturalist*, 139, 125–53.

Endler, J. A. 1993. Some general comments on the evolution and design of animal communication systems. *Philosophical Transactions of the Royal Society B: Biological Sciences*, 340, 215–25.

Endler, J. A. & Basolo, A. L. 1998. Sensory ecology, receiver biases and sexual selection. *Trends in Ecology & Evolution*, 13, 415–20.

Endler, J. A., Basolo, A. L., Glowacki, S. & Zerr, J. 2001. Variation in response to artificial selection for light sensitivity in guppies (*Poecilia reticulata*). *American Naturalist*, 158, 36–48.

Endler, J. A., Westcott, D. A., Madden, J. R. & Robson, T. 2005. Animal visual systems and the evolution of color patterns: Sensory processing illuminates signal evolution. *Evolution*, 59, 1795–818.

Endler, J. A. & Day, L. B. 2006. Ornament colour selection, visual contrast and the shape of colour preference functions in great bowerbirds, *Chlamydera nuchalis*. *Animal Behaviour*, 72, 1405–16.

Enquist, M. & Arak, A. 1993. Selection of exaggerated male traits by female aesthetic senses. *Nature*, 361, 446–8.

Ewert, J. P. & Hock, F. 1972. Movement-sensitive neurones in the toad's retina. *Experimental Brain Research*, 16, 41–59.

Ewert, J. P., Borchers, H. W. & Wiersheim, A. v. 1978. Question of prey feature detection in the toad's *Bufo bufo* (L.) visual system: A correlation analysis. *Journal of Comparative Physiology A*, 126, 43–7.

Fain, G. L., Hardie, R. & Laughlin, S. B. 2010. Phototransduction and the evolution of photoreceptors. *Current Biology*, 20, R114–24.

Fan, C.-M., Yang, E.-C. & Tso, I.-M. 2009. Hunting efficiency and predation risk shapes the color-associated foraging traits of a predator. *Behavioral Ecology*, 20, 808–16.

Fatouros, N. E., Huigens, M. E., van Loon, J. J. A., Dicke, M. & Hilker, M. 2005. Chemical communication—butterfly anti-aphrodisiac lures parasitic wasps. *Nature*, 433, 704.

Fatouros, N. E., Dicke, M., Mumm, R., Meiners, T. & Hilker, M. 2008. Foraging behavior of egg parasitoids exploiting chemical information. *Behavioral Ecology*, 19, 677–89.

Faucher, K., Parmentier, E., Becco, C., Vandewalle, N. & Vandewalle, P. 2010. Fish lateral system is required for accurate control of shoaling behaviour. *Animal Behaviour*, 79, 679–87.

Feng, A. S., Narins, P. M., Xu, C.-H., Lin, W.-Y., Yu, Z.-L., Qiu, Q., Xu, Z.-M. & Shen, J.-X. 2006. Ultrasonic comunication in frogs. *Nature*, 440, 333–6.

Feng, A. S. 2011. Neural mechanisms of target ranging in FM bats: physiological evidence from bats and frogs. *Journal of Comparative Physiology A*, 197, 595–603.

Fernandez, A. A. & Morris, M. R. 2007. Sexual selection and trichromatic color vision in primates: statistical support for the preexisting-bias hypothesis. *American Naturalist*, 170, 10–20.

Fernandez, H. R. & Tsuji, F. I. 1976. Photopigment and spectral sensitivity in the bioluminescent fish *Porichthys notatus*. *Marine Biology*, 34, 101–7.

Fernandez-Juricic, E., O'Rourke, C. & Pitlik, T. 2010. Visual coverage and scanning behavior in two corvid species: American crow and Western scrub jay. *Journal of Comparative Physiology A*, 196, 879–88.

Fernandez-Juricic, E., Gall, M. D., Dolan, T., O'Rourke, C., Thomas, S. & Lynch, J. R. 2011. Visual systems and vigilance behaviour of two ground-foraging avian prey species: white-crowned sparrows and California towhees. *Animal Behaviour*, 81, 705–13.

Fernández-Juricic, E., Erichsen, J. T. & Kacelnik, A. 2004. Visual perception and social foraging in birds. *Trends in Ecology & Evolution*, 19, 25–31.

Fernández-Juricic, E., Gall, M. D., Dolan, T., Tisdale, V. & Martin, G. 2008. The visual fields of two ground-foraging birds, house finches and house sparrows, allow for simultaneous foraging and anti-predator vigilance. *Ibis*, 150, 779–87.

Fernández-Juricic, E., Moore, B. A., Doppler, M., Freeman, J., Blackwell, B. F., Lima, S. L. & DeVault, T. L. 2011. Testing the terrain hypothesis: Canada geese see their world laterally and obliquely. *Brain, Behavior and Evolution*, 77, 147–58.

Fernández-Juricic, E. 2012. Sensory basis of vigilance behavior in birds: synthesis and future prospects. *Behavioral Processes*, 89, 143–52.

Feulner, P. G., Plath, M., Engelmann, J., Kirschbaum, F. & Tiedemann, R. 2009a. Magic trait electric organ discharge (EOD): Dual function of electric signals promotes speciation in African weakly electric fish. *Communicative and Integrative Biology*, 2, 329–31.

Feulner, P. G. D., Kirschbaum, F. & Tiedemann, R. 2008. Adaptive radiation in the Congo River: An ecological speciation scenario for African weakly electric fish (Teleostei; Mormyridae; *Campylomormyrus*). *Journal of Physiology*, 102, 340–6.

Feulner, P. G. D., Plath, M., Engelmann, J., Kirschbaum, F. & Tiedemann, R. 2009b. Electrifying love: electric fish use species-specific discharge for mate recognition. *Biology Letters*, 5, 225–8.

Fisher, H. S., Wong, B. B. M. & Rosenthal, G. G. 2006. Alteration of the chemical environment disrupts communication in a freshwater fish. *Proceedings of the Royal Society B: Biological Sciences*, 273, 1187–93.

Fisher, R. A. 1930. *The Genetical Theory of Natural Selection*. Oxford: Oxford University Press.

Flower, T. 2011. Fork-tailed drongos use deceptive mimicked alarm calls to steal food. *Proceedings of the Royal Society of London B: Biological Sciences*, 278, 1548–55.

Foitzik, S., DeHeer, C. J., Hunjan, D. N. & Herbers, J. M. 2001. Coevolution in host–parasite systems: behavioural strategies of slave–making ants and their hosts. *Proceedings of the Royal Society B: Biological Sciences*, 268, 1139–46.

Font, E. & Carazo, P. 2010. Animals in translation: why there is meaning (but probably no message) in animal communication. *Animal Behaviour*, 80, e1–6.

Ford, E. B. 1940. Polymorphism and taxonomy. In: Huxley, J. *The New Systematics*. Oxford: Clarendon Press. 493–513.

Forrest, T. G., Read, M. P., Farris, H. E. & Hoy, R. R. 1997. A tympanal hearing organ in scarab beetles. *Journal of Experimental Biology*, 200, 601–6.

Forsman, J. T. & Seppänen, J.-T. 2011. Learning what (not) to do: testing rejection and copying of simulated heterospecific behavioural traits. *Animal Behaviour*, 81, 879–83.

Franks, N. R., Pratt, S. C., Mallon, E. B., Britton, N. F. & Sumpter, D. J. T. 2002. Information flow, opinion polling and collective intelligence in house-hunting social insects. *Philosophical Transactions of the Royal Society B: Biological Sciences*, 357, 1567–83.

Franks, N. R., Dornhaus, A., Fitzsimmons, J. P. & Stevens, M. 2003a. Speed versus accuracy in collective decision making. *Proceedings of the Royal Society B: Biological Sciences*, 270, 2457–63.

Franks, N. R., Mallon, E. B., Bray, H. E., Hamilton, M. J. & Mischler, T. C. 2003b. Strategies for choosing between alternatives with different attributes: exemplified by house-hunting ants. *Animal Behaviour*, 65, 215–23.

Franks, N. R., Dechaume-Moncharmont, F.-X., Hanmore, E. & Reynolds, J. K. 2009. Speed versus accuracy in decision-making ants: expediting politics and policy implementation. *Philosophical Transactions of the Royal Society B: Biological Sciences*, 364, 845–52.

Fraser, S., Callahan, A., Klassen, D. & Sherratt, T. N. 2007. Empirical tests of the role of disruptive coloration in reducing detectability. *Proceedings of the Royal Society B: Biological Sciences*, 274, 1325–31.

Frederiksen, R. & Warrant, E. J. 2008. Visual sensitivity in the crepuscular owl butterfly *Caligo memnon* and the diurnal blue morpho *Morpho peleides*: a clue to explain the evolution of nocturnal apposition eyes? *Journal of Experimental Biology*, 211, 844–51.

Frens, M. A. & Van Opstal, A. J. 1998. Visual-auditory interactions modulate saccade-related activity in monkey superior colliculus. *Brain Research Bulletin*, 46, 211–24.

Fried, S. I. & Masland, R. H. 2007. Image processing: how the retina detects the direction of image motion. *Current Biology*, 17, R63–6.

Frye, M. A., Tarsitano, M. & Dickinson, M. H. 2003. Odor localization requires visual feedback during free flight in *Drosophila melanogaster*. *Journal of Experimental Biology*, 206, 843–55.

Frye, M. A. & Dickinson, M. H. 2004. Motor output reflects the linear superposition of visual and olfactory inputs in *Drosophila*. *Journal of Experimental Biology*, 207, 123–31.

Fugère, V. & Krahe, R. 2010. Electric signals and species recognition in the wave-type gymnotiform fish *Apteronotus leptorhynchus*. *Journal of Experimental Biology*, 213, 225–36.

Fullard, J. H. 1994. Auditory changes in noctuid moths endemic to a bat-free habitat. *Journal of Evolutionary Biology*, 7, 435–45.

Fullard, J. H., Simmons, J. A. & Saillant, P. A. 1994. Jamming bat echolocation: the dogbane tiger moth *Cycnia tenera* times its clicks to the terminal attack calls of the big brown bat *Eptesicus fuscus. Journal of Experimental Biology*, 194, 285–98.

Fuller, R. A., Warren, P. H. & Gaston, K. J. 2007. Daytime noise predicts nocturnal singing in urban robins. *Biology Letters*, 3, 368–70.

Fuller, R. C. 2002. Lighting environment predicts the relative abundance of male colour morphs in bluefin killifish (*Lucania goodei*) populations. *Proceedings of the Royal Society B: Biological Sciences*, 269, 1457–65.

Fuller, R. C., Houle, D. & Travis, J. 2005. Sensory bias as an explanation for the evolution of mate preferences. *American Naturalist*, 166, 437–46.

Fuller, R. C. & Noa, L. A. 2010. Female mating preferences, lighting environment, and a test of the sensory bias hypothesis in the bluefin killifish. *Animal Behaviour*, 80, 23–35.

Fuller, R. C., Noa, L. A. & Strellner, R. S. 2010. Teasing apart the many effects of lighting environment on opsin expression and foraging preference in Bluefin killifish. *American Naturalist*, 176, 1–13.

Funk, W. C., Cannatella, D. C. & Ryan, M. J. 2009. Genetic divergence is more tightly related to call variation than landscape features in the Amazonian frogs *Physalaemus petersi* and *P. freibergi. Journal of Evolutionary Biology*, 22, 1839–53.

Fuxjager, M. J., Eastwood, B. S. & Lohmann, K. J. 2011. Orientation of hatchling loggerhead sea turtles to regional magnetic fields along a transoceanic migratory pathway. *Journal of Experimental Biology*, 214, 2504–8.

Gall, M. D. & Fernandez-Juricic, E. 2010. Visual fields, eye movements, and scanning behavior of a sit-and-wait predator, the black phoebe (*Sayornis nigricans*). *Journal of Comparative Physiology A*, 196, 15–22.

Galligan, T. H. & Kleindorfer, S. 2008. Support for the nest mimicry hypothesis in yellow-rumped thornbills *Acanthiza chrysorrhoa. Ibis*, 150, 550–7.

Gan, W., Liu, F., Zhang, Z. & Li, D. 2010. Predator perception of detritus and eggsac decorations spun by orb-web spiders *Cyclosa octotuberculata*: do they function to camouflage the spiders? *Current Zoology*, 56, 379–87.

Gandon, S., Buckling, A., Decaestecker, E. & Day, T. 2008. Host-parasite coevolution and patterns of adaptation across time and space. *Journal of Evolutionary Biology*, 21, 1861–6.

Gandon, S. & Day, T. 2009. Evolutionary epidemiology and the dynamics of adaptation. *Evolution*, 63, 826–38.

Garcia, C. M. & Lemus, Y. S. 2012. Foraging costs drive female resistance to a sensory trap. *Proceedings of the Royal Society B: Biological Sciences*, 279, 2262–8.

Gardiner, J. M. & Atema, J. 2010. The function of bilateral odor arrival time differences in olfactory orientation of sharks. *Current Biology*, 20, 1187–91.

Gaskett, A. C. & Herberstein, M. E. 2010. Colour mimicry and sexual deception by tongue orchids (*Cryptostylis*). *Naturwissenschaften*, 97, 97–102.

Gaskett, A. C. 2011. Orchid pollination by sexual deception: pollinator perspectives. *Biological Reviews*, 86, 33–75.

Gavassa, S., Silva, A. C., Gonzalez, E. & Stoddard, P. K. 2012. Signal modulation as a mechanism for handicap disposal. *Animal Behaviour*, 83, 935–44.

Gegear, R. J., Foley, L. E., Casselman, A. & Reppert, S. M. 2010. Animal cryptochromes mediate magnetoreception by an unconventional photochemical mechanism. *Nature*, 463, 804–7.

Gemeno, C., Yeargan, K. V., & Haynes, K. F. 2000. Aggressive chemical mimicry by the bolas spider *Mastophora hutchinsoni*: identification and quantification of a major prey's sex pheromone components in the spider's volatile emissions. *Journal of Chemical Ecology*, 26, 1235–43.

Geng, J., Liang, D., Jiang, K. & Zhang, P. 2011. Molecular evolution of the infrared sensory gene TRPA1 in snakes and implications for functional studies. *PLoS ONE*, 6, e28644.

Getz, W. M. & Akers, R. P. 1997. Response of American cockroach (*Periplaneta americana*) olfactory receptors to select alcohol odorants and their binary combinations. *Journal of Comparative Physiology A*, 180, 701–9.

Gilbert, L. E. 1976. Postmating female odor in *Heliconius* butterflies: a male-contributed antiaphrodisiac? *Science*, 193, 419–20.

Gillam, E. H. & McCracken, G. F. 2007. Variability in the echolocation of *Tadarida brasiliensis*: effects of geography and local acoustic environment. *Animal Behaviour*, 74, 277–86.

Gillooly, J. F. & Ophir, A. G. 2010. The energetic basis of acoustic communication. *Proceedings of the Royal Society B: Biological Sciences*, 277, 1325–31.

Gilman, R. T., Nuismer, S. L. & Jhwueng, D. C. 2012. Coevolution in multidimensional trait space favours escape from parasites and pathogens. *Nature*, 483, 328–30.

Girard, M. B., Kasumovic, M. M. & Elias, D. O. 2011. Multimodal courtship in the peacock spider, Maratus volans (O.P.-Cambridge, 1874). *PLoS ONE*, 6, e25390.

Gloag, R., Fiorini, V. D., Reboreda, J. C. & Kacelnik, A. 2012. Brood parasite eggs enhance egg survivorship in a multiply parasitized host. *Proceedings of the Royal Society B: Biological Sciences*, 279, 1831–9.

Godin, J.-G. J. & McDonough, H. E. 2003. Predator preference for brightly colored males in the guppy: a viability cost for a sexually selected trait. *Behavioral Ecology*, 14, 194–200.

Goerlitz, H. R. & Siemers, B. M. 2007. Sensory ecology of prey rustling sounds: acoustical features and their classification by wild grey mouse lemurs. *Functional Ecology*, 21, 143–54.

Goerlitz, H. R., ter Hofstede, H. M., Zeale, M. R. K., Jones, G. & Holderied, M. W. 2010. An aerial-hawking bat uses stealth echolocation to counter moth hearing. *Current Biology*, 20, 1568–72.

Goerlitz, H. R., Genzel, D. & Wiegrebe, L. 2012. Bats' avoidance of real and virtual objects: implications for the sonar coding of object size. *Behavioural Processes*, 89, 61–7.

Goldsmith, T. H. 1990. Optimization, constraint, and history in the evolution of eyes. *Quarterly Review of Biology*, 65, 281–322.

Gomez, D. & Théry, M. 2004. Influence of ambient light on the evolution of colour signals: comparative analysis of a Neotropical rainforest bird community. *Ecology Letters*, 7, 279–84.

Gomez, D. & Théry, M. 2007. Simultaneous crypsis and conspicuousness in color patterns: comparative analysis of a neotropical rainforest bird community. *American Naturalist*, 169, S42–61.

Gomez, D., Richardson, C., Lengagne, T., Plenet, S., Joly, P., Léna, J.-P. & Théry, M. 2009. The role of nocturnal vision in mate choice: females prefer conspicuous males in the European tree frog (*Hyla arborea*). *Proceedings of the Royal Society B: Biological Sciences*, 276, 2351–8.

Gómez, P. & Buckling, A. 2011. Bacteria-phage antagonistic coevolution in soil. *Science*, 332, 106–9.

Gordon, S. D. & Uetz, G. W. 2011. Multimodal communication of wolf spiders on different substrates: evidence for behavioural plasticity. *Animal Behaviour*, 81, 367–75.

Goris, R. C. 2011. Infrared organs of snakes: an integral part of vision. *Journal of Herpetology*, 45, 2–14.

Gould, E., McShea, W. & Grand, T. 1993. Function of the star in the star-nosed mole, *Condylura cristata*. *Journal of Mammology*, 74, 108–16.

Gould, J. L., Kirschvink, J. L. & Deffeyes, K. S. 1978. Bees have magnetic remanence. *Science*, 201, 1026–8.

Gould, J. L. 1998. Sensory bases of navigation. *Current Biology*, 8, R731–8.

Gould, J. L., Elliott, S. L., Masters, C. M. & Mukerji, J. 1999. Female preferences in a fish genus without female mate choice. *Current Biology*, 9, 497–500.

Gould, J. L. 2008. Animal navigation: the evolution of magnetic orientation. *Current Biology*, 18, R482–4.

Gould, J. L. 2010. Magnetoreception. *Current Biology*, 20, R431–5.

Gould, J. L. & Gould, C. G. 2012. *The map sense*. Princeton, New Jersey: Princeton University Press.

Gould, S. J. & Lewontin, R. C. 1979. The spandrels of San Marco and the Panglossian paradigm: a critique of the adaptionist programme. *Proceedings of the Royal Society B: Biological Sciences*, 205, 581–98.

Gould, S. J. & Vrba, E. S. 1982. Exaptation—a missing term in the science of form. *Paleobiology*, 8, 4–15.

Gould, S. J. 1991. Red Wings in the Sunset. In: *Bully for Brontosaurus*, pp. 209–28. London: Penguin Books.

Goyret, J. 2010. Look and touch: mulitmodal sensory control of flower inspection movements in the nocturnal hawkmoth *Manduca sexta*. *Journal of Experimental Biology*, 213, 3676–82.

Gracheva, E. O., Ingolia, N. T., Kelly, Y. M., Cordero-Morales, J. F., Hollopeter, G., Chesler, A. T., Sanchez, E. E., Perez, J. C., Weissman, J. S. & Julius, D. 2010. Molecular basis of infrared detection by snakes. *Nature*, 464, 1006–11.

Grafen, A. 1990. Biological signals as handicaps. *Journal of Theoretical Biology*, 144, 517–46.

Grant, J. B. 2007. Ontogenetic colour change and the evolution of aposematism: a case study in panic moth caterpillars. *Journal of Animal Ecology*, 76, 439–47.

Gray, S. M., Dill, L. M., Tantu, F. Y., Loew, E. R., Herder, F. & McKinnon, J. S. 2008. Environment-contingent sexual selection in a colour polymorphic fish. *Proceedings of the Royal Society B: Biological Sciences*, 275, 1785–91.

Greenfield, M. D. 2002. *Signalers and Receivers: Mechanisms and Evolution of Arthropod Communication*. Oxford: Oxford University Press.

Greenwood, A. K. 2010. Sensory evolution: Picking up good vibrations. *Current Biology*, 20, R801–2.

Gregory, J. E., Iggo, A., McIntyre, A. K. & Proske, U. 1989. Responses of electroreceptors in the snout of the echidna. *Journal of Physiology*, 414, 521–38.

Greiner, B. 2006. Adaptations for nocernal vision in insect apposition eyes. *International Review of Cytology*, 250, 1–46.

Gridi-Papp, M., Rand, A. S. & Ryan, M. J. 2006. Animal communication: complex call production in the túngara frog. *Nature*, 441, 38.

Grothe, B., Pecka, M. & McAlpine, D. 2010. Mechanisms of sound localization in mammals. *Physiological Reviews*, 90, 983–1012.

Grusch, M., Barth, F. G. & Eguchi, E. 1997. Fine structural correlates of sensitivity in the eyes of the ctenid spider, *Cupiennius salei* Keys. *Tissue Cell*, 29, 421–30.

Gudger, E. W. 1946. The angler-fish, *Lophius piscatorus* it *americanus*; use of the lure in fishing. *American Naturalist*, 79, 542–8.

Guedes, R. N. C., Matheson, S. M., Frei, B., Smith, M. L. & Yack, J. E. 2012. Vibration detection and discrimination in the masked birch caterpillar (*Drepana arcuata*). *Journal of Comparative Physiology A*, 198, 325–35.

Guilford, T. 1986. How do warning colours work? Conspicuousness may reduce recognition errors in experienced predators. *Animal Behaviour*, 34, 286–8.

Guilford, T. & Dawkins, M. S. 1987. Search images not proven: a reappraisal of recent evidence. *Animal Behaviour*, 35, 1838–45.

Guilford, T. & Dawkins, M. S. 1991. Receiver psychology and the evolution of animal signals. *Animal Behaviour*, 42, 1–14.

Guillemain, M., Martin, G. R. & Fritz, H. 2002. Feeding methods, visual fields and vigilance in dabbling ducks (Anatidae). *Functional Ecology*, 16, 522–9.

Haddock, S. H. D., Moline, M. A. & Case, J. F. 2010. Bioluminescence in the sea. *Annual Review of Marine Science*, 2, 443–93.

Hafernik, J. & Saul-Gershenz, L. S. 2000. Beetle larvae cooperate to mimic bees. *Nature*, 405, 35.

Haff, T. M. & Magrath, R. D. 2010. Vulnerable but not helpless: nestlings are fine-tuned to cues of approaching danger. *Animal Behaviour*, 79, 487–96.

Haff, T. M. & Magrath, R. D. 2011. Calling at a cost: elevated nestling calling attracts predators to active nests. *Biology Letters*, 7, 493–5.

Hagedorn, M. & Zelick, R. 1989. Relative dominance among males is expressed in the electric organ discharge characteristics of a weakly electric fish. *Animal Behaviour*, 38, 520–5.

Hailman, J. P. 1977. *Optical Signals: Animal Communication and Light*. London, UK: Indiana University Press.

Hall, M. I. & Ross, C. F. 2007. Eye shape and activity pattern in birds. *Journal of Zoology*, 271, 437–44.

Hallem, E. A. & Carlson, J. R. 2006. Coding of odors by a receptor repertoire. *Cell*, 125, 143–60.

Hamilton, W. D. & Zuk, M. 1982. Heritable true fitness and bright birds: a role for parasites? *Science*, 218, 384–87.

Hanifin, C. T., Brodie, E. D., Jr & Brodie, E. D., III. 2008. Phenotypic mismatches reveal escape from arms-race coevolution. *PLoS Biology*, 6, e60.

Hanika, S. & Kramer, B. 1999. Electric organ discharges of mormyrid fish as a possible cue for predatory catfish. *Naturwissenschaften*, 86, 286–8.

Hanika, S. & Kramer, B. 2000. Electrosensory prey detection in the African sharptooth catfish, *Clarias gariepinus* (Clariidae), of a weakly electric mormyrid fish, the bulldog (*Marcusenius macrolepidotus*). *Behavioral Ecology and Sociobiology*, 48, 218–28.

Hanika, S. & Kramer, B. 2008. Plasticity of electric organ discharge waveform in the South African Bulldog fish, *Marcusenius pongolensis*: a trade-off between male attractiveness and predator avoidance? *Frontiers in Zoology*, 5, 7.

Hanlon, R. T. 2007. Cephalopod dynamic camouflage. *Current Biology*, 17, 400–4.

Hanlon, R. T., Chiao, C.-C., Mäthger, L. M., Barbosa, A., Buresch, K. C. & Chubb, C. 2009. Cephalopod dynamic camouflage: bridging the continuum between background matching and disruptive coloration. *Philosophical Transactions of the Royal Society B: Biological Sciences*, 364, 429–37.

Hansen, D. M., Van der Niet, T. & Johnson, S. D. 2012. Floral signposts: testing the significance of visual 'nectar guides' for pollinator behaviour and plant fitness. *Proceedings of the Royal Society B: Biological Sciences*, 279, 634–9.

Hansson, B. S. & Stensmyr, M. C. 2011. Evolution of insect olfaction. *Neuron*, 72, 698–711.

Happ, G. M. 1969. Multiple sex pheromones of the mealworm beetle, *Tenebrio molitor* L. *Nature*, 222, 180–1.

Harland, D. P. & Jackson, R. R. 2000. Cues by which *Portia fimbriata*, an araneophagic jumping spider, distinguishes jumping-spider prey from other prey. *Journal of Experimental Biology*, 203, 3485–94.

Harland, D. P. & Jackson, R. R. 2002. Influence of cues from the anterior medial eyes of virtual prey on *Portia fimbriata*, an araneophagic jumping spider. *Journal of Experimental Biology*, 205, 1861–8.

Hart, N. S., Partridge, J. C., Cuthill, I. C. & Bennett, A. T. D. 2000. Visual pigments, oil droplets, ocular media and cone photoreceptor distribution in two species of passerine: The blue tit (*Parus caeruleus* L.) and the blackbird (*Turdus merula* L.). *Journal of Comparative Physiology A*, 186, 375–87.

Hart, N. S. 2001a. The visual ecology of avian photoreceptors. *Progress in Retinal and Eye Research*, 20, 675–703.

Hart, N. S. 2001b. Variations in cone photoreceptor abundance and the visual ecology of birds. *Journal of Comparative Physiology A*, 187, 685–97.

Hart, N. S. 2002. Vision in the peafowl (Aves: *Pavo cristatus*). *Journal of Experimental Biology*, 205, 3925–35.

Hart, N. S. & Hunt, D. M. 2007. Avian visual pigments: characteristics, spectral tuning, and evolution. *American Naturalist*, 169, S7–26.

Hartley, D. J. 1992. Stabilization of perceived echo amplitudes in echolocating bats. I. Echo detection and automatic gain control in the big brown bat, *Eptesicus fuscus*, and the fishing bat, *Noctilio leporinus*. *Journal of the Aucoustical Society of America*, 91, 1120–32.

Hartline, H. K., Wagner, H. G. & Ratliff, F. 1956. Inhibition in the eye of Limulus. *Journal of General Physiology*, 39, 651–73.

Hartline, H. K. & Ratliff, F. 1957. Inhibitory interaction of receptor units in the eye of *Limulus*. *Journal of General Physiology*, 40, 357–76.

Hartline, P. H., Kass, L. & Loop, M. S. 1978. Merging of modalities in the optic tectum: infrared and visual integration in rattlesnakes. *Science*, 199, 1225–9.

Hassanali, A., Nyandat, E., Obenchain, F. A., Otieno, D. A. & Galun, R. 1989. Humidity effects on response of *Argas persicus* (Oken) to guanine, an assembly pheromone of ticks. *Journal of Chemical Ecology*, 15, 791–7.

Hasselquist, D. & Bensch, S. 2008. Daily energy expenditure of singing reed warblers *Acrocephalus arundiaceus*. *Journal of Avian Biology*, 39, 384–8.

Hasson, O. 1994. Cheating signals. *Journal of Theoretical Biology*, 167, 223–38.

Håstad, O., Victorsson, J. & Ödeen, A. 2005. Differences in color vision make passerines less conspicuous in the eyes of their predators. *Proceedings of the National Academy of Sciences of the USA*, 102, 6391–4.

Haynes, K. F., Gemeno, C., Yeargan, K. V., Millar, J. G. & Johnson, K. M. 2002. Aggressive chemical mimicry of moth pheromones by a bolas spider: how does this specialist predator attract more than one species of prey? *Chemoecology*, 12, 99–105.

Hebets, E. A. & Uetz, G. W. 1999. Female responses to isolated signals from multimodal male courtship displays in the wolf spider genus *Schizocosa* (Araneae: Lycosidae). *Animal Behaviour*, 57, 865–72.

Hebets, E. A. & Uetz, G. W. 2000. Leg ornamentation and the efficacy of courtship display in four species of wolf spider (Araneae: Lycosidae). *Behavioral Ecology & Sociobiology*, 47, 280–6.

Hebets, E. A. 2005. Attention-altering interaction in the multimodal courtship display of the wolf spider *Schizocosa uetzi*. *Behavioral Ecology*, 16, 75–82.

Hebets, E. A. & Papaj, D. R. 2005. Complex signal function: developing a framework of testable hypotheses. *Behavioral Ecology and Sociobiology*, 57, 197–214.

Hebets, E. A. 2008. Seismic signal dominance in the multimodal courtship display of the wolf spider *Schizocosa stridulans* Stratton 1991. *Behavioral Ecology*, 19, 1250–7.

Hebets, E. A. 2011. Current status and future directions of research in complex signaling. *Current Zoology*, 57, 1–5.

Hedwig, B. & Meyer, J. 1994. Auditory information processing in strdulating grasshoppers: tympanic membrane vibration and neurophysiology. *Journal of Comparative Physiology A*, 174, 121–31.

Heiligenberg, W. 1991. The Jamming Avoidance Response (JAR) of the electric fish, *Eigenmannia*: computational rules and their neuronal implementation. In: *Seminars in Neuroscience*, pp. 3–18. London: Saunders.

Heiling, A. M., Herberstein, M. E. & Chittka, L. 2003. Crabspiders manipulate flower signals. *Nature*, 421, 334.

Heiling, A. M. & Herberstein, M. E. 2004. Predator-prey coevolution: Australian native bees avoid their spider predators. *Biology Letters*, 271, S196–8.

Heiling, A. M., Cheng, K., Chittka, L., Goeth, A. & Herberstein, M. E. 2005. The role of UV in crab spider signals: effects on perception by prey and predators. *Journal of Experimental Biology*, 208, 3925–31.

Heinrich, M., Warmbold, A., Hoffmann, S., Firzlaff, U. & Wiegrebe, L. 2011. The sonar aperture and its neural representation in bats. *Journal of Neuroscience*, 31, 15618–27.

Hellinger, J. & Hoffmann, K. P. 2009. Magnetic field perception in the rainbow trout, *Oncorhynchus mykiss*. *Journal of Comparative Physiology A*, 195, 873–9.

Hempel de Ibarra, N. & Vorobyev, M. 2009. Flower patterns are adapted for detection by bees. *Journal of Comparative Physiology A*, 195, 319–23.

Henry, K. S. & Lucas, J. R. 2008. Coevolution of auditory sensitivity and temporal resolution with acoustic signal space in three songbirds. *Animal Behaviour*, 76, 1659–71.

Henry, K. S. & Lucas, J. R. 2010. Habitat-related differences in the frequency selectivity of auditory filters in songbirds. *Functional Ecology*, 24, 614–24.

Henry, K. S., Gall, M. D., Bidelman, G. M. & Lucas, J. R. 2011. Songbirds tradeoff auditory frequency resolution and temporal resolution. *Journal of Comparative Physiology A*, 197, 351–9.

Henze, M. J. & Labhart, T. 2007. Haze, clouds and limited sky visibility: polarotactic orientation of crickets under difficult stimulus conditions. *Journal of Experimental Biology*, 210, 3266–76.

Herberstein, M. E., Craig, C. L., Coddington, J. A. & Elgar, M. A. 2000. The functional significance of silk decorations of orb-web spiders: a critical review of the empirical evidence. *Biological Reviews*, 75, 649–69.

Herberstein, M. E., Heiling, A. M. & Cheng, K. 2009. Evidence for UV-based sensory exploitation in Australian but not European crab spiders. *Evolutionary Ecology*, 23, 621–34.

Herring, P. J. & Cope, C. 2005. Red bioluminescence in fishes: on the suborbital photophores of *Malacosteus*, *Pachystomias*, and *Aristostomias*. *Marine Biology*, 148, 383–94.

Higham, J. P., Semple, S., MacLarnon, A., Heistermann, M. & Ross, C. 2009. Female reproductive signaling, and male mating behavior, in the olive baboon. *Hormones and Behavior*, 55, 60–7.

Higham, J. P., Brent, L. J. N., Dubuc, C., Accamando, A. K., Engelhardt, A., Gerald, M. S., Heistermann, M. & Stevens, M. 2010. Color signal information content and the eye of the beholder: a case study in the rhesus macaque. *Behavioral Ecology*, 21, 739–46.

Higham, J. P., Hughes, K. D., Brent, L. J. N., Dubuc, C., Engelhardt, A., Heistermann, M., Maestripieri, D., Santos, L. R. & Stevens, M. 2011. Familiarity affects the assessment of female facial signals of fertility by free-ranging male rhesus macaques. *Proceedings of the Royal Society B: Biological Sciences*, 278, 3452–8.

Hildebrand, J. G. & Shepherd, G. M. 1997. Mechanisms of olfactory discrimination: converging evidence for common principles across phyla. *Annual Review of Neuroscience*, 20, 595–631.

Hill, P. S. M. 2008. *Vibrational Communication in Animals*. Cambridge: Harvard University Press.

Hill, P. S. M. 2009. How do animals use substrate-borne vibrations as an information source? *Naturwissenschaften*, 96, 1355–71.

Hiramatsu, C., Melin, A. D., Aureli, F., Schaffner, C. M., Vorobyev, M., Matsumoto, Y. & Kawamura, S. 2008. Importance of achromatic contrast in short-range fruit foraging of primates. *PLoS ONE*, 3, e3356.

Hofmann, C. M., O'Quin, K. E., Marshall, N. J. & Carleton, K. L. 2010. The relationship between lens transmission and opsin gene expression in cichlids from Lake Malawi. *Vision Research*, 50, 357–63.

Hogg, C., Neveu, M., Stokkan, K.-A., Folkow, L., Cottrill, P., Douglas, R. H., Hunt, D. M. & Jeffery, G. 2011. Arctic reindeer extend their visual range into the ultraviolet. *Journal of Experimental Biology*, 214, 2014–19.

Holland, R. A., Thorup, K., Vonhof, M. J., Cochran, W. W. & Wikelski, M. 2006. Navigation: bat orientation using Earth's magnetic field. *Nature*, 444, 702.

Hölldobler, B. & Wilson, E. O. 1994. *Journey to the ants: a story of scientific exploration*. Cambridge, MA and London, England: The Belknap Press of Harvard University Press.

Homberg, U., Heinze, S., Pfeiffer, K., Kinoshita, M. & el Jundi, B. 2011. Central neural coding of sky polarization in insects. *Philosophical Transactions of the Royal Society B: Biological Sciences*, 366, 680–7.

Honkavaara, J., Åberg, H. & Viitala, J. 2008. Do house mice us UV cues when foraging? *Journal of Ethology*, 26, 339–45.

Hoover, J. P. & Robinson, S. K. 2007. Retaliatory mafia behaviour by a parasitic cowbird favors host acceptance of parasitic eggs. *Proceedings of the National Academy of Sciences of the USA*, 104, 4479–83.

Hopkins, C. D. 2010. A biological function for electroreception in sharks and rays. *Journal of Experimental Biology*, 213, 1005–7.

Hornstein, E. P., O'Carrol, D. C., Anderson, J. C. & Laughlin, S. B. 2000. Sexual dimorphism matches photoreceptor performance to behavioural requirements. *Proceedings of the Royal Society B: Biological Sciences*, 267, 2111–17.

Horváth, G. & Varjú, D. 2004. *Polarized Light in Animal Vision—Polarization Patterns in Nature*. New York: Springer-Verlag.

Hoy, R. R., Nolen, T. & Brodfuehrer, P. 1989. The neuroethology of acoustic startle and escape in flying insects. *Journal of Experimental Biology*, 146, 287–306.

Hoy, R. R. & Robert, D. 1996. Tympanal hearing in insects. *Annual Review of Entomology*, 41, 433–50.

Hristov, N. & Conner, W. E. 2005. Sound strategy: acoustic aposematism in the bat–tiger moth arms race. *Naturwissenschaften*, 92, 164–9.

Huang, J.-N., Cheng, R.-C., Li, D. & Tso, I.-M. 2011. Salticid predation as one potential driving force of ant mimicry in jumping spiders. *Proceedings of the Royal Society B: Biological Sciences*, 278, 1356–64.

Huey, R. B. 1977. Natural selection for juvenile lizards mimicking noxious beetles. *Science*, 193, 201–2.

Hughes, K. A. 2010. More than one way to blanch a lizard. *Proceedings of the National Academy of Sciences of the USA*, 107, 1815–16.

Hunt, D. M., Carvalho, L. S., Cowing, J. A. & Davies, W. L. 2009. Evolution and spectral tuning of visual pigments in birds and mammals. *Philosophical Transactions of the Royal Society B: Biological Sciences*, 364, 2941–55.

Hurd, P. L., Wachtmeister, C.-A. & Enquist, M. 1995. Darwin's principle of antithesis revisited: a role for perceptual biases in the evolution of intraspecific signals. *Proceedings of the Royal Society of London B: Biological Sciences*, 259, 201–5.

Immonen, E. & Ritchie, M. G. 2011. The genomic response to courtship song stimulation in female *Drosophila melanogaster*. *Proceedings of the Royal Society B: Biological Sciences*, 279, 1359–65.

Ingalls, V. 1993. Startle and habituation responses of blue jays (*Cyanocitta cristata*) in a laboratory simulation of anti-predator defenses of catocala moths (Lepidoptera: Noctuidae). *Behaviour*, 126, 77–96.

Isles, A. R., Baum, M. J., Ma, D., Keverne, E. B. & Allen, N. D. 2001. Urinary odour preferences in mice. *Nature*, 409, 783–4.

Jacklyn, P. M. 1992. 'Magnetic' termite mound surfaces are oriented to suit wind and shade conditions. *Oecologia*, 91, 385–95.

Jackson, R. R. & Wilcox, R. S. 1990. Aggressive mimicry, prey-specific predatory behaviour and predator-recognition in the predator-prey interactions of *Portia fimbriata* and *Euryattus* sp., jumping spiders from Queensland. *Behavioral Ecology and Sociobiology*, 26, 111–19.

Jacobs, G. H. 1996. Primate photopigments and primate color vision. *Proceedings of the National Academy of Sciences of the USA*, 93, 577–81.

Jacobs, G. H. & Deegan, J. F., 2nd. 1999. Uniformity of colour vision in Old World monkeys. *Proceedings of the Royal Society B: Biological Sciences*, 266, 2023–8.

Jacobs, G. H. 2009. Evolution of colour vision in mammals. *Philosophical Transactions of the Royal Society B: Biological Sciences*, 364, 2957–67.

Jaffe, K., Mirás, B. & Cabrera, A. 2007. Mate selection in the moth *Neoleucinodes elegantalis*: evidence for a supernormal chemical stimulus in sexual attraction. *Animal Behaviour*, 73, 727–34.

Janzen, D. H. 1980. When is it coevolution? *Evolution*, 34, 611–12.

Jarvis, J., Bohn, K. M., Tressler, J. & Smotherman, M. 2010. A mechanism for the antiphonal echolocation by free-tailed bats. *Animal Behaviour*, 79, 787–96.

Jersakova, J., Johnson, S. D. & Kindlmann, P. 2006. Mechanisms and evolution of deceptive pollination in orchids. *Biological Reviews*, 81, 219–35.

Jetz, W., Rowe, C. & Guilford, T. 2001. Non-warning odors trigger innate color aversions-as long as they are novel. *Behavioral Ecology*, 12, 134–9.

Jiggins, C., Naisbit, R. E., Coe, R. L. & Mallet, J. 2001. Reproductive isolation caused by colour pattern mimicry. *Nature*, 411, 302–5.

Jiggins, C., Estrada, C. & Rodrigues, A. 2004. Mimicry and the evolution of premating isolation in *Heliconius melpomene* Linnaeus. *Journal of Evolutionary Biology*, 17, 680–91.

Jiggins, C. 2008. Ecological speciation in mimetic butterflies. *BioScience*, 58, 541–8.

Johnsen, S. 2001. Hidden in plain sight: the ecology and physiology of organismal transparency. *Biology Bulletin*, 201, 301–18.

Johnsen, S. & Lohmann, K. J. 2005. The physics and neurobiology of magnetoreception. *Nature Reviews Neuroscience*, 6, 703–12.

Johnsen, S., Marshall, N. J. & Widder, E. A. 2011. Polarization sensitivity as a contrast enhancer in pelagic predators: lessons from in situ polarization imaging of transparent zooplankton. *Philosophical Transactions of the Royal Society B: Biological Sciences*, 366, 655–70.

Johnsen, S. 2012. *The Optics of Life: A Biologists Guide to Light in Nature*. Princeton: Princeton University Press.

Jones, G. 2005. Echolocation. *Current Biology*, 15, R484–8.

Jones, G. & Teeling, E. C. 2006. The evolution of echolocation in bats. *Trends in Ecology & Evolution*, 21, 149–56.

Jones, G. & Holderied, M. W. 2007. Bat echolocation calls: adaptation and convergent evolution. *Proceedings of the Royal Society B: Biological Sciences*, 274, 905–12.

Jones, G. 2010. Molecular evolution: gene convergence in echolocating mammals. *Current Biology*, 20, R62–4.

Jones, G. & Siemers, B. M. 2010. The communicative potential of bat echolocation pulses. *Journal of Comparative Physiology A*, 197, 447–57.

Jones, K. J. & Hill, W. L. 2001. Auditory perception of hawks and owls for passerine bird calls. *Ethology*, 107, 717–26.

Joron, M. & Mallet, J. 1998. Diversity in mimicry: paradox or paradigm? *Trends in Ecology and Evolution*, 13, 461–6.

Joron, M., Jiggins, C., Papanicolaou, A. & McMillan, W. O. 2006. *Heliconius* wing patterns: an evo-devo model for understanding phenotypic diversity. *Heredity*, 97, 157–67.

Julian, D., Crampton, W. G. R., Wohlgemuth, S. E. & Albert, J. S. 2003. Oxygen consumption in weakly electric Neotropical fishes. *Oecologia*, 137, 502–11.

Kaib, M., Heinze, J. & Ortius, D. 1993. Cuticular hydrocarbon profiles in the slave-making ant *Harpagoxenus sublaevis* and its hosts. *Naturwissenschaften*, 80, 281–5.

Kalberer, N. M., Reisenman, C. E. & Hildebrand, J. G. 2010. Male moths bearing transplanted female antennae express characteristically female behaviour and central neural activity. *Journal of Experimental Biology*, 213, 1272–80.

Kalko, E. & Schnitzler, H. U. 1993. Plasticity in echolocation signals of European pipistrelle bats in search flight: implications for habitat use and prey detection. *Behavioral Ecology and Sociobiology*, 33, 415–28.

Kalmijn, A. J. 1971. The electric sense of sharks and rays. *Journal of Experimental Biology*, 55, 371–83.

Kandel, E. R., Schwartz, J. H. & Jessell, T. M. 1995. *Essentials of Neural Science and Behavior, International Edition*. London: McGraw-Hill.

Kaplan, E. & Barlow, R. B., Jr 1980. Circadian clock in *Limulus* brain increases response and decreases noise of retinal photoreceptors. *Nature*, 286, 393–5.

Kaupp, U. B. 2010. Olfactory signalling in vertebrates and insects: differences and commonalities. *Nature Reviews Neuroscience*, 11, 188–200.

Kawata, M., Shoji, A., Kawamura, S. & Seehausen, O. 2007. A genetically explicit model of speciation by sensory drive within a continuous population in aquatic environments. *BMC Evolutionary Biology*, 7, 99.

Kekäläinen, J., Huuskonen, H., Kiviniemi, V. & Taskinen, J. 2010. Visual conditions and habitat shape the coloration of the Eurasian perch (*Perca fluviatilis* L.): a trade-off between camouflage and communication? *Biological Journal of the Linnean Society*, 99, 47–59.

Kelber, A., Vorobyev, M. & Osorio, D. 2003. Animal colour vision—behavioural tests and physiological concepts. *Biological Reviews*, 78, 81–118.

Kelber, A. & Roth, L. S. V. 2006. Nocturnal colour vision—not as rare as we might think. *Journal of Experimental Biology*, 209, 781–8.

Kelber, A. & Osorio, D. 2010. From spectral information to animal colour vision: experiments and concepts. *Proceedings of the Royal Society B: Biological Sciences*, 277, 1617–25.

Kelley, J. L., Phillips, B., Cummins, G. H. & Shand, J. 2012. Changes in the visual environment affect colour signal brightness and shoaling behaviour in freshwater fish. *Animal Behaviour*, 83, 783–91.

Kelley, L. A., Coe, R. L., Madden, J. R. & Healy, S. D. 2008. Vocal mimicry in songbirds. *Animal Behaviour*, 76, 521–8.

Kelley, L. A. & Healy, S. D. 2011. Vocal mimicry. *Current Biology*, 21, R9–10.

Kenward, B., Wachtmeister, C. A., Ghirlanda, S. & Enquist, M. 2004. Spots and stripes: the evolution of repetition in visual signal form. *Journal of Theoretical Biology*, 230, 407–19.

Kick, S. A. & Simmons, J. A. 1984. Automatic gain control in the bat's sonar receiver and the neuroethology of echolocation. *Journal of Neuroscience*, 4, 2725–37.

Kilner, R. M., Noble, D. G. & Davies, N. B. 1999. Signals of need in parent-offspring communication and their exploitation by the common cuckoo. *Nature*, 397, 667–72.

Kilner, R. M. & Langmore, N. E. 2011. Cuckoos versus hosts in insects and birds: adaptations, counter-adaptations and outcomes. *Biological Reviews*, 86, 836–52.

King, A. J. & Palmer, A. R. 1985. Integration of visual and auditory information in bimodal neurones in the guinea-pig superior colliculus. *Experimental Brain Research*, 60, 492–500.

Kingdon, J., Agwanda, B., Kinnaird, M., O'Brien, T., Holland, C., Gheysens, T., Boulet-Audet, M. & Vollrath, F. 2012. A poisonous surprise under the coat of the African crested rat. *Proceedings of the Royal Society B: Biological Sciences*, 279, 675–80.

Kingston, T. & Rossiter, S. J. 2004. Harmonic-hopping in Wallacea's bats. *Nature*, 429, 654–7.

Kirkpatrick, M. & Price, T. 2008. In sight of speciation. *Nature*, 455, 601–2.

Kirschel, A. N. G., Blumstein, D. T., Cohen, R. E., Buermann, W., Smith, T. B. & Slabbekoorn, H. 2009. Birdsong tuned to the environment: green hylia song varies with elevation, tree cover, and noise. *Behavioral Ecology*, 20, 1089–95.

Kirschvink, J. L. & Gould, J. L. 1981. Biogenic magnetite as a basis for magnetic field detection in animals. *Biosystems*, 13, 181–201.

Kleindorfer, S., Hoi, H. & Fessl, B. 1996. Alarm calls and chick reactions in the moustached warbler, *Acrocephalus melanopogon*. *Animal Behaviour*, 51, 1199–206.

Klump, G. M., Kretzschmar, E. & Curio, E. 1986. The hearing of an avian predator and its prey. *Behavioral Ecology and Sociobiology*, 18, 317–23.

Knudsen, D. P. & Gentner, T. Q. 2010. Mechanisms of song perception in oscine birds. *Brain & Language*, 115, 59–68.

Knudsen, E. I. & Konishi, M. 1978a. Center-surround organization of auditory receptive fields in the owl. *Science*, 202, 778–80.

Knudsen, E. I. & Konishi, M. 1978b. A neural map of auditory space in the owl. *Science*, 200, 795–7.

Knudsen, E. I., Blasdel, G. G. & Konishi, M. 1979. Sound localisation by the barn owl (*Tyto alba*) measured with the search coil technique. *Journal of Comparative Physiology A*, 133, 1–11.

Knudsen, E. I. & Konishi, M. 1979. Mechanisms of sound localisation in the barn owl (*Tyto alba*). *Journal of Comparative Physiology A*, 133, 13–21.

Kojima, W., Takanashi, T. & Ishikawa, Y. 2012. Vibratory communication in the soil: pupal signals deter larval intrusion in a group-living beetle *Trypoxylus dichotoma*. *Behavioral Ecology and Sociobiology*, 66, 171–9.

Kokko, H., Brooks, R., Jennions, M. D. & Morley, J. 2003. The evolution of mate choice and mating biases. *Proceedings of the Royal Society B: Biological Sciences*, 270, 653–64.

Kokko, H., Jennions, M. D. & Brooks, R. 2006. Unifying and testing models of sexual selection. *Annual Review of Ecology, Evolution, and Systematics*, 37, 43–66.

Konishi, M. 1970. Comparative neurophysiological studies of hearning and vocalizations in songbirds. *Zeitschrift für vergleichende Physiologie*, 66, 257–72.

Konishi, M. 1973. How the owl tracks its prey: experiments with trained barn owls reveal how their acute sense of hearing enables them to catch prey in the dark. *American Scientist*, 61, 414–24.

Köppl, C., Gleich, O. & Manley, G. A. 1993. An auditory fovea in the barn owl cochlea. *Journal of Comparative Physiology A*, 171, 695–704.

Köppl, C. 2011. Birds—same thing, but different? Convergent evolution in the avian and mammalian auditory systems provides informative comparative models. *Hearing Research*, 273, 65–71.

Koshitaka, H., Kinoshita, M., Vorobyev, M. & Arikawa, K. 2008. Tetrachromacy in a butterfly that has eight varieties of spectral receptors. *Proceedings of the Royal Society B: Biological Sciences*, 275, 947–54.

Kössl, M. & Vater, M. 1985. The cochlear frequency map of the mustache bat, *Pteronotus parnellii*. *Journal of Comparative Physiology A*, 157, 687–97.

Kotiaho, J. S., Alatalo, R. V., Mappes, J., Nielsen, M. G., Parri, S. & Rivero, A. 1998a. Energetic costs of size and sexual signalling in a wolf spider. *Proceedings of the Royal Society B: Biological Sciences*, 265, 2203–9.

Kotiaho, J. S., Alatalo, R. V., Mappes, J. & Parri, S. 1998b. Male mating success and risk of predation in a wolf spider: a balance between sexual and natural selection? *Journal of Animal Ecology*, 67, 287–91.

Kraaijeveld, K., Kraaijeveld-Smit, F. J. L. & Maan, M. E. 2011. Sexual selection and speciation: the comparative evidence revisited. *Biological Reviews*, 86, 367–77.

Kraft, P., Evangelista, C., Dacke, M., Labhart, T. & Srinivasan, M. V. 2011. Honeybee navigation: following routes using polarized-light cues. *Philosophical Transactions of the Royal Society B: Biological Sciences*, 366, 703–8.

Krakauer, D. C. & Johnstone, R. A. 1995. The evolution of exploitation and honesty in animal communication: a model using artificial neural networks. *Philosophical Transactions of the Royal Society B: Biological Sciences*, 348, 355–61.

Kramer, B. & Kuhn, B. 1993. Electric signaling and impedance matching in a variable environment. *Naturwissenschaften*, 80, 43–6.

Kramer, G. 1953. Wird die Sonnenhöhe bei der Heimfindeorientierung verwertet? *Journal für Ornithologie*, 94, 201–19.

Krekelberg, B. 2008. Motion detection mechanisms. In: *The Senses: A Comprehensive Reference* (Ed. by Basbaum, A.), pp. 133–54. Oxford: Elsevier Inc.

Kreuter, K., Twele, R., Francke, W. & Ayasse, M. 2010. Specialist *Bombus vestalis* and generalist *Bombus bohemicus* use different odour cues to find their host *Bombus terrestris*. *Animal Behaviour*, 80, 297–302.

Kronforst, M. R., Young, L. G., Kapan, D. D., McNeely, C., O'Neill, R. J. & Gilbert, L. E. 2006. Linkage of butterfly mate preference and wing color preference cue at the genomic location of *wingless*. *Proceedings of the National Academy of Sciences of the USA*, 103, 6575–80.

Kronforst, M. R., Young, L. G. & Gilbert, L. E. 2007. Reinforcement of mate preference among hybridizing *Heliconius* butterflies. *Journal of Evolutionary Biology*, 20, 278–85.

Krüger, O., Sorenson, M. D. & Davies, N. B. 2009. Does coevolution promote species richness in parasitic cuckoos? *Proceedings of the Royal Society B: Biological Sciences*, 276, 3871–79.

Krüger, O. 2007. Cuckoos, cowbirds and hosts: adaptations, trade-offs and constraints. *Philosophical Transactions of the Royal Society B: Biological Sciences*, 362, 1873–86.

Krüger, O. 2011. Brood parasitism selects for no defence in a cuckoo host. *Proceedings of the Royal Society B: Biological Sciences*, 278, 2777–83.

Kuffler, S. W. 1953. Discharge patterns and functional organisation of mammalian retina. *Journal of Physiology*, 16, 37–68.

Kukuk, P. 1985. Evidence for an antiaphrodisiac in the sweat bee *Lasioglossum zephyrum* (Dialictus). *Science*, 227, 656–7.

Kulahci, I. G., Dornhaus, A. & Papaj, D. R. 2008. Multimodal signals enhance decision making in foraging bumble-bees. *Proceedings of the Royal Society B: Biological Sciences*, 275, 797–802.

Kürten, L., Schmidt, U. & Schafer, K. 1984. Warm and cold receptors in the nose of the vampire bat *Desmodus rotundus*. *Naturwissenschaften*, 71, 327–8.

Kyhn, L. A., Jensen, F. H., Beedholm, K., Tougaard, J., Hansen, M. & Madsen, P. T. 2010. Echolocation in sympatric Peale's dolphins (*Lagenorhynchus australis*) and Commerson's dolphins (*Cephalorhynchus commersonii*) producing narrow-band high-frequency clicks. *Journal of Experimental Biology*, 213, 1940–9.

Labhart, T. 1980. Specialized photoreceptors at the dorsal rim of the honeybee's compound eye: polarizational and angular sensitivity. *Journal of Comparative Physiology A*, 141, 19–30.

Labhart, T. 1988. Polarization-opponent interneurons in the insect visual system. *Nature*, 331, 435–7.

Labhart, T. & Meyer, E. P. 2002. Neural mechanisms in insect navigation: polarization compass and odometer. *Current Opinion in Neurobiology*, 12, 707–14.

Lahti, D. C. 2005. Evolution of bird eggs in the absence of cuckoo parasitism. *Proceedings of the National Academy of Sciences of the USA*, 102, 18057–62.

Lall, A. B., Chapman, R. M., Trouth, C. O. & Holloway, J. A. 1980a. Spectral mechanisms of the compound eye in the firefly *Photinus pyralis* (Coleoptera: Lampryridae). *Journal of Comparative Physiology A*, 135, 21–7.

Lall, A. B., Seliger, H. H., Biggley, W. H. & Lloyd, J. E. 1980b. Ecology of colors of firefly bioluminescence. *Science*, 210, 560–2.

Lall, A. B. 1981a. Electroretinogram and the spectral sensitivity of the compound eyes in the firefly *Photuris versicolor* (Coleoptera-Lampyridae): a corresondence between green sensitivity and species bioluminescence emission. *Journal of Insect Physiology*, 27, 461–8.

Lall, A. B. 1981b. Vision tuned to species bioluminescence emission in firefly *Photinus pyralis*. *Journal of Experimental Zoology*, 216, 317–9.

Lall, A. B., Lord, E. T. & Trouth, C. O. 1982. Vision in the firefly *Photurius lucicrescens* (Coleoptera: Lampyridae): spectral sensitivity and selective adaptation in the compound eye. *Journal of Comparative Physiology A*, 147, 195–200.

Lall, A. B., Strother, G. K., Cronin, T. W. & Seliger, H. H. 1988. Modification of spectral sensitivities by screening pigments in the compound eyes of twilight-active fireflies (Coleoptera: Lampyridae). *Journal of Comparative Physiology A*, 162, 23–33.

Lall, A. B., Ventura, D. S., Bechara, E. J., de Souza, J. M., Colepicolo-Neto, P. & Viviani, V. R. 2000. Spectral correspondence between visual spectral sensitivity and bioluminescence emission spectra in the click beetle *Pyrophorus punctatissimus* (Coleoptera: Elateridae). *Journal of Insect Physiology*, 46, 1137–41.

Lall, A. B. & Worthy, K. M. 2000. Action spectra of the female's response in the firefly *Photinus pyralis* (Coleoptera: Lampyridae): evidence for an achromatic detection of the bioluminescent optical signal. *Journal of Insect Physiology*, 46, 965–8.

Lall, A. B., Cronin, T. W., Carvalho, A. A., de Souza, J. M., Barros, M. P., Stevani, C. V., Bechara, E. J., Ventura, D. F., Viviani, V. R. & Hill, A. A. 2010. Vision in click beetles

(Coleoptera: Elateridae): pigments and spectral correspondence between visual sensitivity and species bioluminescence emission. *Journal of Comparative Physiology A*, 196, 629–38.

Lamb, T. D. 2009. Evolution of vertebrate retinal photoreception. *Philosophical Transactions of the Royal Society B: Biological Sciences*, 364, 2911–24.

Lamb, T. D., Arendt, D. & Collin, S. P. 2009. The evolution of phototransduction and eyes. *Philosophical Transactions of the Royal Society B: Biological Sciences*, 364, 2791–3.

Land, M. F. & Nilsson, D.-E. 2012. *Animal Eyes, Second Edition*. Oxford: Oxford University Press.

Langmore, N. E., Hunt, S. & Kilner, R. M. 2003. Escalation of a coevolutionary arms race through host rejection of brood parasitic young. *Nature*, 422, 157–60.

Langmore, N. E. & Kilner, R. M. 2009. Why do Horsfield's bronze-cuckoo *Chalcites basalis* eggs mimic those of their hosts? *Behavioral Ecology and Sociobiology*, 63, 1127–31.

Langmore, N. E., Stevens, M., Maurer, G. & Kilner, R. M. 2009. Are dark cuckoo eggs cryptic in host nests? *Animal Behaviour*, 78, 461–8.

Langmore, N. E. & Kilner, R. M. 2010. The coevolutionary arms race between Horsfield's bronze-cuckoos and superb fairy-wrens. *Emu*, 110, 32–8.

Langmore, N. E., Stevens, M., Maurer, G., Heinsohn, R., Hall, M. L., Peters, A. & Kilner, R. M. 2011. Visual mimicry of host nestlings by cuckoos. *Proceedings of the Royal Society B: Biological Sciences*, 278, 2455–63.

Langridge, K., Broom, M. & Osorio, D. 2007. Selective signalling by cuttlefish to predators. *Current Biology*, 17, R1044–5.

Laughlin, S. B. & Weckström, M. 1993. Fast and solw photoreceptors—a comparative study of the functional diversity of coding and conductances in the Diptera. *Journal of Comparative Physiology A*, 172, 5933–5609.

Laughlin, S. B. 1996. Matched filtering by a photoreceptor membrane. *Vision Research*, 36, 1529–41.

Laughlin, S. B., de Ruyter van Steveninck, R. R. & Anderson, J. C. 1998. The metabolic cost of neural information. *Nature Neuroscience*, 1, 36–41.

Laughlin, S. B. 2001. Energy as a constraint on the coding and processing of sensory information. *Current Opinion in Neurobiology*, 11, 475–80.

Laughlin, S. B. 2011. Energy, information, and the work of the brain. In: *Work meets life* (Ed. by Levin, R. A., Laughlin, S. B., De la Rocha, C. L. & Blackwell, A. F.), pp. 39–67. Cambridge Mass: MIT Press.

Laumann, R. A., Cokl, A., Lopes, A. P. S., Fereira, J. B. C. & Moraes, M. C. B. 2011. Silent singers are not safe: selective response of a parasitoid to substrate-borne vibratory signals of stink bugs. *Animal Behaviour*, 82, 1175–83.

Lawrence, E. S. & Allen, J. A. 1983. On the term 'search image'. *Oikos*, 40, 313–14.

Lazure, L. & Fenton, M. B. 2010. High duty cycle echolocation and prey detection by bats. *Journal of Experimental Biology*, 214, 1131–7.

Lee, S. & Zhou, Z. J. 2006. The synaptic mechanism of direction selectivity in distal processes of starburst amacrine cells. *Neuron*, 51, 787–99.

Lehmann, G. U. C., Heller, K.-G. & Lehmann, A. W. 2001. Male bushcrickets favoured by parasitoid flies when acoustically more attractive for conspecific females (Orthoptera: Phanopteridae/Diptera: Tachinidae). *Entomologia Generalis*, 25, 135–40.

Lemmon, E. M. 2009. Diversification of conspecific signals in sympatry: geographic overlap drives multidimensional reproductive character displacement in frogs. *Evolution*, 63, 1155–70.

Lent, D. D., Graham, P. & Collett, T. S. 2010. Image-matching during ant navigation occurs through saccade-like body turns controlled by learned visual features. *Proceedings of the National Academy of Sciences of the USA*, 107, 16348–53.

Leonard, A. S., Dornhaus, A. & Papaj, D. R. 2011. Flowers help bees cope with uncertainty: signal detection and the function of floral complexity. *Journal of Experimental Biology*, 214, 113–21.

Lerner, A., Sabbah, S., Erlick, C. & Shashar, N. 2011. Navigation by light polarization in clear and turbid waters. *Philosophical Transactions of the Royal Society B: Biological Sciences*, 366, 671–9.

Lewis, S. M. & Cratsley, C. K. 2008. Flash signal evolution, mate choice, and predation in fireflies. *Annual Review of Entomology*, 53, 293–321.

Lewkiewicz, D. A. & Zuk, M. 2004. Latency to resume calling after disturbance in the field cricket, *Teleogryllus oceanicus*, corresponds to population-level differences in parasitism risk. *Behavioral Ecology and Sociobiology*, 55, 569–73.

Li, G., Wang, J., Rossiter, S. J., Jones, G. & Zhang, S. 2007. Accelerated *FoxP2* evolution in echolocating bats. *PLoS ONE*, 2, e900.

Li, G., Wang, J., Rossiter, S. J., Jones, G., Cotton, J. A. & Zhang, S. 2008a. The hearing gene *Prestin* reunites echolocating bats. *Proceedings of the National Academy of Sciences of the USA*, 105, 13959–64.

Li, J., Lim, M. L. M., Zhang, Z., Liu, Q., Liu, F., Chen, J. & Li, D. 2008b. Sexual dichromatism and male color morph in ultraviolet-B reflectance in two populations of the jumping spider *Phintella vittata* (Araneae: Salticidae) from tropical China. *Biological Journal of the Linnean Society*, 94, 7–20.

Li, J., Zhang, Z., Liu, F., Liu, Q., Gan, W., Chen, J., Lim, M. L. & Li, D. 2008c. UVB-based mate-choice cues used by females of the jumping spider *Phintella vittata*. *Current Biology*, 18, 699–703.

Li, Y., Liu, Z., Shi, P. & Zhang, J. 2010. The hearing gene *Prestin* unites echolocating bats and whales. *Current Biology*, 20, R55–66.

Liang, W., Yang, C., Stokke, B. G., Antonov, A., Fossoy, F., Vikan, J. R., Moksnes, A., Roskaft, E., Shykoff, J. A., Moller, A. P. & Takasu, F. 2012. Modelling the maintenance of egg polymorphism in avian brood parasites and their hosts. *Journal of Evolutionary Biology*, 25, 916–29.

Lim, M. L. M. & Li, D. 2006. Extreme ultraviolet sexual dimorphism in jumping spiders (Araneae: Salticidae). *Biological Journal of the Linnean Society*, 89, 397–406.

Lim, M. L. M., Land, M. F. & Li, D. 2007. Sex-specific UV and fluorescence signals in jumping spiders. *Science*, 315, 481.

Lim, M. L. M., Li, J. & Li, D. 2008. Effect of UV-reflecting markings on female mate-choice decisions in *Cosmophasis umbratica*, a jumping spider from Singapore. *Behavioral Ecology*, 19, 61–6.

Lind, O. & Kelber, A. 2009. The intensity threshold of colour vision in two species of parrot. *Journal of Experimental Biology*, 212, 3693–9.

Lindström, L., Rowe, C. & Guilford, T. 2001. Pyrazine odour makes visually conspicuous prey aversive. *Proceedings of the Royal Society B: Biological Sciences*, 268, 159–62.

Linnenschmidt, M., Beedholm, K., Wahlberg, M., Hojer-Kristensen, J. & Nachtigall, P. E. 2012. Keeping returns optimal: gain control exerted through sensitivity adjustments in the harbour porpoise auditory system. *Proceedings of the Royal Society B: Biological Sciences*, 279, 2237–45.

Lissmann, H. W. 1951. Continuous Electrical Signals from the Tail of a Fish, *Gymnarchus niloticus* Cuv. *Nature*, 167, 201–2.

Liu, Y., Cotton, J. A., Shen, B., Han, X., Rossiter, S. J. & Zhang, S. 2010. Convergent sequence evolution between echolocating bats and dolphins. *Current Biology*, 20, R53–4.

Lloyd, J. E. & Wing, S. R. 1983. Nocturnal aerial predation of fireflies by light-seeking fireflies. *Science*, 222, 634–5.

Lohmann, K. J., Salmon, M. & Wyneken, J. 1990. Functional autonomy of land and sea orientation systems in sea turtle hatchlings. *Biology Bulletin*, 179, 214–18.

Lohmann, K. J. 1991. Magnetic orientation by hatchling loggerhead sea turtles (*Caretta caretta*). *Journal of Experimental Biology*, 155, 37–49.

Lohmann, K. J., Swartz, A. & Lohmann, C. M. 1995. Perception of ocean wave direction by sea turtles. *Journal of Experimental Biology*, 198, 1079–85.

Lohmann, K. J. & Lohmann, C. M. 1996a. Orientation and open-sea navigation in sea turtles. *Journal of Experimental Biology*, 199, 73–81.

Lohmann, K. J. & Lohmann, C. M. F. 1996b. Detection of magnetic field intensity by sea turtles. *Nature*, 380, 59–61.

Lohmann, K. J., Lohmann, C. M., Ehrhart, L. M., Bagley, D. A. & Swing, T. 2004. Geomagnetic map used in sea-turtle navigation. *Nature*, 428, 909–10.

Lohmann, K. J., Lohmann, C. M. & Putman, N. F. 2007. Magnetic maps in animals: nature's GPS. *Journal of Experimental Biology*, 210, 3697–705.

Lohmann, K. J., Lohmann, C. M. & Endres, C. S. 2008a. The sensory ecology of ocean navigation. *Journal of Experimental Biology*, 211, 1719–28.

Lohmann, K. J., Luschi, P. & Hays, G. C. 2008b. Goal navigation and island-finding in sea turtles. *Journal of Experimental Marine Biology and Ecology*, 356, 83–95.

Longbauer, J. W. R., Payne, K. B., Charif, R. A., Rapaport, L. & Osborn, F. 1991. African elephants respond to distant playbacks of low-frequency conspecific calls. *Journal of Experimental Biology*, 157, 35–46.

Lubbock, J. 1882. *Ants, bees, and wasps: A record of observations on the habits of the social hymenoptera. Second Edition.* London: Kegan Paul, Trench, & Co.

Lugli, M. 2010. Sounds of shallow water fishes pitch within the quiet window of the habitat ambient noise. *Journal of Comparative Physiology A*, 196, 439–51.

Lythgoe, J. N. & Hemmings, C. C. 1967. Polarized light and underwater vision. *Nature*, 213, 893–4.

Lythgoe, J. N. 1979. *The Ecology of Vision.* Oxford: Clarendon Press.

Maan, M. E., Hofker, K. D., van Alphen, J. J. M. & Seehausen, O. 2006. Sensory drive in cichlid speciation. *American Naturalist*, 167, 947–54.

Maan, M. E. & Cummings, M. E. 2008. Female preferences for aposematic signal components in a polymorphic poison frog. *Evolution*, 62, 2334–45.

Maan, M. E. & Cummings, M. E. 2009. Sexual dimorphism and directional sexual selection on aposematic signals in a poison frog. *Proceedings of the National Academy of Sciences of the USA*, 106, 19072–7.

Maan, M. E. & Seehausen, O. 2010. Mechanisms of species divergence through visual adaptation and sexual selection. *Current Zoology*, 56, 285–99.

Maan, M. E. & Seehausen, O. 2011. Ecology, sexual selection and speciation. *Ecology Letters*, 14, 591–602.

Maan, M. E. & Cummings, M. E. 2012. Poison frog colors are honest signals of toxicity, particularly for bird predators. *American Naturalist*, 179, E1–14.

Machnik, P. & Kramer, B. 2008. Female choice by electric pulse duration: attractiveness of the males' communication signal assessed by female bulldog fish, *Marcusenius pongolensis* (Mormyridae, Teleostei). *Journal of Experimental Biology*, 211, 1969–77.

Machnik, P. & Kramer, B. 2010. Novel electrosensory advertising during diurnal resting period in male snout-fish, *Marcusenius altisambesi* (Mormyridae, Teleostei). *Journal of Ethology*, 29, 131–42.

Madden, J. R., Kilner, R. M. & Davies, N. B. 2005. Nestling responses to adult food and alarm calls: 1. Species-specific responses in two cowbird hosts. *Animal Behaviour*, 70, 619–27.

Maeda, K., Henbest, K. B., Cintolesi, F., Kuprov, I., Rodgers, C. T., Liddell, P. A., Gust, D., Timmel, C. R. & Hore, P. J. 2008. Chemical compass model of avian magnetoreception. *Nature*, 453, 387–90.

Magrath, R. D., Pitcher, B. J. & Dalzill, A. H. 2007. How to be fed but not eaten: nestling responses to parental food calls and the sound of predator footsteps. *Animal Behaviour*, 74, 1117–29.

Mallet, J. & Barton, N. H. 1989. Strong natural selection in a warning-color hybrid zone. *Evolution*, 43, 421–31.

Mallon, E. B. & Franks, N. R. 2000. Ants estimate area using Buffon's needle. *Proceedings of the Royal Society B: Biological Sciences*, 267, 765–70.

Mallon, E. B., Pratt, S. C. & Franks, N. R. 2001. Individual and collective decision-making during nest site selection by the ant *Leptothorax albipennis*. *Behavioral Ecology and Sociobiology*, 50, 352–9.

Manceau, M., Domingues, V. S., Mallarino, R. & Hoekstra, H. E. 2011. The developmental role of agouti in color pattern evolution. *Science*, 331, 1062–5.

Mänd, T., Tammaru, T. & Mappes, J. 2007. Size dependent predation risk in cryptic and conspicuous insects. *Evolutionary Ecology*, 21, 485–98.

Mappes, J., Alatalo, R. V., Kotiaho, J. S. & Parri, S. 1996. Viability costs of condition-dependent sexual male display in a drumming wolf spiders. *Proceedings of the Royal Society B: Biological Sciences*, 263, 785–9.

Marchetti, K. 1993. Dark habitats and bright birds illustrate the role of the environment in species divergence. *Nature*, 362, 149–52.

Marhold, S. & Wiltschko, W. 1997. A magnetic polarity compass for direction finding in a subterranean mammal. *Naturwissenschaften*, 84, 421–3.

Markham, M. R., McAnelly, M. L., Stoddard, P. K. & Zakon, H. H. 2009. Circadian and social cues regulate ion channel trafficking. *PLoS Biology*, 7, e1000203.

Marler, P. 1957. Specific distinctiveness in the communication signals of birds. *Behaviour*, 11, 13–39.

Marler, P. 1961. The logical analysis of animal communication. *Journal of Theoretical Biology*, 1, 295–317.

Marler, P. 1967. Animal communication signals: we are beginning to understand how the structure of animal signals relates to the function they serve. *Science*, 157, 769–74.

Marler, P. R. 1955. Characteristics of some animal calls. *Nature*, 176, 6–8.

Marples, N. M., van Veelen, W. & Brakefield, P. M. 1994. The relative importance of colour, taste and smell in the protection of an aposematic insect, *Coccinella septempunctata*. *Animal Behaviour*, 48, 967–74.

Marr, D. & Hildreth, E. 1980. Theory of edge detection. *Proceedings of the Royal Society B: Biological Sciences*, 207, 187–217.

Marrow, P., Law, R. & Cannings, C. 1992. The coevolution of predator-prey interactions: ESSS and Red Queen dynamics. *Proceedings of the Royal Society B: Biological Sciences*, 250, 133–41.

Marshall, J. & Cronin, T. W. 2011. Polarisation vision. *Current Biology*, 21, R101–5.

Marshall, N. J. 1988. A unique colour and polarization vision system in mantis shrimp. *Nature*, 333, 557–60.

Marshall, N. J., Cronin, T. W., Shashar, N. & Land, M. 1999. Behavioural evidence for polarisation vision in stomatopods reveals a potential channel for communication. *Current Biology*, 9, 755–8.

Marshall, N. J. 2000. Communication and camouflage with the same 'bright' colours in reef fishes. *Philosophical Transactions of the Royal Society B: Biological Sciences*, 355, 1243–8.

Martin, S. J., Carruthers, J., Williams, P. & Drijfhout, F. P. 2010. Host specific social parasites (*Psithyrus*) indicate chemical recognition system in bumblebees. *Journal of Chemical Ecology*, 36, 855–63.

Martin, S. J., Helantera, H. & Drijfhout, F. P. 2011. Is parasite pressure a driver of chemical cue diversity in ants? *Proceedings of the Royal Society B: Biological Sciences*, 278, 496–503.

Masse, N. Y., Turner, G. C. & Jefferis, G. S. X. E. 2009. Olfactory information processing in *Drosophila*. *Current Biology*, 19, R700–13.

Masson, C. & Mustaparta, H. 1990. Chemical information processing in the olfactory system of insects. *Physiological Reviews*, 70, 199–245.

Masta, S. E. & Maddison, W. P. 2002. Sexual selection driving diversification in jumping spiders. *Proceedings of the National Academy of Sciences of the USA*, 99, 4442–7.

Mäthger, L. M., Shashar, N. & Hanlon, R. T. 2009. Do cephalopods communicate using polarized light reflections from their skin? *Journal of Experimental Biology*, 212, 2133–40.

Maynard Smith, J. & Harper, D. 2003. *Animal Signals*. Oxford: Oxford University Press.

Mayr, E. 1942. *Systematics and the origin of species* New York: Columbia University Press.

McGraw, K. J. 2006. Chapter 5: Mechanics of Carotenoid-Based Coloration. In: *Bird Coloration. Volume I: Mechanisms and Measurements* (Ed. by Hill, G. E. & McGraw, K. J.), pp. 177–242. Cambridge MA: Harvard University Press.

McNaught, M. K. & Owens, I. P. F. 2002. Interspecific variation in plumage colour among birds: species recognition or light environment? *Journal of Experimental Biology*, 15, 505–14.

McNett, G., Luan, L. & Cocroft, R. 2010. Wind-induced noise alters signaler and receiver behavior in vibrational communication. *Behavioral Ecology and Sociobiology*, 64, 2043–51.

Melin, A. D., Fedigan, L. M., Hiramatsu, C., Sendall, C. L. & Kawamura, S. 2007. Effects of colour vision phenotype on insect capture by a free-ranging population of white-faced capuchins, *Cebus capucinus*. *Animal Behaviour*, 73, 205–14.

Melin, A. D., Fedigan, L. M., Young, H. C. & Kawamura, S. 2010. Can color vision variation explain sex differences in invertebrate foraging by capuchin monkeys? *Current Zoology*, 56, 300–12.

Mendelson, T. C. & Shaw, K. L. 2005. Rapid speciation in an arthropod. *Nature*, 433, 375.

Merilaita, S. & Lind, J. 2005. Background-matching and disruptive coloration, and the evolution of cryptic coloration. *Proceedings of the Royal Society B: Biological Sciences*, 272, 665–70.

Merrill, R. M., Gompert, Z., Dembeck, L. M., Kronforst, M. R., McMillan, W. O. & Jiggins, C. D. 2011a. Mate preference across the speciation continuum in a clade of mimetic butterflies. *Evolution*, 65, 1489–500.

Merrill, R. M., Van Schooten, B., Scott, J. A. & Jiggins, C. D. 2011b. Pervasive genetic associations between traits causing reproductive isolation in *Heliconius* butterflies. *Proceedings of the Royal Society B: Biological Sciences*, 278, 511–18.

Merrill, R. M., Wallbank, R. W. R., Bull, W. Salazar, P. A., Mallet, J., Stevens, M. & Jiggins, C. D. In Press. Disruptive ecological selection on a mating cue. *Proceedings of the Royal Society B.*

Miller, L. A. 1991. Arctiid moth clicks can degrade the accuracy of range difference discrimination in echolocating big brown bats, *Eptesicus fuscus*. *Journal of Comparative Physiology A*, 168, 571–9.

Miller, L. A. & Surlykke, A. 2001. How some insects detect and avoid being eaten by bats: Tactics and countertactics of prey and predator. *BioScience*, 51, 570–81.

Møhl, B. & Miller, L. A. 1976. Ultrasonic clicks produced by the peacock butterfly: a possible bat-repellent mechanism. *Journal of Experimental Biology*, 64, 639–44.

Moody, M. F. & Parriss, J. R. 1961. Discrimination of polarised light by Octopus: a behavioral and morphological study. *Zeitschrift für vergleichende Physiologie*, 44, 268–91.

Moritz, R. F. A., Kirchner, W. H. & Crewe, R. M. 1991. Chemical camouflage of the Death's Head hawkmoth (*Acherontia atropos* L.) in honeybee colonies. *Naturwissenschaften*, 78, 179–82.

Morrone, M. C. & Owens, R. A. 1987. Feature detection from local energy. *Pattern Recognition Letters*, 6, 303–13.

Morrone, M. C. & Burr, D. C. 1988. Feature detection in human vision: a phase-dependent energy model. *Proceedings of the Royal Society B: Biological Sciences*, 235, 221–45.

Morrongiello, J. R., Bond, N. R., Crook, D. A. & Wong, B. B. M. 2010. Nuptial coloration varies with ambient light environment in a freshwater fish. *Journal of Evolutionary Biology*, 23, 2718–25.

Mougi, A. & Iwasa, Y. 2010. Evolution towards oscillation or stability in a predator–prey system. *Proceedings of the Royal Society B: Biological Sciences*, 277, 3163–71.

Mueller, J. C., Pulido, F. & Kempenaers, B. 2011. Identification of a gene associated with avian migratory behaviour. *Proceedings of the Royal Society B: Biological Sciences*, 278, 2848–56.

Mugford, S. T., Mallon, E. B. & Franks, N. R. 2001. The accuracy of Buffon's needle: a rule of thumb used by ants to estimate area. *Behavioral Ecology*, 12, 655–8.

Muheim, R., Moore, F. R. & Phillips, J. B. 2006. Calibration of magnetic and celestial compass cues in migratory birds—a review of cue-conflict experiments. *Journal of Experimental Biology*, 209, 2–17.

Muheim, R., Åkesson, S. & Phillips, J. B. 2007. Magnetic compass of migratory savannah sparrows is calibrated by skylight polarisation at sunrise and sunset. *Journal of Ornithology*, 148, S485–94.

Mullen, P. & Pohland, G. 2008. Studies on UV reflection in feathers of some 1000 bird species: are UV peaks in feathers correlated with violet-sensitive and ultraviolet-sensitive cones? *Ibis*, 150, 59–68.

Müller, M. & Wehner, R. 2010. Path integration provides a scaffold for landmark learning in desert ants. *Current Biology*, 20, 1368–71.

Munoz, N. E. & Blumstein, D. T. 2012. Multisensory perception in uncertain environments. *Behavioral Ecology*, 23, 457–62.

Musolf, K., Hoffmann, F. & Penn, D. J. 2010. Ultrasonic courtship vocalizations in wild house mice, *Mus musculus musculus*. *Animal Behaviour*, 79, 757–64.

Nachman, M. W., Hoekstra, H. E. & D'Agostino, S. L. 2003. The genetic basis of adaptive melanism in pocket mice. *Proceedings of the National Academy of Sciences of the USA*, 100, 5268–73.

Nachtigall, P. E. & Supin, A. Y. 2008. A false killer whale adjusts its hearing when it echolocates. *Journal of Experimental Biology*, 211, 1714–18.

Nakano, R., Ishikawa, Y., Tatsuki, S., Surlykke, A., Skals, N. & Takanashi, T. 2006. Ultrasonic courtship song in the Asian corn borer moth *Ostrinia furnacalis*. *Naturwissenschaften*, 93, 292–6.

Nakano, R., Skals, N., Takanashi, T., Surlykke, A., Koike, T., Yoshida, K., Maruyama, H., Tatsuki, S. & Ishikawa, Y. 2008. Moths produce extremely quiet ultrasonic courtship songs by rubbing specialized scales. *Proceedings of the National Academy of Sciences of the USA*, 105, 11812–17.

Nakano, R., Ishikawa, Y., Tatsuki, S., Skals, N., Surlykke, A. & Takanashi, T. 2009a. Private ultrasonic whispering in moths. *Communicative and Integrative Biology*, 2, 123–6.

Nakano, R., Takanashi, T., Fujii, T., Skals, N., Surlykke, A. & Ishikawa, Y. 2009b. Moths are not silent but whisper ultrasonic courtship songs. *Journal of Experimental Biology*, 212, 4072–8.

Nakano, R., Takanashi, T., Skals, N., Surlykke, A. & Ishikawa, Y. 2010. To females of a noctuid moth, male courtship songs are nothing more than bat echolocation calls. *Biology Letters*, 6, 582–4.

Nakata, K. 2009. To be or not to be conspicuous: the effect of prey availability and predator risk on spider's wed decoration building. *Animal Behaviour*, 78, 1255–60.

Narendra, A., Reid, S. F., Greiner, B., Peters, R. A., Hemmi, J. M., Ribi, W. A. & Zeil, J. 2011. Caste-specific visual adaptations to distinct daily activity schedules in Australian *Myrmecia* ants. *Proceedings of the Royal Society B: Biological Sciences*, 278, 1141–9.

Narins, P. M., Hödl, W. & Grabul, D. S. 2003. Bimodal signal requisite for agonistic behavior in a dart-poison frog, *Epipedobates femoralis*. *Proceedings of the National Academy of Sciences of the USA*, 100, 577–80.

Narins, P. M., Grabul, D. S., Soma, K. K., Gaucher, P. & Hödl, W. 2005. Cross-modal integration in a dart-poison frog. *Proceedings of the National Academy of Sciences of the USA*, 102, 2425–9.

Naug, D. & Arathi, H. S. 2007. Receiver bias for exaggerated signals in honeybees and its implications for the evolution of floral displays. *Biology Letters*, 3, 635–7.

Nawroth, J. C., Greer, C. A., Chen, W. R., Laughlin, S. B. & Shepherd, G. M. 2007. An energy budget for the olfactory glomerulus. *Journal of Neuroscience*, 27, 9790–800.

Nelson, M. E. & MacIver, M. A. 2006. Sensory acquisition in active sensing systems. *Journal of Comparative Physiology A*, 192, 573–86.

Nelson, X. J. & Jackson, R. R. 2006. Compound mimicry and trading predators by the males of sexually dimorphic Batesian mimics. *Proceedings of the Royal Society B: Biological Sciences*, 273, 367–72.

Nĕmec, P., Altmann, J., Marhold, S., Burda, H. & Oelschläger, H. H. 2001. Neuroanatomy of magnetoreception: the superior colliculus involved in magnetic orientation in a mammal. *Science*, 294, 366–8.

Nemerov, A. 1997. Vanishing Americans: Abbott Thayer, Theodore Roosevelt, and the attraction of camouflage. *American Art*, 11, 50–81.

Neubauer, R. L. 1999. Super-normal length song preferences of female zebra finches (*Taeniopygia guttata*) and a theory of the evolution of bird song. *Evolutionary Ecology*, 13, 365–80.

Neuweiler, G. 1984. Foraging, echolocation and audition in bats. *Naturwissenschaften*, 71, 446–55.

Neuweiler, G. 1989. Foraging ecology and audition in echolocating bats. *Trends in Ecology & Evolution*, 4, 160–6.

Neuweiler, G. 1990. Auditory adaptations for prey capture in echolocating bats. *Physiological Reviews*, 70, 615–40.

Newman, E. A. & Hartline, P. H. 1982. The infrared 'vision' of snakes. *Scientific American*, 46, 116–24.

Newton, I. 1718. *Opticks, or, A treatise of the reflections, refractions, inflections and colours of light. The second edition, with additions*. London Printed for W. and J. Innys.

Nießner, C., Denzau, S., Gross, J. C., Peichl, L., Bischof, H.-J., Fleissner, G., Wiltschko, W. & Wiltschko, R. 2011. Avian ultraviolet/violet cones identified as probable magnetoreceptors. *PLoS ONE*, 6, e20091.

Nilsson, D.-E. & Warrant, E. J. 1999. Visual discrimination: seeing the third quality of light. *Current Biology*, 9, R535–7.

Nilsson, D.-E. 2009. The evolution of eyes and visually guided behaviour. *Philosophical Transactions of the Royal Society B: Biological Sciences*, 364, 2833–47.

Niskanen, M. & Mappes, J. 2005. Significance of the dorsal zigzag pattern of *Vipera latastei gaditana* against avian predators. *Journal of Animal Ecology*, 74, 1091–01.

Niven, J. E., Vahasoyrinki, M. & Juusola, M. 2003. *Shaker* K(+)-channels are predicted to reduce the metabolic cost of neural information in *Drosophila* photoreceptors. *Proceedings of the Royal Society B: Biological Sciences*, 270 Suppl 1, S58–61.

Niven, J. E. & Laughlin, S. B. 2008. Energy limitation as a selective pressure on the evolution of sensory systems. *Journal of Experimental Biology*, 211, 1792–804.

Noble, D. G., Davies, N. B., Hartley, I. R. & McRae, S. B. 1999. The red gape of the nestling cuckoo (*Cuculus canorus*) is not a supernormal stimulus for three common cuckoo hosts. *Behaviour*, 136, 759–77.

Nokelainen, O., Hegna, R. H., Reudler, J. H., Lindstedt, C. & Mappes, J. 2012. Trade-off between warning signal efficacy and mating success in the wood tiger moth. *Proceedings of the Royal Society B: Biological Sciences*, 1727, 257–65.

Noonan, B. P. & Comeault, A. A. 2009. The role of predator selection on polymorphic aposematic poison frogs. *Biology Letters*, 5, 51–4.

Norris, K. S. & Harvey, G. W. 1974. Sound transmission in the porpoise head. *Journal of the Aucoustical Society of America*, 56, 659–64.

Nosil, P. & Crespi, B. J. 2006. Experimental evidence that predation promotes divergence in adaptive radiation. *Proceedings of the National Academy of Sciences of the USA*, 103, 9090–5.

O'Quin, K. E., Hofmann, C. M., Hofmann, H. A. & Carleton, K. L. 2010. Parallel evolution of opsin gene expression in African cichlid fishes. *Molecular Biology and Evolution*, 27, 2839–54.

O'Rourke, C. T., Hall, M. I., Pitlik, T. & Fernandez-Juricic, E. 2010a. Hawk eyes I: diurnal raptors differ in visual fields and degree of eye movement. *PLoS ONE*, 5, e12802.

O'Rourke, C. T., Pitlik, T., Hoover, M. & Fernandez-Juricic, E. 2010b. Hawk eyes II: diurnal raptors differ in head movement strategies when scanning from perches. *PLoS ONE*, 5, e12169.

O'Connell-Rodwell, C., Wood, J., Rodwell, T., Puria, S., Partan, S., Keefe, R., Shriver, D., Arnason, B. & Hart, L. 2006. Wild elephant (*Loxodonta africana*) breeding herds respond to artificially transmitted seismic stimuli. *Behavioral Ecology and Sociobiology*, 59, 842–50.

Ödeen, A. & Håstad, O. 2003. Complex distribution of avian color vision systems revealed by sequencing the SWS1 opsin from total DNA. *Molecular Biology and Evolution*, 20, 855–61.

Ödeen, A. & Håstad, O. 2010. Pollinating birds differ in spectral sensitivity. *Journal of Comparative Physiology A*, 196, 91–6.

Ödeen, A., Håstad, O. & Alström, P. 2011. Evolution of ultraviolet vision in the largest avian radiation—the passerines. *BMC Evolutionary Biology*, 11, 313.

Ödeen, A., Pruett-Jones, S., Driskell, A. C., Armenta, J. K. & Håstad, O. 2012. Multiple shifts between violet and ultraviolet vision in a family of passerine birds with associated changes in plumage coloration. *Proceedings of the Royal Society B: Biological Sciences*, 279, 1269–76.

Okamura, J. Y. & Strausfeld, N. J. 2007. Visual system of Calliphorid flies: motion- and orientation-sensitive visual interneurons supplying dorsal optic glomeruli. *Journal of Comparative Neurology*, 500, 189–208.

Okanoya, K. & Dooling, R. J. 1990. Detection of gaps in noise by budgerigars (*Melopsittacus undulatus*) and zebra finches (*Poephila guttata*). *Hearing Research*, 50, 185–92.

Olofsson, M., Vallin, A., Jakobsson, S. & Wiklund, C. 2011. Winter predation on two species of hibernating butterflies: monitoring rodent attacks with infrared cameras. *Animal Behaviour*, 81, 529–34.

Olofsson, M., Jakobsson, S. & Wiklund, C. 2012. Auditory defence in the peacock butterfly (*Inachis io*) against mice (*Apodemus flavicollis* and *A. sylvaticus*). *Behavioral Ecology and Sociobiology*, 66, 209–15.

Ophir, A. G., Schrader, S. B. & Gillooly, J. F. 2010. Energetic cost of calling: general constraints and species-specific differences. *Journal of Experimental Biology*, 23, 1564–69.

Ord, T. J., Peters, R. A., Clucas, B. & Stamps, J. A. 2007. Lizards speed up visual displays in noisy motion habitats. *Proceedings of the Royal Society B: Biological Sciences*, 274, 1057–62.

Ord, T. J. & Stamps, J. A. 2008. Alert signals enhance animal communication in 'noisy' environments. *Proceedings of the National Academy of Sciences of the USA*, 105, 18830–5.

Osorio, D. & Srinivasan, M. V. 1991. Camouflage by edge enhancement in animal coloration patterns and its implications for visual mechanisms. *Proceedings of the Royal Society B: Biological Sciences*, 244, 81–5.

Osorio, D. & Vorobyev, M. 1996. Colour vision as an adaptation to frugivory in primates. *Proceedings of the Royal Society B: Biological Sciences*, 263, 593–9.

Osorio, D., Vorobyev, M. & Jones, C. D. 1999. Colour vision of domestic chicks. *Journal of Experimental Biology*, 202, 2951–9.

Osorio, D., Smith, A. C., Vorobyev, M. & Buchanan-Smith, H. M. 2004. Detection of fruit and the selection of primate visual pigments for color vision. *American Naturalist*, 164, 696–708.

Osorio, D. & Vorobyev, M. 2005. Photoreceptor spectral sensitivities in terrestrial animals: adaptations for luminance and colour vision. *Proceedings of the Royal Society B: Biological Sciences*, 272, 1745–52.

Owren, M., Rendall, D. & Ryan, M. J. 2010. Redefining animal signaling: influence versus information in communication. *Biology and Philosophy*, 25, 755–80.

Ozaki, M., Wada-Katsumata, A., Fujikawa, K., Iwasaki, M., Yokohari, F., Satoji, Y., Nisimura, T. & Yamaoka, R. 2005. Ant nestmate and non-nestmate discrimination by a chemosensory sensillum. *Science*, 309, 311–14.

Page, R. A. & Ryan, M. J. 2008. The effect of signal complexity on localization performance in bats that localize frog calls. *Animal Behaviour*, 76, 761–9.

Pamminger, T., Scharf, I., Pennings, P. S. & Foitzik, S. 2011. Increased host aggression as an induced defense against slave-making ants. *Behavioral Ecology*, 22, 255–60.

Panhuis, T. M., Butlin, R., Zuk, M. & Tregenza, T. 2001. Sexual selection and speciation. *Trends in Ecology & Evolution*, 16, 364–71.

Parker, A. 2003. *In the Blink of an Eye: How Vision Kick-Started the Big Bang of Evolution* London: Free Press.

Párraga, C. A., Troscianko, T. & Tolhurst, D. J. 2002. Spatiochromatic properties of natural images and human vision. *Current Biology*, 12, 483–7.

Partan, S., Fulmer, A. G., Gounard, M. A. M. & Redmond, J. E. 2010. Multimodal alarm behavior in urban and rural gray squirrels studied by means of observation and a mechanical robot. *Current Zoology*, 56, 313–26.

Partan, S. R. & Marler, P. R. 1999. Communication goes multimodal. *Science*, 283, 1272–3.

Partan, S. R. & Marler, P. R. 2005. Issues in the classification of multimodal signals. *American Naturalist*, 166, 231–45.

Partan, S. R., Larco, C. P. & Owens, M. J. 2009. Wild tree squirrels respond with multisensory enhancement to conspecific robot alarm behaviour. *Animal Behaviour*, 77, 1127–35.

Partridge, J. C., Shand, J., Archer, S. N., Lythgoe, J. N. & Groningen-Luyben, W. A. H. M. 1989. Interspecific variation in the visual pigments of deep-sea fishes. *Journal of Comparative Physiology A*, 164, 513–29.

Partridge, J. C. & Douglas, R. H. 1995. Far-red sensitivity of dragon fish. *Nature*, 375, 21–2.

Patten, M. A., Rotenberry, J. T. & Zuk, M. 2004. Habitat selection, acoustic adaptation, and the evolution of reproductive isolation. *Evolution*, 58, 2144–55.

Patullo, B. W. & Macmillan, D. L. 2010. Making sense of electrical sense in crayfish. *Journal of Experimental Biology*, 213, 651–7.

Payne, R. S. & Drury, W. H. 1958. *Tyto alba*, Part II. *Natural History N.Y.*, 67, 316–23.

Payne, R. S. 1971. Acoustic location of prey by barn owls (*Tyto alba*). *Journal of Experimental Biology*, 54, 535–73.

Pearn, S. M., Bennett, A. T. D. & Cuthill, I. C. 2001. Ultraviolet vision, fluorescence and mate choice in a parrot, the budgerigar *Melopsittacus undulatus*. *Proceedings of the Royal Society B: Biological Sciences*, 268, 2273–9.

Pellissier, L., Wassef, J., Bilat, J., Brazzola, G., Buri, P., Colliard, C., Fournier, B., Hausser, J., Yannic, G. & Perrin, N. 2011. Adaptive colour polymorphism of *Acrida ungarica* H. (Orthoptera: Acrididae) in a spatially heterogeneous environment. *Acta Oecologica*, 37, 93–8.

Peña, J. L., Viete, S., Albeck, Y. & Konishi, M. 1996. Tolerance to sound intensity of binaural coincidence detection in the nucleus laminaris of the owl. *Journal of Neuroscience*, 16, 7046–54.

Peters, R. C., Eeuwes, L. B. & Bretschneider, F. 2007. On the electrodetection threshold of aquatic vertebrates with ampullary or mucous gland electroreceptor organs. *Biological Reviews*, 82, 361–73.

Pettigrew, J. D. 1999. Electroreception in monotremes. *Journal of Experimental Biology*, 202, 1447–54.

Pfennig, K. S. & Pfennig, D. W. 2009. Character displacement:ecological and reproductive responses to a common evolutionary problem. *Quarterly Review of Biology*, 84, 253–76.

Phillips, J. B. & Borland, S. C. 1992. Behavioural evidence for use of a light-dependent magnetoreception mechanism by a vertebrate. *Nature*, 359, 142–4.

Pieprzyk, A. R., Weiner, W. W. & Chamberlain, S. C. 2003. Mechanisms controlling the sensitivity of the *Limulus* lateral eye in natural lighting. *Journal of Comparative Physiology A*, 189, 643–53.

Pietrewicz, A. T. & Kamil, A. C. 1977. Visual detection of cryptic prey by blue jays (*Cyanocitta cristata*). *Science*, 195, 580–2.

Pietsch, T. W. & Grobecker, D. B. 1978. The compleat angler: aggressive mimicry in an antennariid anglerfish. *Science*, 201, 369–70.

Pignatelli, V., Champ, C., Marshall, J. & Vorobyev, M. 2010. Double cones are used for colour discrimination in the reef fish, *Rhinecanthus aculeatus*. *Biology Letters*, 6, 537–9.

Planqué, R., Dechaume-Moncharmont, F.-X., Franks, N. R., Kovacs, T. & Marshall, J. A. R. 2007. Why do house-hunting ants recruit in both directions? *Naturwissenschaften*, 94, 911–18.

Plotnick, R. E., Dornbos, S. Q. & Chen, J. 2010. Information landscapes and sensory ecology of the Cambrian Radiation. *Paleobiology*, 36, 303–17.

Podos, J. 2001. Correlated evolution of morphology and vocal signal structure in Darwin's finches. *Nature*, 409, 185–8.

Pollak, G. D. & Bodenhamer, R. D. 1981. Specialized characteristics of single units in inferior colliculus of mustache bat: frequency representation, tuning, and discharge patterns. *Journal of Neurophysiology*, 46, 605–20.

Porter, M. L., Blasic, J. R., Bok, M. J., Cameron, E. G., Pringle, T., Cronin, T. W. & Robinson, P. R. 2012. Shedding new light on opsin evolution. *Proceedings of the Royal Society B: Biological Sciences*, 279, 3–14.

Pothmann, L., Wilkens, L. A. & Hofmann, M. H. 2012. Two modes of information processing in the electrosensory system of the paddlefish (*Polyodon spathula*). *Journal of Comparative Physiology A*, 198, 1–10.

Pratt, S. C., Mallon, E. B., Sumpter, D. J. T. & Franks, N. R. 2002. Quorum sensing, recruitment, and collective decision-making during colony emigration by the ant *Leptothorax albipennis*. *Behavioral Ecology and Sociobiology*, 52, 117–27.

Prinz, K. & Wiltschko, W. 1992. Migratory orientation of pied flycatchers: interaction of stellar and magnetic information during ontogeny. *Animal Behaviour*, 44, 539–45.

Proctor, H. C. 1991. Courtship in the water mite *Neumania papillator*: males capitalize on female adoptions for predation. *Animal Behaviour*, 42, 589–98.

Proctor, H. C. 1992. Sensory exploitation and the evolution of male mating behaviour: a cladistic test using water mites (Acari: *Parasitengona*). *Animal Behaviour*, 44, 745–52.

Pye, J. D. 2010. The distribution of circularly polarized light reflection in the Scarabaeoidea (Coleoptera). *Biological Journal of the Linnean Society*, 100, 585–96.

Quinn, T. P. 1980. Evidence for celestial and magnetic compsss orientation in lake migrating sockeye salmon fry. *Journal of Comparative Physiology A*, 137, 243–8.

Quirici, V. & Costa, F. G. 2007. Seismic sexual signal design of two sympatric burrowing tarantula spiders from meadows of Uruguay: *Eupalaestrus weijenberghi* and *Acanthoscurria suina* (Araneae, Theraphosidae). *Journal of Arachnology*, 35, 38–45.

Qvarnström, A., Haavie, J., Sæther, S. A., Eriksson, D. & Pärt, T. 2006. Song similarity predicts hybridization in flycatchers. *Journal of Evolutionary Biology*, 19, 1202–9.

Radford, A. N., Bell, M. B. V., Hollén, L. I. & Ridley, A. R. 2011. Singing for your supper: sentinel calling by kleptoparasites can mitigate the cost to victims. *Evolution*, 65, 900–6.

Raffa, K. F., Hobson, K. R., LaFontaine, S. & Aukema, B. H. 2007. Can chemical communication be cryptic? Adaptations by herbivores to natural enemies exploiting prey semiochemistry. *Oecologia*, 153, 1009–19.

Raine, N. E. & Chittka, L. 2007. The adaptive significance of a sensory bias in a foraging context: floral colour preferences in the bumblebee *Bombus terrestris*. *PLoS ONE*, 2, e556.

Ralls, K. 1967. Auditory sensitivity in mice, *Peromyscus* and *Mus musculus*. *Animal Behaviour*, 15, 123–8.

Ramsier, M. A., Cunningham, A. J., Moritz, G. L., Finneran, J. J., Williams, C. V., Ong, P. S., Gursky-Doyen, S. L. & Dominy, N. J. 2012. Primate communication in the pure ultrasound. *Biology Letters*, 8, 508–11.

Ratcliffe, J. M. & Nydam, M. L. 2008. Multimodal warning signals for a multiple predator world. *Nature*, 455, 96–100.

Rauschecker, J. P. 1995. Compensatory plasticity and sensory substitution in the cerebral cortex. *Trends in Neurosciences*, 18, 36–43.

Reardon, E. E., Parisi, A., Krahe, R. & Chapman, L. J. 2011. Energetic constraints on electric signalling in wave-type weakly electric fishes. *Journal of Experimental Biology*, 214, 4141–50.

Reby, D. & McComb, K. 2003. Anatomical constraints generate honesty: acoustic cues to age and weight in the roars of red deer stags. *Animal Behaviour*, 65, 519–30.

Reches, A. & Gutfreund, Y. 2009. Auditory and multisensory responses in the tectofugal pathway of the barn owl. *Journal of Neuroscience*, 29, 9602–13.

Reches, A., Netser, S. & Gutfreund, Y. 2010. Interactions between stimulus-specific adaptation and visual auditory integration in the forebrain of the barn owl. *Journal of Neuroscience*, 30, 6991–8.

Regnier, F. E. & Wilson, E. O. 1971. Chemical communication and 'propaganda' in slave-maker ants. *Science*, 172, 267–9.

Reid, S. F., Narendra, A., Hemmi, J. M. & Zeil, J. 2011. Polarised skylight and the landmark panorama provide night-active bull ants with compass information during route following. *Journal of Experimental Biology*, 214, 363–70.

Reinhold, K., Greenfield, M. D., Jang, Y. & Broce, A. 1998. Energetic cost of sexual attractiveness: ultrasonic advertisement in wax moths. *Animal Behaviour*, 55, 905–13.

Reisert, J. & Restrepo, D. 2009. Moecular tuning of odorant receptors and its implication for odor signal processing. *Chemical Senses*, 34, 535–45.

Ren, T. 2002. Longitudinal pattern of basilar membrane vibration in the sensitive cochlea. *Proceedings of the National Academy of Sciences of the USA*, 99, 17101–6.

Ren, Z.-X., Li, D.-Z., Bernhardt, P. & Wang, H. 2011. Flowers of *Cypripedium fargesii* (Orchidaceae) fool flat-footed flies (Platypezidae) by faking fungus-infected foliage. *Proceedings of the National Academy of Sciences of the USA*, 108, 7478–80.

Rendall, D., Owren, M. J. & Ryan, M. J. 2009. What do animal signals mean? *Animal Behaviour*, 78, 233–40.

Reynolds, R. G. & Fitzpatrick, B. M. 2007. Assortative mating in poison-dart frogs based on an ecologically important trait. *Evolution*, 61, 2253–9.

Riba-Hernandez, P., Stoner, K. E. & Osorio, D. 2004. Effect of polymorphic colour vision for fruit detection in the spider monkey *Ateles geoffroyi*, and its implications for the maintenance of polymorphic colour vision in platyrrhine monkeys. *Journal of Experimental Biology*, 207, 2465–70.

Richards, D. G. 1981. Alerting and message components in songs of Rufous-Sided Towhees. *Behaviour*, 76, 223–49.

Richards-Zawacki, C. L. & Cummings, M. E. 2011. Intraspecific reproductive character displacement in a polymorphic poison dart frog, *Dendrobates pumilio*. *Evolution*, 65, 259–67.

Ripmeester, E. A. P., Mulder, M. & Slabbekoorn, H. 2010. Habitat-dependent acoustic divergence affects playback response in urban and forest populations of the European blackbird. *Behavioral Ecology*, 21, 876–83.

Ritchie, M. G. 2007. Sexual selection and speciation. *Annual Review of Ecology, Evolution and Systematics*, 38, 79–102.

Ritz, T., Adem, S. & Schulten, K. 2000. A model for photoreceptor-based magnetoreception in birds. *Biophysical Journal*, 78, 707–18.

Robert, D., Amoroso, J. & Hoy, R. R. 1992. The evolutionary convergence of hearing in a parasitoid fly and its cricket host. *Science*, 258, 1135–7.

Roberts, J. A., Taylor, P. W. & Uetz, G. W. 2007. Consequences of complex signaling: predator detection of multimodal cues. *Behavioral Ecology*, 18, 236–40.

Roberts, N. W., Chiou, T. H., Marshall, N. J. & Cronin, T. W. 2009. A biological quarter-wave retarder with excellent achromaticity in the visible wavelength region. *Nature Photonics*, 3, 641–4.

Roberts, S. C., Gosling, L. M., Thornton, E. A. & McClung, J. 2001. Scent-marking by male mice under the risk of predation. *Behavioral Ecology*, 12, 698–705.

Roberts, S. C. & Gosling, L. M. 2003. Genetic similarity and quality interact in mate choice decisions by female mice. *Nature Genetics*, 35, 103–6.

Robinson, E. J. H., Franks, N. R., Ellis, S., Okuda, S. & Marshall, J. A. R. 2011. A simple threshold rule is sufficient to explain sophisticated collective decision-making. *PLoS ONE*, 6, e19981.

Roeder, K. D. 1967. Turning tendency of moths exposed to ultrasound while in stationary flight. *Journal of Insect Physiology*, 13, 873–88.

Rosenblum, E. B., Hoekstra, H. E. & Nachman, M. W. 2004. Adaptive reptile color variation and the evolution of the MC1R gene. *Evolution*, 58, 1794–808.

Rosenblum, E. B. 2006. Convergent evolution and divergent selection: Lizards at the White Sands ecotone. *American Naturalist*, 167, 1–15.

Rosenblum, E. B., Römpler, H., Schöneberg, T. & Hoekstra, H. E. 2010. Molecular and functional basis of phenotypic convergence in white lizards at White Sands. *Proceedings of the National Academy of Sciences of the USA*, 107, 2113–17.

Rosenthal, G. G., Flores Martinez, T. Y., García de León, F. J. & Ryan, M. J. 2001. Shared preferences by predators and females for male ornaments in swordtails. *American Naturalist*, 158, 146–54.

Rota, J. & Wagner, D. L. 2006. Predator mimicry: metalmark moths mimic their jumping spider predators. *PLoS ONE*, 1, e45.

Rothschild, M. 1975. Remarks on Carotenoids in the Evolution of Signals. In: *Coevolution in Animals and Plants* (Ed. by Gilbert, L. E. & Raven, P. H.), pp. 20–52. Austin, Texas: University of Texas Press.

Rothstein, S. I. 1990. A model system for coevolution: avian brood parasitism. *Annual Review of Ecology, Evolution and Systematics*, 21, 481–508.

Rowe, C. & Guilford, T. 1996. Hidden colour aversions in domestic chicks triggered by pyrazine odours of insect warning displays. *Nature*, 383, 520–2.

Rowe, C. 1999a. One signal or two? *Science*, 284, 741.

Rowe, C. 1999b. Receiver psychology and the evolution of multicomponent signals. *Animal Behaviour*, 58, 921–31.

Rowe, C. & Guilford, T. 1999a. The evolution of multimodal warning displays. *Evolutionary Ecology*, 13, 655–71.

Rowe, C. & Guilford, T. 1999b. Novelty effects in a multimodal warning signal. *Animal Behaviour*, 57, 341–6.

Rowe, C. 2002. Sound improves visual discrimination learning in avian predators. *Proceedings of the Royal Society B: Biological Sciences*, 269, 1353–7.

Rowe, C. & Skelhorn, J. 2004. Avian psychology and communication. *Proceedings of the Royal Society B: Biological Sciences*, 271, 1435–42.

Rowe, M. P., Coss, R. G. & Owings, D. H. 1986. Rattlesnake rattles and burrowing owl hisses: A case of acoustic batesian mimicry. *Ethology*, 72, 53–71.

Rowland, W. J. 1989. Mate choice and the supernormality effect in female sticklebacks (*Gasterosteus aculeatus*). *Behavioral Ecology and Sociobiology*, 24, 433–8.

Royer, L. & McNeil, J. N. 1993. Effect of relative humidity conditions on responsiveness of European corn borer (*Ostrinia nubilalis*) males to female sex pheromone in a wind tunnel. *Journal of Chemical Ecology*, 19, 61–9.

Ruggero, M. A. 1992. Responses to sound of the basilar membrane of the mammalian cochlea. *Current Biology*, 2, 449–56.

Rundle, H. D. & Nosil, P. 2005. Ecological speciation. *Ecology Letters*, 8, 336–52.

Rundus, A. S., Owings, D. H., Joshi, S. S., Chinn, E. & Giannini, N. 2007. Ground squirrels use an infrared signal to deter rattlesnake predation. *Proceedings of the National Academy of Sciences of the USA*, 104, 14372–6.

Rundus, A. S., Santer, R. D. & Hebets, E. A. 2010. Multimodal courtship efficacy of *Schizocosa retrorsa* wolf spiders: implications of an additional signal modality. *Behavioral Ecology*, 21, 701–7.

Rundus, A. S., Sullivan-Beckers, L., Wilgers, D. & Hebets, E. A. 2011. Females are choosy in the dark: context-dependent reliance on courtship components and its impact on fitness. *Evolution*, 65, 268–82.

Ruxton, G. D., Sherratt, T. N. & Speed, M. P. 2004. *Avoiding Attack*. Oxford: Oxford University Press.

Ruxton, G. D. 2009. Non-visual crypsis: a review of the empirical evidence for camouflage to senses other than vision. *Philosophical Transactions of the Royal Society B: Biological Sciences*, 364, 549–57.

Ruxton, G. D. & Schaefer, H. M. 2011. Resolving current disagreements and ambiguities in the terminology of animal communication. *Journal of Experimental Biology*, 24, 2574–85.

Ryan, M. J., Tuttle, M. D. & Rand, A. S. 1982. Bat predation and sexual advertisement in a neotropical anuran. *American Naturalist*, 119, 136–9.

Ryan, M. J., Tuttle, M. D. & Barclay, R. M. R. 1983. Behavioral responses of the frog-eating bat, *Trachops cirrhosus*, to sonic frequencies. *Journal of Comparative Physiology A*, 150, 413–18.

Ryan, M. J. & Brenowitz, E. A. 1985. The role of body size, phylogeny, and ambient noise in the evolution of bird song. *American Naturalist*, 126, 87–100.

Ryan, M. J. 1990. Sexual selection, sensory systems and sensory exploitation. *Oxford Surveys in Evolutionary Biology*, 7, 157–95.

Ryan, M. J., Cocroft, R. & Wilcznski, W. 1990a. The role of environmental selection in intraspecific divergence of mate recognition signals in the cricket frog, *Acris crepitans*. *Evolution*, 44, 1869–72.

Ryan, M. J., Fox, J. H., Wilczynski, W. & Rand, A. S. 1990b. Sexual selection for sensory exploitation in the frog *Physalaemus pustulosus*. *Nature*, 343, 66–7.

Ryan, M. J. & Rand, A. S. 1990. The sensory bias of sexual selection for complex calls in the tungara frog, *Physalaemus pustulosus* (sexual selection for sensory exploitation). *Evolution*, 44, 305–14.

Ryan, M. J. & Keddy-Hector, A. 1992. Directional patterns of female mate choice and the role of sensory biases. *American Naturalist*, 139, S4–35.

Ryan, M. J. & Rand, A. S. 1993. Sexual selection and signal evolution: the ghost of biases past. *Philosophical Transactions of the Royal Society B: Biological Sciences*, 340, 187–95.

Ryan, M. J. 2001. Food, song and speciation. *Nature*, 409, 139–40.

Ryan, M. J. & Cummings, M. E. 2005. Animal signals and the overlooked costs of efficacy. *Evolution*, 59, 1160–1.

Ryan, M. J., Bernal, X. E. & Rand, A. S. 2010. Female mate choice and the potential for ornament evolution in túngara frogs *Physalaemus pustulosus*. *Current Zoology*, 56, 343–57.

Sabbah, S., Laria, R. L., Gray, S. M. & Hawryshyn, C. W. 2010. Functional diversity in the color vision of cichlid fishes. *BMC Biology*, 8, 133.

Safer, A. B. & Grace, M. S. 2004. Infrared imaging in vipers: differential responses of crotaline and viperine snakes to paired thermal targets. *Behavioral Brain Research*, 154, 55–61.

Saito, A., Mikami, A., Kawamura, S., Ueno, Y., Hiramatsu, C., Widayati, K. A., Suryobroto, B., Teramoto, M., Mori, Y., Nagano, K., Fujita, K., Kuroshima, H. & Hasegawa, T. 2005. Advantage of dichromats over trichromats in discrimination of color-camouflaged stimuli in nonhuman primates. *American Journal of Primatology*, 67, 425–36.

Salazar, V. L. & Stoddard, P. K. 2008. Sex differences in energetic costs explain sexual dimorphism in the circadian rhythm modulation of the electrocommunication signal of the gymnotiform fish *Brachyhypopomus pinnicaudatus*. *Journal of Experimental Biology*, 211, 1012–20.

Sass, H. 1983. Production, release and effectiveness of two female sex pheromone components of *Periplaneta americana*. *Journal of Comparative Physiology A*, 152, 309–17.

Sato, N. J., Tokue, K., Noske, R. A., Mikami, O. K. & Ueda, K. 2010. Evicting cuckoo nestlings from the nest: a new anti-parasite behaviour. *Biology Letters*, 6, 67–9.

Satou, M. & Shiraishi, A. 1991. Local motion processing in the optic tectum of the Japanese toad, *Bufo japonicus*. *Journal of Comparative Physiology A*, 169, 569–89.

Saul-Gershenz, L. S. & Millar, J. G. 2006. Phoretic nest parasites use sexual deception to obtain transport to their host's nest. *Proceedings of the National Academy of Sciences of the USA*, 103, 14039–44.

Schaefer, H. M. & Stobbe, N. 2006. Disruptive coloration provides camouflage independent of background matching. *Proceedings of the Royal Society B: Biological Sciences*, 273, 2427–32.

Schaefer, H. M. & Ruxton, G. D. 2009. Deception in plants: mimicry or perceptual exploitation? *Trends in Ecology & Evolution*, 24, 676–85.

Schaefer, H. M. 2010. Chapter 1: Visual communication: evolution, ecology, and functional mechanisms. In: *Animal Behaviour: Evolution and Mechanisms* (Ed. by Kappeler, P. M.), pp. 3–28. London: Springer Berlin Heidelberg.

Schaefer, H. M. & Ruxton, G. D. 2011. *Plant-Animal Communication*. Oxford: Oxford University Press.

Scheich, H., Langner, G., Tidemann, C., Coles, R. B. & Guppy, A. 1986. Electroreception and electrolocation in platypus. *Nature*, 319, 401–2.

Schenkman, B. N. & Nilsson, M. E. 2010. Human echolocation: Blind and sighted persons' ability to detect sounds recorded in the presence of a reflecting object. *Perception*, 39, 483–501.

Schiestl, F. P., Peakall, R., Mant, J. G., Ibarra, F., Schulz, C., Franke, S. & Francke, W. 2003. The chemistry of sexual deception in an orchid-wasp pollination system. *Science*, 302, 437–8.

Schluter, D. 2001. Ecology and the origin of species. *Trends in Ecology & Evolution*, 16, 372–80.

Schmitt, D. E. & Esch, H. E. 1993. Magnetic orientation of honeybees in the laboratory. *Naturwissenschaften*, 80, 41–3.

Schmitz, L. & Motani, R. 2010. Morphological differences between the eyeballs of nocturnal and diurnal amniotes revisited from optical perspectives of visual environments. *Vision Research*, 50, 936–46.

Schnitzler, H. U., Suga, N. & Simmons, J. A. 1976. Peripheral auditory tuning for fine frequency analysis by the CF-FM bat, *Rhinolophus ferrumequinum*; III. Chochlear microphonics and auditory nerve responses. *Journal of Comparative Physiology A*, 106, 99–110.

Schnitzler, H. U. & Denzinger, A. 2011. Auditory fovea and Doppler shift compensation: adaptations for flutter detection in echolocating bats using CF-FM signals. *Journal of Comparative Physiology A*, 197, 541–59.

Schöneich, S. & Hedwig, B. 2010. Hyperacute directional hearing and phonotactic steering in the cricket (*Gryllus bimaculatus* deGeer). *PLoS ONE*, 5, e15141.

Schuchmann, M. & Siemers, B. M. 2010a. Variability in echolocation call intensity in a community of horseshoe bats: a role for resource partitioning or communication? *PLoS ONE*, 5, e12842.

Schuchmann, M. & Siemers, B. M. 2010b. Behavioral evidence for community-wide species discrimination from echolocation calls in bats. *American Naturalist*, 176, 72–82.

Schuller, G. & Pollack, G. 1979. Disproportionate frequency representation in the inferior colliculus of Doppler-compensating greater horseshoe bats: evidence for an acoustic fovea. *Journal of Comparative Physiology A*, 132, 47–54.

Scott-Samuel, N. E., Baddeley, R., Palmer, C. E. & Cuthill, I. C. 2011. Dazzle camouflage affects speed perception. *PLoS ONE*, 6, e20233.

Seddon, N. 2005. Ecological adaptation and species recognition drives vocal evolution in neotropical suboscine birds. *Evolution*, 59, 200–15.

Seddon, N. & Tobias, J. A. 2010. Character displacement from the receiver's perspective: species and mate recognition despite convergent signals in suboscine birds. *Proceedings of the Royal Society B: Biological Sciences*, 277, 2475–83.

Seehausen, O., van Alphen, J. J. M. & Witte, F. 1997. Cichlid fish diversity threatened by eutrophication that curbs sexual selection. *Science*, 277, 1808–11.

Seehausen, O., Terai, Y., Magalhaes, I. S., Carleton, K. L., Mrosso, H. D. J., Miyagi, R., van der Sluijs, I., Schneider, M. V., Maan, M. E., Tachida, H., Imai, H. & Okada, N. 2008. Speciation through sensory drive in cichlid fish. *Nature*, 455, 620–6.

Selzer, R. 1984. On the specificities of antennal olfactory receptor cells of *Periplaneta americana*. *Chemical Senses*, 8, 375–95.

Seppänen, J.-T. & Forsman, J. T. 2007. Interspecific social learning: novel preference can be acquired from a competing species. *Current Biology*, 17, 1248–52.

Seppänen, J.-T., Forsman, J. T., Mönkkönen, M., Krams, I. & Salmi, T. 2011. New behavioural trait adopted or rejected by observing heterospecific tutor fitness. *Proceedings of the Royal Society B: Biological Sciences*, 278, 1736–41.

Servedio, M. R., Van Doorn, G. S., Kopp, M., Frame, A. M. & Nosil, P. 2011. Magic traits in speciation: 'magic' but not rare? *Trends in Ecology & Evolution*, 26, 389–97.

Seyfarth, R. M., Cheney, D. L., Bergman, T., Fischer, J., Zuberbühler, K. & Hammerschmidt, K. 2010. The central importance of information in studies of animal communication. *Animal Behaviour*, 80, 3–8.

Shannon, C. E. & Weaver, W. 1949. *A Mathematical Model of Communication*. Urbana, IL: University of Illinois Press.

Shashar, N. & Cronin, T. W. 1996. Polarization contrast vision in octopus. *Journal of Experimental Biology*, 199, 999–1004.

Shashar, N., Rutledge, P. S. & Cronin, T. W. 1996. Polarization vision in cuttlefish—a concealed communication channel? . *Journal of Experimental Biology*, 199, 2077–84.

Shashar, N., Hanlon, R. T. & Petz, A. D. 1998. Polarization vision helps detect transparent prey. *Nature*, 393, 222–3.

Shashar, N., Hagan, R., Boal, J. G. & Hanlon, R. T. 2000. Cuttlefish use polarization sensitivity in predation on silvery fish. *Vision Research*, 40, 71–5.

Sherratt, T. N. & Beatty, C. D. 2003. The evolution of warning signals as reliable indicators of prey defence. *American Naturalist*, 162, 377–89.

Shichida, Y. & Matsuyama, T. 2009. Evolution of opsins and phototransduction. *Philosophical Transactions of the Royal Society B: Biological Sciences*, 364, 2881–95.

Siebeck, U. E., Parker, A. N., Sprenger, D., Mäthger, L. M. & Wallis, G. 2010. A species of reef fish that uses ultraviolet patterns for covert face recognition. *Current Biology*, 20, 407–10.

Siemers, B. M., Schauermann, G., Turni, H. & von Merten, S. 2009. Why do shrews twitter? Communication or simple echo-based orientation. *Biology Letters*, 5, 593–6.

Siemers, B. M. & Schaub, A. 2011. Hunting at the highway: traffic noise reduces foraging efficiency in acoustic predators. *Proceedings of the Royal Society B: Biological Sciences*, 278, 1646–52.

Silveira, H. C., Oliveira, P. S. & Trigo, J. R. 2012. Attracting predators without falling prey: chemical camouflage protects honeydew-producing treehoppers from ant predation. *American Naturalist*, 175, 261–8.

Simmons, P. J. & Young, D. 2010. *Nerve Cells and Animal Behaviour*. Cambridge: Cambridge University Press.

Simon, R., Holderied, M. W., Koch, C. U. & von Helversen, O. 2011. Floral acoustics: conspicuous echoes of a dish-shaped leaf attract bat pollinators. *Science*, 333, 631–3.

Simon, V. B. 2007. Not all signals are equal: male brown anole lizards (*Anolis sagrei*) selectively decrease pushup frequency following a simulated predatory attack. *Ethology*, 113, 793–801.

Sivalinghem, S., Kasumovic, M. M., Mason, A. C., Andrade, M. C. B. & Elias, D. O. 2010. Vibratory communication in the jumping spider *Phidippus clarus*: polyandry, male courtship signals, and mating success. *Behavioral Ecology*, 21, 1308–14.

Skelhorn, J., Rowland, H. M. & Ruxton, G. D. 2010a. The evolution and ecology of masquerade. *Biological Journal of the Linnean Society*, 99, 1–8.

Skelhorn, J., Rowland, H. M., Speed, M. P. & Ruxton, G. D. 2010b. Masquerade: Camouflage without crypsis. *Science*, 327, 51.

Skelhorn, J. & Ruxton, G. D. 2010. Predators are less likely to misclassify masquerading prey when their models are present. *Biology Letters*, 6, 597–9.

Skelhorn, J., Rowland, H. M., Delf, J., Speed, M. P. & Ruxton, G. D. 2011. Density-dependent predation influences the evolution and behavior of masquerading prey. *Proceedings of the National Academy of Sciences of the USA*, 108, 6532–6.

Slabbekoorn, H. & Peet, M. 2003. Birds sing at a higher pitch in urban noise. *Nature*, 424, 267.

Slabbekoorn, H. & den Boer-Visser, A. 2006. Cities change the songs of birds. *Current Biology*, 16, 2326–31.

Sloan, J. L. & Hare, J. F. 2004. Monotony and the information content of Richardson's ground squirrel (*Spermophilus richardsonii*) repeated calls: tonic communication or signal certainty? *Ethology*, 110, 147–56.

Sloan, J. L., Wilson, D. R. & Hare, J. F. 2005. Functional morphology of Richardson's ground squirrel, *Spermophilus richardsonii*, alarm calls: the meaning of chirps, whistles and chucks. *Animal Behaviour*, 70, 937–44.

Smadja, C. M. & Butlin, R. K. 2011. A framework for comparing processes of speciation in the presence of gene flow. *Molecular Ecology*, 20, 5123–40.

Smith, A. C., Buchanan-Smith, H. M., Surridge, A. K., Osorio, D. & Mundy, N. I. 2003. The effect of colour vision status on the detection and selection of fruits by tamarins (*Saguinus* spp.). *Journal of Experimental Biology*, 206, 3159–65.

Smith, A. C., Surridge, A. K., Prescott, M. J., Osorio, D., Mundy, N. I. & Buchanan-Smith, H. M. 2012. Effect of colour vision status on insect prey capture efficiency of captive and wild tamarins (*Saguinus* app.). *Animal Behaviour*, 83, 479–86.

Smith, B. H. & Getz, W. M. 1994. Nonpheremonal olfactory processing in insects. *Annual Review of Entomology*, 39, 351–75.

Snowden, R., Thompson, P. & Troscianko, T. 2006. *Basic Vision: An Introduction to Visual Perception*. Oxford: Oxford University Press.

Sobel, J. M., Chen, G. F., Watt, L. R. & Schemske, D. W. 2009. The biology of speciation. *Evolution*, 64, 295–315.

Soler, M., Soler, J. J., Martínez, J. G. & Møller, A. P. 1995. Magpie host manipulation by Great Spotted Cuckoos: Evidence for an avian mafia? *Evolution*, 49, 770–5.

Somanathan, H., Kelber, A., Borges, R., Wallén, R. & Warrant, E. J. 2009. Visual ecology of Indian carpenter bees II: adaptations of eyes and ocelli to nocturnal and diurnal lifestyles. *Journal of Comparative Physiology A*, 195, 571–83.

Speck, J. & Barth, F. G. 1982. Vibration sensitivity of the pretarsal slit sensilla in the spider leg. *Journal of Comparative Physiology A*, 148, 187–94.

Spottiswoode, C. N. & Stevens, M. 2010. Visual modeling shows that avian host parents use multiple visual cues in rejecting parasitic eggs. *Proceedings of the National Academy of Sciences of the USA*, 107, 8672–6.

Spottiswoode, C. N. & Stevens, M. 2011. How to evade a coevolving brood parasite: egg discrimination versus egg variability as host defences. *Proceedings of the Royal Society of London B: Biological Sciences*, 278, 3566–73.

Spottiswoode, C. N. & Stevens, M. 2012. Host-parasite arms races and rapid changes in bird egg appearance. *American Naturalist*, 179, 633–48.

Srinivasan, M. V., Zhang, S. W. & Witney, K. 1994. Visual discrimination of pattern orientation by honeybees: performance and implications for 'cortical' processing. *Philosophical Transactions of the Royal Society B: Biological Sciences*, 343, 199–210.

Stanford, T. R., Quessy, S. & Stein, B. E. 2005. Evaluating the operations underlying multisensory integration in the cat superior colliculus. *Journal of Neuroscience*, 25, 6499–508.

Stapput, K., Thalau, P., Wiltschko, R. & Wiltschko, W. 2008. Orientation of birds in total darkness. *Current Biology*, 18, 602–6.

Stapput, K., Gunturkun, O., Hoffmann, K. P., Wiltschko, R. & Wiltschko, W. 2010. Magnetoreception of directional information in birds requires nondegraded vision. *Current Biology*, 20, 1259–62.

Stauffer, H.-P. & Semlitsch, R. 1993. Effects of visual, chemical and tactile cues of fish on the behavioural responses of tadpoles. *Animal Behaviour*, 46, 355–64.

Steiner, C. C., Weber, J. N. & Hoekstra, H. E. 2007. Adaptive variation in beach mice produced by two interacting pigmentation genes. *PLoS Biology*, 5, e219.

Stevens, M. 2005. The role of eyespots as anti-predator mechanisms, principally demonstrated in the Lepidoptera. *Biological Reviews*, 80, 573–88.

Stevens, M. & Cuthill, I. C. 2006. Disruptive coloration, crypsis and edge detection in early visual processing. *Proceedings of the Royal Society B: Biological Sciences*, 273, 2141–7.

Stevens, M., Cuthill, I. C., Windsor, A. M. M. & Walker, H. J. 2006. Disruptive contrast in animal camouflage. *Proceedings of the Royal Society B: Biological Sciences*, 273, 2433–8.

Stevens, M. 2007. Predator perception and the interrelation between different forms of protective coloration. *Proceedings of the Royal Society B: Biological Sciences*, 274, 1457–64.

Stevens, M. & Cuthill, I. C. 2007. Hidden messages: Are ultraviolet signals a special channel in avian communication? *BioScience*, 57, 501–7.

Stevens, M., Hopkins, E., Hinde, W., Adcock, A., Connelly, Y., Troscianko, T. & Cuthill, I. C. 2007. Field experiments on the effectiveness of 'eyespots' as predator deterrents. *Animal Behaviour*, 74, 1215–27.

Stevens, M., Hardman, C. J. & Stubbins, C. L. 2008a. Conspicuousness, not eye mimicry, makes 'eyespots' effective anti-predator signals. *Behavioral Ecology*, 19, 525–31.

Stevens, M., Yule, D. H. & Ruxton, G. D. 2008b. Dazzle coloration and prey movement. *Proceedings of the Royal Society B: Biological Sciences*, 275, 2639–43.

Stevens, M. & Merilaita, S. 2009a. Defining disruptive coloration and distinguishing its functions. *Philosophical Transactions of the Royal Society B: Biological Sciences*, 364, 481–8.

Stevens, M. & Merilaita, S. 2009b. Introduction. Animal camouflage: current issues and new perspectives. *Philosophical Transactions of the Royal Society B: Biological Sciences*, 364, 423–7.

Stevens, M. & Merilaita, S. 2011a. Chapter 1: Animal Camouflage: An Introduction. In: *Animal Camouflage: From*

Mechanisms & Function (Ed. by Stevens, M. & Merilaita, S.), pp. 1–16. Cambridge: Cambridge University Press.

Stevens, M. & Merilaita, S. 2011b. *Animal Camouflage: From Mechanisms & Function*. Cambridge: Cambridge University Press.

Stevens, M., Searle, W. T. L., Seymour, J. E., Marshall, K. L. A. & Ruxton, G. D. 2011. Motion dazzle and camouflage as distinct anti-predator defenses. *BMC Biology*, 9, 81.

Stevens, M. & Ruxton, G. D. 2012. Linking the evolution and form of warning coloration in nature. *Proceedings of the Royal Society B: Biological Sciences*, 279, 417–26.

Stewart, F. J., Baker, D. A. & Webb, B. 2010. A model of visual-olfactory integration for odour localisation in free-flying fruit flies. *Journal of Experimental Biology*, 213, 1886–900.

Stoddard, M. C. & Stevens, M. 2010. Pattern mimicry of host eggs by the common cuckoo, as seen through a bird's eye. *Proceedings of the Royal Society B: Biological Sciences*, 277, 1387–93.

Stoddard, M. C. & Stevens, M. 2011. Avian vision and the evolution of egg color mimicry in the common cuckoo. *Evolution*, 65, 2004–13.

Stoddard, M. C. & Prum, R. O. 2012. How colorful are birds? Evolution of the avian plumage color gamut. *Behavioral Ecology*, 22, 1042–52.

Stoddard, P. K. 1999. Predation enhances complexity in the evolution of electric fish signals. *Nature*, 400, 254–6.

Stoddard, P. K. 2002. The evolutionary origins of electric signal complexity. *Journal of Physiology*, 96, 485–91.

Stoddard, P. K. & Markham, M. R. 2008. Signal cloaking by electric fish. *BioScience*, 58, 415–25.

Stoddard, P. K. & Salazar, V. L. 2011. Energetic cost of communication. *Journal of Experimental Biology*, 214, 200–5.

Stökl, J., Brodmann, J., Dafni, A., Ayasse, M. & Hansson, B. S. 2011. Smells like aphids: orchid flowers mimic aphid alarm pheromones to attract hoverflies for pollination. *Proceedings of the Royal Society B: Biological Sciences*, 278, 1216–22.

Stowe, M. K., Tumlinson, J. H. & Heath, R. R. 1987. Chemical mimicry: Bolas spiders emit components of moth prey species pheromones. *Science*, 236, 964–6.

Stowe, M. K., Turlings, T. C. J., Loughrin, J. H., Lewis, W. J. & Tumlinson, J. H. 1995. The chemistry of eavesdropping, alarm, and deceit. *Proceedings of the National Academy of Sciences of the USA*, 92, 23–8.

Stowe, S. 1980. Rapid synthesis of photoreceptor membrane and assembly of new microvilli in a crab at dusk. *Cell & Tissue Research*, 211, 419–40.

Stowe, S. 1981. Effects of illumination changes on rhabdom synthesis in a crab. *Journal of Comparative Physiology A*, 142, 19–25.

Stratton, G. E. & Uetz, G. W. 1983. Communication via substratum-coupled stridulation and reproductive iso-

lation in wolf spiders (Araneae: Lycosidae). *Animal Behaviour*, 31, 164–72.

Stroeymeyt, N., Giurfa, M. & Franks, N. R. 2010. Improving decision speed, accuracy and group cohesion through early information gathering in house-hunting ants. *PLoS ONE*, 5, e13059.

Stuart-Fox, D., Moussalli, A., Marshall, N. J. & Owens, I. P. F. 2003. Conspicuous males suffer higher predation risk: visual modelling and experimental evidence from lizards. *Animal Behaviour*, 66, 541–50.

Stuart-Fox, D., Moussalli, A. & Whiting, M. J. 2007. Natural selection on social signals: signal efficacy and the evolution of chameleon display coloration. *American Naturalist*, 170, 916–30.

Stuart-Fox, D. & Moussalli, A. 2008. Selection for social signalling drives the evolution of chameleon colour change. *PLoS Biology*, 6, e25.

Stuart-Fox, D., Moussalli, A. & Whiting, M. J. 2008. Predator-specific camouflage in chameleons. *Biology Letters*, 4, 326–9.

Stuart-Fox, D. & Moussalli, A. 2009. Camouflage, communication and thermoregulation: lessons from colour changing organisms. *Philosophical Transactions of the Royal Society B: Biological Sciences*, 364, 463–70.

Stynoski, J. L. & Noble, V. R. 2012. To beg or to freeze: multimodal sensory integration directs behaviour in a tadpole. *Behavioral Ecology and Sociobiology*, 66, 191–9.

Su, C. Y., Menuz, K. & Carlson, J. R. 2009. Olfactory perception: receptors, cells, and circuits. *Cell*, 139, 45–59.

Suga, N. & Jen, P. H.-S. 1975. Peripheral control of acoustic signals in the auditory system of echolocating bats. *Journal of Experimental Biology*, 62, 277–311.

Suga, N., Simmons, J. A. & Jen, P. H.-S. 1975. Peripheral specialization for fine analysis of doppler-shifted echoes in the auditory system of the 'CF-FM' bat *Pteronotus parnellii*. *Journal of Experimental Biology*, 63, 161–92.

Suga, N., Neuweiler, G. & Möller, J. 1976. Peripheral auditory tuning for fine frequency analysis by the CF-FM bat, *Rhinolophus ferrumequinum*; IV: Properties of peripheral auditory neurons. *Journal of Comparative Physiology A*, 106, 111–25.

Suga, N. & Jen, P. H.-S. 1977. Further studies on the peripheral auditory system of 'CF-FM' bats specialized for fine frequency analysis of doppler-shifted echoes. *Journal of Experimental Biology*, 69, 207–32.

Sullivan, J. P., Lavoué, S. & Hopkins, C. D. 2002. Discovery and phylogenetic analysis of a riverine species flock of African electric fishes (Mormyridae: Teleostei). *Evolution*, 56, 597–616.

Sullivan, W. E., 3rd. 1982a. Neural representation of target distance in auditory cortex of the echolocating bat *Myotis lucifugus*. *Journal of Neurophysiology*, 48, 1011–32.

Sullivan, W. E., 3rd. 1982b. Possible neural mechanisms of target distance coding in auditory system of the echolocating bat *Myotis lucifugus*. *Journal of Neurophysiology*, 48, 1033–47.

Summers, K., Symula, R., Clough, M. & Cronin, T. W. 1999. Visual mate choice in poison frogs. *Proceedings of the Royal Society B: Biological Sciences*, 266, 2141–5.

Summers, K. & Clough, M. 2001. The evolution of coloration and toxicity in the poison frog family (Dendrobatidae). *Proceedings of the National Academy of Sciences of the USA*, 98, 6227–32.

Sumner, P. & Mollon, J. D. 2000. Catarrhine photopigments are optimized for detecting targets against a foliage background. *Journal of Experimental Biology*, 203, 1963–86.

Surlykke, A. & Miller, L. A. 1985. The influence of arctiid moth clicks on bat echolocation; jamming or warning? *Journal of Comparative Physiology A*, 156, 831–43.

Surlykke, A. & Skals, N. 1998. Sonic hearing in a diurnal geometrid moth, *Archiearis parthenias*, temporally isolated from bats. *Naturwissenschaften*, 85, 36–7.

Surlykke, A. & Kalko, E. K. 2008. Echolocating bats cry out loud to detect their prey. *PLoS ONE*, 3, e2036.

Surlykke, A., Boel Pedersen, S. & Jakobsen, L. 2009. Echolocating bats emit a highly directional sonar sound beam in the field. *Proceedings of the Royal Society B: Biological Sciences*, 276, 853–60.

Swan, D. C. & Hare, J. F. 2008. The first cut is the deepest: primary syllables of Richardson's ground squirrel, *Spermophilus richardsonii*, repeated calls alert receivers. *Animal Behaviour*, 76, 47–54.

Sweeney, A., Jiggins, C. & Johnsen, S. 2003. Polarized light as a butterfly mating signal. *Nature*, 423, 31–2.

Swihart, S. L. 1972. The neural basis of colour vision in the butterfly, *Heliconius erato*. *Journal of Insect Physiology*, 18, 1015–25.

Swynnerton, C. F. M. 1918. Rejections by birds of eggs unlike their own: with remarkson some of the cuckoo problems. *Ibis*, 1918, 127–54.

Számadó, S. 2011. The cost of honesty and the fallacy of the handicap principle. *Animal Behaviour*, 81, 3–10.

Takahashi, N., Kashino, M. & Hironaka, N. 2010. Structure of rat ultrasonic vocalisations and its relevance to behaviour. *PLoS ONE*, 5, e14115.

Takahashi, T. T. 2010. How the owl tracks its prey—II. *Journal of Experimental Biology*, 213, 3399–408.

Takasu, F. 2003. Co-evolutional dynamics of egg appearance in avian brood parasitism. *Evolutionary Ecology Research*, 5, 345–62.

Takasu, F. 2005. A theoretical consideration on co-evolutionary interactions between avian brood parasites and their hosts. *Ornithological Science*, 4, 65–7.

Talbot, C. M. & Marshall, J. 2010. Polarization sensitivity in two species of cuttlefish—*Sepia plangon* (Gray 1849) and *Sepia mestus* (Gray 1849)—demonstrated with polarized optomotor stimuli. *Journal of Experimental Biology*, 213, 3364–70.

Talbot, C. M. & Marshall, J. N. 2011. The retinal topography of three species of coleoid cephalopod: significance for perception of polarized light. *Philosophical Transactions of the Royal Society B: Biological Sciences*, 366, 724–33.

Tan, E. J. & Li, D. 2009. Detritus decorations of an orb-weaving spider, *Cyclosa mulmeinensis* (Thorell): for food or camouflage? *Journal of Experimental Biology*, 212, 1832–9.

Tan, E. J., Seah, S. W. H., Yap, L.-M. Y. L., Goh, P. M., Gan, W., Liu, F. & Li, D. 2010. Why do orb-weaving spiders (*Cyclosa ginnaga*) decorate their webs with silk spirals and plant detritus? *Animal Behaviour*, 79, 179–86.

Tan, S., Amos, W. & Laughlin, S. B. 2005. Captivity selects for smaller eyes. *Current Biology*, 15, R540–2.

Tanaka, K., Morimoto, G., Stevens, M. & Ueda, K. 2011. Rethinking visual supernormal stimuli in cuckoos: visual modeling of host and parasite signals. *Behavioral Ecology*, 22, 1012–19.

Tanaka, K. D. & Ueda, K. 2005. Horsfield's hawk-cuckoo nestlings simulate multiple gapes for begging. *Science*, 308, 653.

Tarsitano, M., Jackson, R. R. & Kirchner, W. H. 2000. Signals and signal choices made by the araneophagic jumping spider *Portia fimbriata* while hunting the orb-weaving web spiders *Zygiella x-notata* and *Zosis geniculatus*. *Ethology*, 106, 595–615.

Taylor, R. C., Klein, B. A., Stein, J. & Ryan, M. J. 2011. Multimodal signal variation in space and time: how important is matching a signal with its signaler? *Journal of Experimental Biology*, 214, 815–20.

Teeling, E. C. 2009. Hear, hear: the convergent evolution of echolocation in bats? *Trends in Ecology & Evolution*, 24, 351–4.

Temple, S. E., Pignatelli, V., Cook, T., How, M. J., Chiou, T. H., Roberts, N. W. & Marshall, N. J. 2012. High-resolution polarisation vision in a cuttlefish. *Current Biology*, 22, R121–2.

ter Hofstede, H. M., Kalko, E. K. & Fullard, J. H. 2010. Auditory-based defence against gleaning bats in neotropical katydids (Orthoptera: Tettigoniidae). *Journal of Comparative Physiology A*, 196, 349–58.

Teramitsu, I. & White, S. A. 2008. Motor learning: the FoxP2 puzzle piece. *Current Biology*, 18, R335–7.

Thaler, L., Arnott, S. R. & Goodale, M. A. 2011. Neural correlates of natural human echolocation in early and late blind echolocation experts. *PLoS ONE*, 6, e20162.

Thayer, G. H. 1909. *Concealing-Coloration in the Animal Kingdom: An Exposition of the Laws of Disguise Through Color and Pattern: Being a Summary of Abbott H. Thayer's Discoveries*. New York: Macmillan.

Théry, M. & Casas, J. 2002. Predator and prey views of spider camouflage. *Nature*, 415, 133.

Théry, M. & Casas, J. 2009. The multiple disguises of spiders: web colour and decorations, body colour and movement. *Philosophical Transactions of the Royal Society B: Biological Sciences*, 364, 471–80.

Thompson, A. B. & Hare, J. F. 2010. Neighbourhood watch: multiple alarm callers communicate directional predator movement in Richardson's ground squirrels, *Spermophilus richardsonii*. *Animal Behaviour*, 80, 269–75.

Tinbergen, L. 1960. The natural control of insects in pine woods I. Factors influencing the intensity of predation by songbirds. *Archives Neerlandaises de Zoologie*, 13, 265–343.

Tinbergen, N. 1948. Social releasers and the experimental method required for their study. *Wilson Bulletin*, 60, 6–51.

Tinbergen, N. 1951. *Study of Instinct*. Oxford: Oxford University Press.

Tisdale, V. & Fernandez-Juricic, E. 2009. Vigilance and predator detection vary between avian species with different visual acuity and coverage. *Behavioral Ecology*, 20, 936–45.

Tobias, J. A., Aben, J., Brumfield, R. T., Derryberry, E. P., Halfwerk, W., Slabbekoorn, H. & Seddon, N. 2010. Song divergence by sensory drive in Amazonian birds. *Evolution*, 64, 2820–39.

Tokue, K. & Ueda, K. 2010. Mangrove gerygones reject and eject little bronze-cuckoo hatchlings from parasitized nests. *Ibis*, 152, 835–9.

Tougaard, J., Casseday, J. H. & Covey, E. 1998. Arctiid moths and bat echolocation: broad-band clicks interfere with neural responses to auditory stimuli in the nuclei of the lateral lemniscus of the big brown bat. *Journal of Comparative Physiology A*, 182, 203–15.

Tovée, M. J. 1995. Ultraviolet photoreceptors in the animal kingdom: Their distribution and function. *Trends in Ecology & Evolution*, 10, 455–60.

Tseng, L. & Tso, I.-M. 2009. A risky defence by a spider using conspicuous decoys resembling itself in appearance. *Animal Behaviour*, 78, 425–31.

Tso, I.-M., Liao, C.-P., Huang, R.-P. & Yang, E.-C. 2006. Function of being colorful in web spiders: attracting prey or camouflaging oneself? *Behavioral Ecology*, 17, 606–13.

Tso, I. M., Tai, P. L., Ku, T. H., Kuo, C. H. & Yang, E. C. 2002. Colour-associated foraging success and population genetic structure in a sit-and-wait predator *Nephila maculata* (Araneae: Tetragnathidae). *Animal Behaviour*, 63, 175–82.

Tso, I. M., Lin, C. W. & Yang, E. C. 2004. Colorful orb-weaving spiders through a bee's eyes. *Journal of Experimental Biology*, 207, 2631–7.

Tullberg, B. S., Merilaita, S. & Wiklund, C. 2005. Aposematism and crypsis combined as a result of distance dependence: functional versatility of the colour pattern in the swallowtail butterfly larva. *Proceedings of the Royal Society B: Biological Sciences*, 272, 1315–21.

Turner, J. R., White, E. M., Collins, M. A., Partridge, J. C. & Douglas, R. H. 2009. Vision in lanternfish (Myctophidae): Adaptations for viewing bioluminescence in the deep-sea. *Deep Sea Research Part I: Oceanographic Research Papers*, 56, 1003–17.

Tuttle, M. D. & Ryan, M. J. 1981. Bat predation and the evolution of frog vocalisations in the neotropics. *Science*, 214, 677–8.

Tuttle, M. D. & Ryan, M. J. 1982. The role of synchronized calling, ambient light, and ambient noise, in anti-bat predator behavior of a treefrog. *Behavioral Ecology and Sociobiology*, 11, 125–31.

Tuttle, M. D., Taft, L. K. & Ryan, M. J. 1982. Evasive behaviour of a frog in response to bat predation. *Animal Behaviour*, 30, 393–7.

Twig, G. & Perlman, I. 2004. Homogeneity and diversity of color-opponent horizontal cells in the turtle retina: Consequences for potential wavelength discrimination. *Journal of Vision*, 4, 403–14.

Uetz, G. W. & Roberts, J. A. 2002. Multisensory cues and multimodal communication in spiders: insights from video/audio playback studies. *Brain, Behavior and Evolution*, 59, 222–30.

Uetz, G. W., Roberts, J. A. & Taylor, P. W. 2009. Multimodal communication and mate choice in wolf spiders: female response to multimodal versus unimodal signals. *Animal Behaviour*, 78, 299–305.

Uy, J. A. C. & Borgia, G. 2000. Sexual selection drives rapid divergence in bowerbird display traits. *Evolution*, 54, 273–8.

Uy, J. A. C. & Endler, J. A. 2004. Modification of the visual background increases the conspicuousness of golden-collared manakin displays. *Behavioral Ecology*, 15, 1003–10.

Vácha, M., Drštková, D. & Pᵁžová, T. 2008. Tenebrio beetles use magnetic inclination compass. *Naturwissenschaften*, 95, 761–5.

Valkonen, J., Niskanen, M., Björklund, M. & Mappes, J. 2011. Disruption or aposematism? Significance of dorsal zigzag pattern of European vipers. *Evolutionary Ecology*, 25, 1047–63.

Vallin, A., Jakobsson, S., Lind, J. & Wiklund, C. 2005. Prey survival by predator intimidation: an experimental study of peacock butterfly defence against blue tits. *Proceedings of the Royal Society of: Biological Sciences*, 272, 1203–7.

Vallin, A., Jakobsson, S. & Wiklund, C. 2007. 'An eye for an eye?' -on the generality of the intimidating quality of eyespots in a butterfly and a hawkmoth. *Behavioral Ecology and Sociobiology*, 61, 1419–24.

Van Deemter, J. H. & Buf, J. M. H. 2000. Simultaneous detection of lines and edges using compound gabor filters. *International Journal of Pattern Recognition and Artificial Intelligence*, 14, 757–77.

van der Sluijs, I., Gray, S. M., Amorim, M., Barber, I., Candolin, U., Hendry, A., Krahe, R., Maan, M. E., Utne-Palm, A., Wagner, H.-J. & Wong, B. B. M. 2011. Communication in troubled waters: responses of fish communication systems to changing environments. *Evolutionary Ecology*, 25, 623–40.

Van Dijk, P., Mason, M. J., Schoffelen, R. L., Narins, P. M. & Meenderink, S. W. 2011. Mechanics of the frog ear. *Hearing Research*, 273, 46–58.

van Doorn, S. G. & Weissing, F. J. 2001. Ecological versus sexual selection models of sympatric speciation: a synthesis. *Selection*, 2, 17–40.

van Doorn, S. G., Dieckmann, U. & Weissing, F. J. 2004. Sympatric speciation by sexual selection: a critical reevaluation. *American Naturalist*, 163, 709–25.

Van Dyke, J. U. & Grace, M. S. 2010. The role of thermal contrast in infrared-based defensive targeting by the copperhead, *Agkistrodon contortrix*. *Animal Behaviour*, 79, 993–9.

van Valen, L. 1973. A new evolutionary law. *Evolutionary Theory*, 1, 1–30.

Vasconcelos, R. O., Fonseca, P. J., Amorim, M. C. & Ladich, F. 2011. Representation of complex vocalizations in the Lusitanian toadfish auditory system: evidence of fine temporal, frequency and amplitude discrimination. *Proceedings of the Royal Society B: Biological Sciences*, 278, 826–34.

Vaughan, F. A. 1983. Startle responses of blue jays to visual stimuli presented during feeding. *Animal Behaviour*, 31, 385–96.

Ventura, D. F., deSouza, J. M., Devoe, R. D. & Zana, Y. 1999. UV responses in the retina of the turtle. *Visual Neuroscience*, 16, 191–204.

Ventura, D. F., Zana, Y., de Souza, J. M. & DeVoe, R. D. 2001. Ultraviolet colour opponency in the turtle retina. *Journal of Experimental Biology*, 204, 2527–34.

Vermeij, G. J. 1994. The evolutionary interaction among species: Selection, escalation and coevolution. *Annual Review of Ecology, Evolution and Systematics*, 25, 219–36.

Verpooten, J. & Nelissen, M. 2010. Sensory exploitation and cultural transmission: the late emergence of iconic representations in human evolution. *Theory in Biosciences*, 129, 211–21.

Vickers, N. J. 2006. Winging it: moth flight behavior and responses of olfactory neurons are shaped by pheromone plume dynamics. *Chemical Senses*, 31, 155–66.

Vignieri, S. N., Larson, J. G. & Hoekstra, H. E. 2010. The selective advantage of crypsis in mice. *Evolution*, 64, 2153–8.

Viitala, J., Korpimaki, E., Palokangas, P. & Koivula, M. 1995. Attraction of kestrels to vole scent marks visible in ultraviolet light. *Nature*, 373, 425–7.

Virant-Doberlet, M., King, R. A., Polajnar, J. & Symondson, W. O. C. 2011. Molecular diagnostics reveal spiders that exploit prey vibrational signals used in sexual communication. *Molecular Ecology*, 20, 2204–16.

Vogel, E. R., Neitz, M. & Dominy, N. J. 2006. Effect of color vision phenotype on the foraging of wild white-faced capuchins, *Cebus capucinus*. *Behavioural Ecology*, 18, 292–7.

von der Emde, G., Schwarz, S., Gomez, L., Budelli, R. & Grant, K. 1998. Electric fish measure distance in the dark. *Nature*, 395, 890–4.

von der Emde, G. 1999. Active electrolocation of objects in weakly electric fish. *Journal of Experimental Biology*, 202, 1205–15.

von Helversen, D. & von Helversen, O. 1999. Acoustic guide in bat-pollinated flower. *Nature*, 398, 759–60.

von Philipsborn, A. & Labhart, T. 1990. A behavioural study of polarization vision in the fly, *Musca domestica*. *Journal of Comparative Physiology A*, 167, 737–43.

Vorobyev, M. 2003. Coloured oil droplets enhance colour discrimination. *Proceedings of the Royal Society B: Biological Sciences*, 270, 1255–61.

Wallace, A. R. 1889. *Darwinism. An Exposition of the Theory of Natural Selection With Some of its Applications*. London: Macmillan & Co.

Wandell, B. A. 1995. *Foundations of Vision*. Sunderland, Massachusetts: Sinauer Associates, Inc.

Wang, I. J. & Shaffer, H. B. 2008. Rapid color evolution in an aposematic species: A phylogenetic analysis of color variation in the strikingly polymorphic strawberry poison-dart frog. *Evolution*, 62, 2742–59.

Wang, I. J. & Summers, K. 2010. Genetic structure is correlated with phenotypic divergence rather than geographic isolation in the highly polymorphic strawberry poison-dart frog. *Molecular Ecology*, 19, 447–58.

Wang, J. H., Jackson, J. K. & Lohmann, K. J. 1998. Perception of wave surge motion by hatchling sea turtles. *Journal of Experimental Marine Biology and Ecology*, 229, 177–86.

Warkentin, K. J., Keeley, A. T. H. & Hare, J. F. 2001. Repetitive calls of juvenile Richardson's ground squirrels (*Spermophilus richardsonii*) communicate response urgency. *Canadian Journal of Zoology*, 79, 569–73.

Warrant, E. J. 2010. Polarisation vision: beetles see circularly polarised light. *Current Biology*, 20, R610–11.

Webster, D. R. & Weissburg, M. J. 2009. The hydrodynamics of chemical cues among aquatic organisms. *Annual Review of Fluid Mechanics*, 41, 73–90.

Weckström, M., Hardie, R. C. & Laughlin, S. B. 1991. Voltage-activated potassium channels in blowfly photoreceptors and their role in light adaptation. *Journal of Physiology*, 440, 635–57.

Wehner, R., Bernard, G. D. & Geiger, E. 1975. Twisted and non-twisted rhabdoms and their significance for polarization detection in the bee. *Journal of Comparative Physiology A*, 104, 225–45.

Wehner, R. 2001. Polarization vision- a uniform sensory capacity? *Journal of Experimental Biology*, 204, 2589–96.

Weindler, P., Wiltschko, R. & Wiltschko, W. 1996. Magnetic information affects the stellar orientation of young bird migrants. *Nature*, 383, 158–60.

Weissing, F. J., Edelaar, P. & van Doorn, G. S. 2011. Adaptive speciation theory: a conceptual review. *Behavioral Ecology and Sociobiology*, 65, 461–80.

Welbergen, J. A. & Davies, N. B. 2009. Strategic variation in mobbing as a front line of defense against brood parasitism. *Current Biology*, 19, 235–40.

Welbergen, J. A. & Davies, N. B. 2011. A parasite in wolf's clothing: hawk mimicry reduces mobbing of cuckoos by hosts. *Behavioral Ecology*, 22, 574–9.

Whitchurch, E. A. & Takahashi, T. T. 2006. Combined auditory and visual stimuli facilitate head saccades in the barn owl (*Tyto alba*). *Journal of Neurophysiology*, 96, 730–45.

Widder, E. A. 1998. A predatory use of counterillumination by the squaloid shark, *Isistius brasiliensis*. *Environmental Biology of Fishes*, 53, 267–73.

Widder, E. A. 2010. Bioluminescence in the ocean: origins of biological, chemical, and ecological diversity. *Science*, 328, 704–8.

Wignall, A. E. & Taylor, P. W. 2009. Alternative predatory tactics of an araneophagic assassin bug (*Stenolemus bituberus*). *Acta Ethologica*, 12, 23–7.

Wignall, A. E., Jackson, R. R., Wilcox, R. S. & Taylor, P. W. 2011. Exploitation of environmental noise by an araneophagic assassin bug. *Animal Behaviour*, 82, 1037–42.

Wignall, A. E. & Taylor, P. W. 2011. Assassin bug uses aggressive mimicry to lure spider prey. *Proceedings of the Royal Society B: Biological Sciences*, 278, 1427–33.

Wiklund, C., Vallin, A., Friberg, M. & Jakobsson, S. 2008. Rodent predation on hibernating peacock and small tortoiseshell butterflies. *Behavioral Ecology and Sociobiology*, 62, 379–89.

Wilder, S. M., DeVito, J., Persons, M. H. & Rypstra, A. L. 2005. The effects of moisture and heat on the efficacy of chemical cues used in predator detection by the wolf spider *Pardosa milvina* (ARANEAE, LYCOSIDAE). *Journal of Arachnology*, 33, 857–61.

Wiley, R. H. 1991. Associations of song properties with habitats for territorial oscine birds of eastern North America. *American Naturalist*, 138, 973–93.

Williams, N. 2010. Winging it. *Current Biology*, 20, 544–5.

Willis, M. A. 2008. Chemical plume tracking behavior in animals and mobile robots. *Navigation*, 55, 127–35.

Willis, M. A., Avondet, J. L. & Zheng, E. 2011. The role of vision in odor-plume tracking by walking and flying insects. *Journal of Experimental Biology*, 214, 4121–32.

Wilson, D. R. & Hare, J. F. 2004. Ground squirrel uses ultrasonic alarms. *Nature*, 430, 523.

Wilson, D. R. & Hare, J. F. 2006. The adaptive utility of Richardson's ground squirrel (*Spermophilus richardsonii*) short-range ultrasonic alarm signals. *Canadian Journal of Zoology*, 84, 1322–30.

Wilts, B. D., Michielsen, K., Kuipers, J., De Raedt, H. & Stavenga, D. G. 2012. Brilliant camouflage: photonic crystals in the diamond weevil, *Entimus imperialis*. *Proceedings of the Royal Society B: Biological Sciences*, 279, 2524–30.

Wiltschko, R., Munro, U., Ford, H. & Wiltschko, W. 2008a. Response to the comments by R. Muheim, S. Åkesson, and J. B. Phillips to our paper 'Contradictory results on the role of polarized light in compass calibration in migratory songbirds'. *Journal of Ornithology*, 149, 663–4.

Wiltschko, R., Munro, U., Ford, H. & Wiltschko, W. 2008b. Contradictory results on the role of polasized light in compass calibration in migratory songbirds. *Journal of Ornithology*, 149, 607–14.

Wiltschko, R. & Wiltschko, W. 2009. Avian navigation. *Auk*, 126, 717–43.

Wiltschko, R. & Wiltschko, W. 2010. Avian magnetic compass: its functional properties and physical basis. *Current Zoology*, 56, 265–76.

Wiltschko, R., Denzau, S., Gehring, D., Thalau, P. & Wiltschko, W. 2011. Magnetic orientation of migratory robins, *Erithacus rubecula*, under long-wavelength light. *Journal of Experimental Biology*, 214, 3096–101.

Wiltschko, W., Weindler, P. & Wiltschko, R. 1998. Interaction of magnetic and celestial cues in the migratory orientation of passerines. *Journal of Avian Biology*, 29, 606–17.

Wiltschko, W., Traudt, J., Gunturkun, O., Prior, H. & Wiltschko, R. 2002. Lateralization of magnetic compass orientation in a migratory bird. *Nature*, 419, 467–70.

Wiltschko, W. & Wiltschko, R. 2005. Magnetic orientation and magnetoreception in birds and other animals. *Journal of Comparative Physiology A*, 191, 675–93.

Wiltschko, W., Munro, U., Ford, H. & Wiltschko, R. 2006. Bird navigation: what type of information does the magnetite-based receptor provide? *Proceedings of the Royal Society B: Biological Sciences*, 273, 2815–20.

Wiltschko, W. & Wiltschko, R. 2007. Magnetoreception in birds: two receptors for two different tasks. *Journal of Ornithology*, 148, S61–76.

Windmill, J. F., Fullard, J. H. & Robert, D. 2007. Mechanics of a 'simple' ear: tympanal vibrations in noctuid moths. *Journal of Experimental Biology*, 210, 2637–48.

Witte, K., Farris, H. E., Ryan, M. J. & Wilczynski, W. 2005. How cricket frog females deal with a noisy world: habitat-related differences in auditory tuning. *Behavioral Ecology*, 16, 571–9.

Wolf, M. C. & Moore, P. A. 2002. Effects of the herbicide metolachlor on the perception of chemical stimuli by *Orconectes rusticus*. *Journal of the North American Benthological Society*, 21, 457–67.

Wood, S. R., Sanderson, K. J. & Evans, C. S. 2000. Perception of terrestrial and aerial alarm calls by honeyeaters and falcons. *Australian Journal of Zoology*, 48, 127–34.

Woolhouse, M. E. J., Webster, J. P., Domingo, E., Charlesworth, B. & Levin, B. R. 2002. Biological and biomedical implications of the co-evolution of pathogens and their hosts. *Nature Genetics*, 32, 569–77.

Wyatt, T. D. 2010. Pheromones and signature mixtures: defining species-wide signals and variable cues for identity in both invertebrates and vertebrates. *Journal of Comparative Physiology A*, 196, 685–700.

Wyatt, T. D. 2013. Perception and response to chemical communication: from receptors to brains, behavior, and development. In: *Pheromones and animal behavior: chemical signals and signature mixtures*. Cambridge: Cambridge University Press.

Wyneken, J., Salmon, M. & Lohmann, K. J. 1990. Orientation by hatchling loggerhead sea turtles *Caretta caretta* L. in a wave tank. *Journal of Experimental Marine Biology and Ecology*, 139, 43–50.

Yack, J. E., Kalko, E. K. & Surlykke, A. 2007. Neuroethology of ultrasonic hearing in nocturnal butterflies (Hedyloidea). *Journal of Comparative Physiology A*, 193, 577–90.

Yager, D. D. & May, M. L. 1990. Ultrasound-triggered, flight-gated evasive maneuvers in the praying mantis *Parasphendale agrionina*. II. Tethered flight. *Journal of Experimental Biology*, 152, 41–58.

Yager, D. D., Cook, A. P., Pearson, D. L. & Spangler, H. G. 2000. A comparative study of ultrasound-triggered behaviour in tiger beetles (Cicindelidae). *Journal of Zoology*, 251, 355–68.

Yang, C., Liang, W., Cai, Y., Shi, S., Takasu, F., Møller, A. P., Antonov, A., Fossøy, F., Moksnes, A., Røskaft, E. & Stokke, B. G. 2010. Coevolution in action: disruptive selection on egg colour in an avian brood parasite and its host. *PLoS ONE*, 5, e10816.

Yang, E. C. & Maddess, T. 1997. Orientation-sensitive neurons in the brain of the honey bee (*Apis mellifera*). *Journal of Insect Physiology*, 43, 329–36.

Yeargan, K. V. 1988. Ecology of a bolas spider, *Mastophora hutchinsoni*: phenology, hunting tactics, and evidence for aggressive chemical mimicry. *Oecologia*, 74, 524–30.

Yoder, J. B. & Nuismer, S. L. 2010. When does coevolution promote diversification? *American Naturalist*, 176, 802–17.

Yokoyama, S., Altun, A. & DeNardo, D. F. 2011. Molecular convergence of infrared vision in snakes. *Molecular Biology and Evolution*, 28, 45–8.

Yoshizawa, M., Goricki, S., Soares, D. & Jeffery, W. R. 2010. Evolution of a behavioral shift mediated by superficial neuromasts helps cavefish find food in darkness. *Current Biology*, 20, 1631–6.

Zahar, Y., Reches, A. & Gutfreund, Y. 2009. Multisensory enhancement in the optic tectum of the barn owl: spike count and spike timing. *Journal of Neurophysiology*, 101, 2380–94.

Zahavi, A. 1975. Mate selection—a selection for a handicap. *Journal of Theoretical Biology*, 53, 205–14.

Zakon, H., Oestreich, J., Tallarovic, S. & Triefenbach, F. 2002. EOD modulations of brown ghost electric fish: JARs, chirps, rises, and dips. *Journal of Physiology*, 96, 451–8.

Zakon, H. H., Lu, Y., Zwickl, D. J. & Hillis, D. M. 2006. Sodium channel genes and the evolution of diversity in communication signals of electric fishes: convergent molecular evolution. *Proceedings of the National Academy of Sciences of the USA*, 103, 3675–80.

Zakon, H. H., Zwickl, D. J., Lu, Y. & Hillis, D. M. 2008. Molecular evolution of communication signals in electric fish. *Journal of Experimental Biology*, 211, 1814–18.

Zheng, J., Shen, W., He, D. Z., Long, K. B., Madison, L. D. & Dallos, P. 2000. Prestin is the motor protein of cochlear outer hair cells. *Nature*, 405, 149–55.

Zuk, M. & Kolluru, G. R. 1998. Exploitation of sexual signals by predators and parasitoids. *The Quarterly Review of Biology*, 73, 415–38.

Zuk, M., Rotenberry, J. T. & Tinghitella, R. M. 2006. Silent night: adaptive disappearance of a sexual signal in a parasitized population of field crickets. *Biology Letters*, 2, 521–4.

Zwislocki, J. J. 1981. Sound analysis in the ear: a history of discoveries. *American Scientist*, 69, 184–92.

Zylinksi, S., How, M. J., Osorio, D., Hanlon, R. T. & Marshall, J. A. R. 2011. To be seen or to hide: visual characteristics of body patterns for camouflage and communication in the Australian giant cuttlefish *Sepia apama*. *American Naturalist*, 177, 681–90.

Zylinski, S. & Johnsen, S. 2011. Mesopelagic cephalopods switch between transparency and pigmentation to optimize camouflage in the deep. *Current Biology*, 21, 1937–41.

Index